Glencoe Science

Reading Essentials for Biology

An Interactive Student Textbook

The McGraw·Hill Companies

Send all inquiries to:
McGraw-Hill Education
8787 Orion Place
Columbus, OH 43240-4027

ISBN: 978-0-07-896099-4
MHID: 0-07-896099-1

Printed in the United States of America.

7 RHR 13

Table of Contents

To the Student

Reading Essentials for Biology takes the stress out of reading, learning, and understanding biology. This book covers important concepts in biology, offers ideas for how to learn the information, and helps you review what you have learned. Understanding biology concepts will help you improve your critical-thinking skills, solve problems effectively, and make useful decisions.

The chapters of *Reading Essentials for Biology* include the following elements.

- **Before You Read** sparks your interest in what you will learn and relates it to your world. The **Main Idea** and **What You'll Learn** statements help focus on the most important concepts in the section.
- **Read to Learn** describes important biology concepts with words and graphics. Next to the text you can find a variety of study tips and ideas for organizing and learning information:
  - **Study Coach** and **Mark the Text** offer tips for getting the main ideas out of the text.
  - **Foldables® Study Organizers** help you divide the information into smaller, easier-to-remember concepts.
  - **Reading Checks** ask questions about key concepts. The questions are placed so you know whether you understand the material.
  - **Think It Over** elements help you consider the material in-depth, giving you an opportunity to use your critical-thinking skills.
  - **Picture This** questions relate to the illustrations used with the text. The questions will help get you actively involved in illustrating the important concepts.
  - **Applying Math** reinforces the connection between math and science.

Dinah Zike's Foldables®

A Foldable is a 3-D, interactive graphic organizer. By using Foldables, you can quickly organize and retain information. Every chapter in ***Reading Essentials for Biology*** includes a Foldable that can be used to organize important ideas in the chapter. Later, the Foldable can be used as a study guide for main ideas and key points in the chapter. Foldables can also be used for a more in-depth investigation of the key terms, concepts, or ideas presented in the chapter.

The Foldables for this book can be created using notebook paper or plain sheets of paper. Some will require scissors to cut the tabs. The Foldables created for this book can be stored in a plastic bag, a box, or sheet protectors in a three-ring binder. By keeping your Foldables organized, you will have a ready study tool. You will also be creating a portfolio of your work.

Your teacher might ask you to make the Foldables found on the Start-Up Activities pages in the Student Edition, in addition to the Foldables you will make for ***Reading Essentials for Biology.*** As you become familiar with Foldables, you might see other opportunities to use Foldables to create additional study tools. Keep together all the Foldables you make for a chapter. Use them as you review the chapter and study for assessments.

chapter 1

The Study of Life

section 1 Introduction to Biology

Before You Read

What does it mean to be alive? On the lines below, list characteristics that you think living things have. Then read the section to learn what you have in common with other living things.

____

____

MAIN Idea

All living things share the characteristics of life.

What You'll Learn

- the definition of biology
- possible benefits from studying biology
- characteristics of living things

Read to Learn

The Science of Life

Biology is the science of life. In biology, you will learn the origins and history of life and once-living things. You will also learn structures, functions, and interactions of living things.

Study Coach

Make Flash Cards Make a flash card for each key term in this section. Write the term on one side of the card. Write the definition on the other side. Use the flash cards to review what you have learned.

What do biologists do?

Biologists make discoveries and look for explanations by performing laboratory and field studies. Some biologists study animals in their natural environment. For example, Jane Goodall's observations helped scientists know how best to protect chimpanzees.

Other biologists research diseases to develop new treatments. Many biologists work to develop new technology. Technology is the application of scientific knowledge to solve human needs and to extend human capabilities. For example, Dr. Charles Drew developed methods to separate blood plasma for transfusions. His research led to blood banks.

Some biologists study genetic engineering of plants. They try to develop plants that can grow in poor soils and resist insects and disease. Environmental biologists try to protect animals and plants from extinction by developing reproductive strategies and ways to protect them.

FOLDABLES™

Summarize Information Make an eight-tab Foldable from a sheet of paper. Label the tabs with the question heads in this section. As you read, summarize the answers under the tabs.

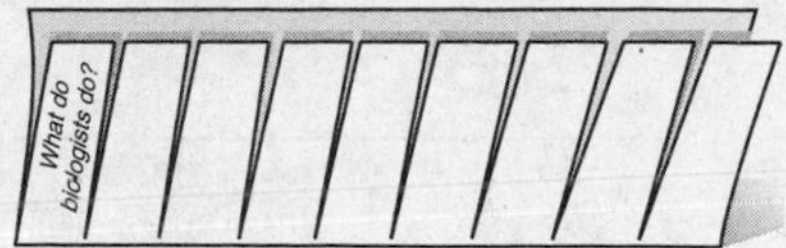

Picture This

1. Highlight each characteristic of life in the table as you read about it in the section. Use the descriptions in the table to review what you have learned.

The Characteristics of Life

From many observations, biologists concluded that all living things have certain characteristics. The characteristics of life are listed in the table below. An **organism** is anything that has or once had all these characteristics.

Characteristic of Life	Description
Made of one or more cells	The cell is the basic unit of life. Some organisms have one cell only. Others have many cells.
Displays organization	The organization of a biological system begins with atoms and molecules. Each organized structure in an organism has a specific function. For example, an anteater's snout is long because it functions as a container for the long tongue.
Grows and develops	Growth results in an increase in mass. Development results in different abilities. For example, a tadpole grows larger and develops into an adult frog.
Reproduces	Organisms reproduce and pass on traits to the next generation. Reproduction must occur for a species to continue to exist.
Responds to stimuli	Reactions to stimuli from inside and outside the body are called responses. For example, a cheetah responds to the need for food by chasing a gazelle. The gazelle responds by running away.
Requires energy	Energy is needed for life processes. Many organisms get energy by taking in food. Other organisms make their own food.
Maintains homeostasis	Homeostasis is the process that keeps conditions inside the bodies of all organisms stable. For example, humans perspire when hot to lower body temperature.
Adaptations evolve over time	Adaptations are inherited changes that occur over time and help the species survive.

What determines a cell's structure?

Cells are the basic units of structure and function in all living things. Some organisms, such as bacteria, are unicellular—they have just one cell. Humans and plants are multicellular—they have many cells. The structure of a cell is related to its function. For example, each cell in a tree's roots has a structure that enables it to take in water from soil.

Reading Check

2. Sequence the levels of organization, from least complex to most complex.

______

______

______

How are living things organized?

Living things display **organization**. This means they are arranged in an orderly way. Each cell is made up of atoms and molecules. Tissues are groups of specialized cells that work together. Tissues are organized into organs, which perform functions such as digestion. Organ systems work together to support an organism. ☑

How does development differ from growth?

Growth adds mass to an organism. Many organisms form new cells and new structures as they grow. **Development** is the process of natural changes that take place during the life of an organism. For example, after baby birds hatch they cannot fly for a few weeks. As they grow, structures develop that give them the ability to fly.

Why is reproduction important to a species?

Reproduction is the production of offspring. If a species is to continue to exist, some members of the species must reproduce. A **species** is a group of organisms that can breed with one another and produce fertile offspring. Without reproduction, a species will become extinct.

Why is the ability to respond to stimuli critical?

An organism's external environment includes all things that surround it, such as air, water, soil, rocks, and other organisms. An organism's internal environment includes all things inside it. A **stimulus** (plural, stimuli) is anything that is part of either environment that causes some reaction by the organism. The reaction to a stimulus is a **response**. For example, a houseplant responds to the stimulus of sunlight coming through a window by growing toward it. The ability to respond to stimuli is important for survival.

How do organisms obtain energy?

Living things need energy to fuel their life functions. Living things get their energy from food. Most plants and some unicellular organisms use light energy from the Sun to make their own food. Organisms that cannot make their own food get energy by consuming other organisms.

Why must an organism maintain homeostasis?

Homeostasis (hoh mee oh STAY sus) is the regulation of an organism's internal conditions to maintain life. If anything upsets an organism's normal state, processes to restore the normal state begin. If homeostasis is not restored, the organism might die. ☑

How do adaptations benefit a species?

An **adaptation** is any inherited characteristic that results from changes to a species over time. Adaptations make the members of a species better able to survive and, therefore, better able to pass their genes to their offspring.

Think it Over

3. **Apply** Give an example of an internal stimulus for a rabbit. Describe an appropriate response to the stimulus.

___

___

___

Reading Check

4. **Summarize** the importance of homeostasis.

___

___

___

chapter 1

The Study of Life

section 2 The Nature of Science

MAIN Idea

Science is the process based on inquiry that seeks to develop explanations.

What You'll Learn

- characteristics of science
- how to distinguish science from pseudoscience
- the importance of the metric system and SI

Before You Read

When you see a headline such as *Alien Baby Found in Campsite,* how do you know whether to believe it? Write your thoughts on the lines below. Then read the section to learn how to tell the difference between science and pseudoscience.

____

____

____

Mark the Text

Restate the Main Point Highlight the main point in each paragraph. Then restate each main point in your own words.

Read to Learn

What is science?

Science is a body of knowledge based on the study of nature and its physical setting. The purpose of science is scientific inquiry—the development of explanations. Scientific inquiry is a creative process as well as a process involving observation and experimentation.

How are scientific theories developed?

A **theory** is an explanation of a natural phenomenon supported by many observations and experiments over time. A scientific explanation combines what is already known about something with many observations and experiments. An explanation becomes a theory only when investigations produce enough evidence to support the idea. On the other hand, a **law** describes the scientific relationships under certain conditions of nature. ☑

A pseudoscience (soo doh SI uhnts) is an area of study that tries to imitate science. Astrology, horoscopes, and psychic reading are pseudosciences. They are not supported by science-based evidence.

Reading Check

1. **Identify** the difference between a theory and a law.

____

____

____

How does science expand knowledge?

Science is guided by research that results in a constant reevaluation of what is known. This reevaluation process leads to new knowledge. It also leads to new questions that require more research. ☑

Reading Check

2. Contrast the role of research in science and pseudoscience.

What happens when scientists disagree?

Scientists welcome debate. Disagreements among scientists often lead to further investigation. Science advances when new discoveries are added to the existing body of knowledge. For example, scientific research has dramatically increased our understanding of metabolic syndrome in athletes.

How do scientists deal with inconsistent data?

When observations or data are not consistent with current understanding, scientists investigate the inconsistencies. For example, some early biologists suggested that bats had traits that were more similar to those of mammals than those of birds, as shown in the figure below. This idea led to further investigation. The new evidence confirmed that bats are more closely related to mammals than to birds.

In pseudoscience, observations that are not consistent with beliefs are ignored.

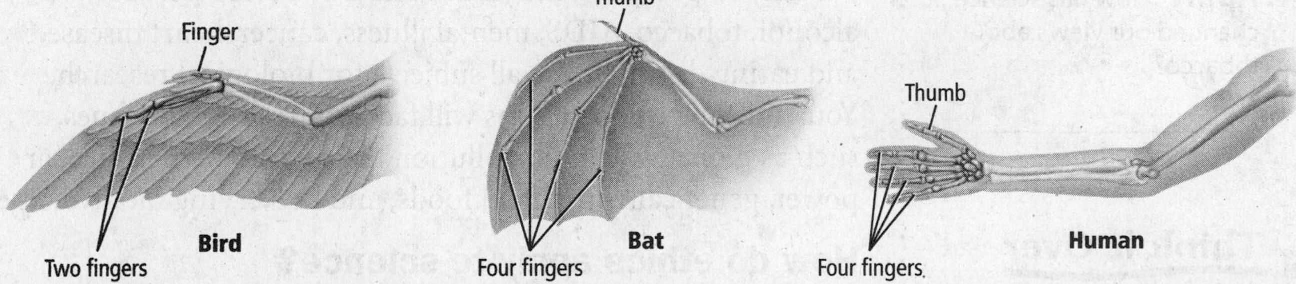

Picture This

3. Compare Are the structures of a bat's wing more like a human arm or a bird's wing? Explain.

How do scientists test claims?

In science, all research follows standard procedures. Conclusions are based on evidence from carefully controlled investigations. Pseudoscientists make claims that cannot be tested. These claims are a mix of facts and opinions.

How are scientific investigations evaluated?

Scientific investigations undergo peer review. **Peer review** in science is a process in which the procedures used during an experiment and the results are evaluated by scientists who are in the same field or are doing similar research.

How do scientists conduct research?

Scientists have several ways to research a topic. They could do experiments, run surveys, analyze work that has already been done, or do a case study. A case study is an intensive study of a group, an incident, or a community. For example, one scientist conducted a case study on college football players and their exercise routines and nutrition habits. The results of the case study showed that athletes may have risk factors associated with heart disease and diabetes.

Science in Everyday Life

Science is not limited to the laboratory. It is all around you. Many popular television shows about crime are based on **forensics**—the field that uses science to investigate crime. The media is filled with information on medical advances, new scientific discoveries, and new technologies.

Why is science literacy important?

To evaluate the vast amount of information available in print, online, and on television, you must be science literate. To be science literate, you need to combine a basic understanding of science and its process with reasoning and thinking skills.

Many important issues today relate to biology. Drugs, alcohol, tobacco, AIDS, mental illness, cancer, heart disease, and eating disorders are all subjects for biological research. You and future generations will face environmental issues, such as global warming, pollution, use of fossil fuels, nuclear power, genetically modified foods, and preserving biodiversity.

How do ethics apply to science?

Many scientific inquiries involve ethics. **Ethics** are a set of moral principles or values. Ethical issues are involved in the study of cloning, genetic engineering, euthanasia (yoo thuh NAY zhuh), and cryonics (kri AH niks). Euthanasia is permitting death for reasons of mercy. Cryonics is freezing a dead organism with the hope of reviving it in the future.

Scientists provide information about new discoveries and technology. As a scientifically literate adult, you will be able to participate in discussions about important issues. You will have the opportunity to support policies that reflect your values.

Think it Over

4. Apply How has science changed our views about tobacco?

____

Think it Over

5. Summarize Complete the following sentence: I need to be science literate in order to . . .

____

____

____

____

____

____

The Study of Life

section 3 Methods of Science

Before You Read

Suppose you want to identify a bird that visits the feeder in your yard. On the lines below, describe some methods you might use to identify the bird. Then read the section to learn the methods scientists use to gather information and answer questions.

MAIN Idea

Biologists use specific methods when conducting research.

What You'll Learn

- the difference between an observation and an inference
- how a control, an independent variable, and a dependent variable differ

Read to Learn

Ask a Question

Scientific inquiry begins with observation. **Observation** is a direct method of gathering information in an orderly way. It often involves recording information. For example, if you want to identify a bird, you observe it. You note how it behaves and what it eats. You might draw or photograph it.

Scientific inquiry involves asking questions and using information from reliable sources. By combining information from other sources with your observations of the bird, you could start making logical conclusions. This process is called making **inferences**, or inferring. For example, if you saw a photo of a bird that was similar to your bird, you might infer that your bird was related to the bird in the photo.

Biologists work in many settings. They work in the field. They work in laboratories, universities, and museums. No matter where they work, all biologists use similar methods to gather information and to answer questions. These methods are an organized series of events called **scientific methods**. Throughout the process, biologists continue to observe and make inferences.

Study Coach

Make an Outline Make an outline of the information you learn in this section. Start with the headings. Include the boldface terms.

Think it Over

1. **Explain** how inferences relate to observation.

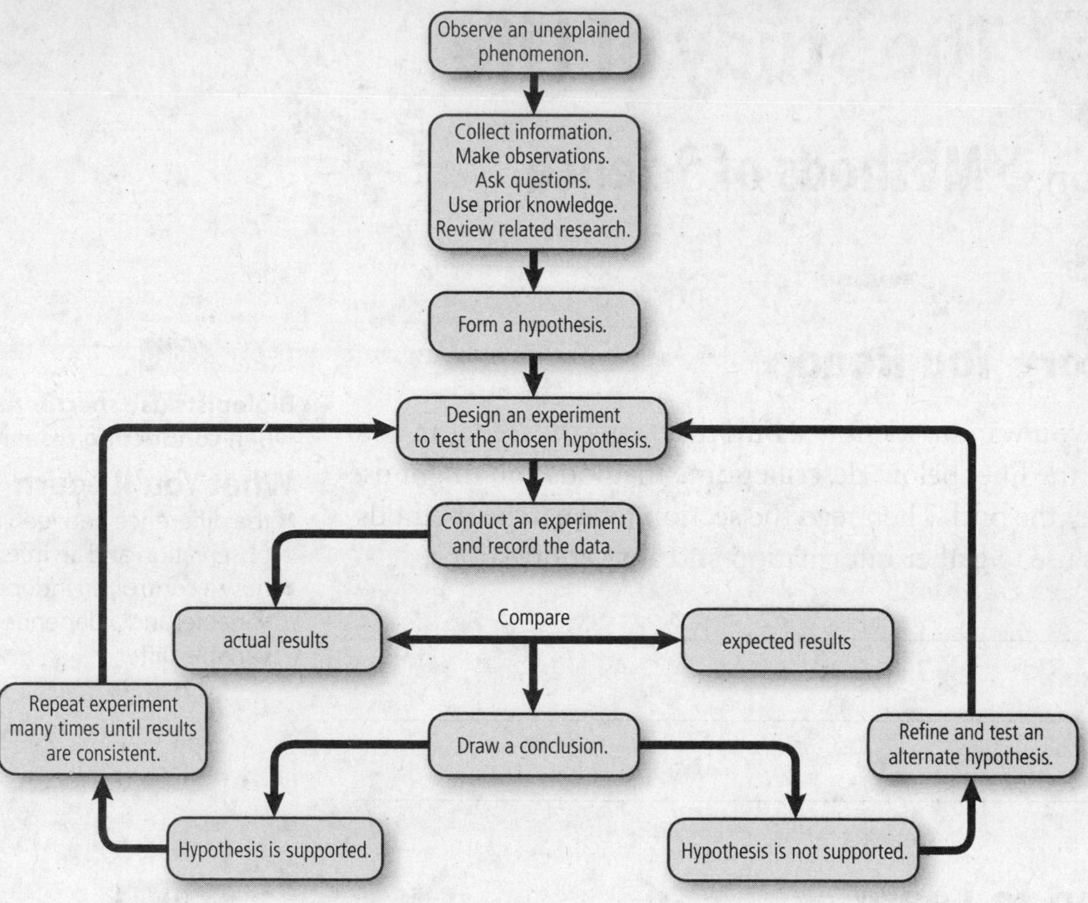

Form a Hypothesis

The figure below shows the sequence of events in scientific methods. Scientists use the information they gather from observation and other sources to form a hypothesis. A **hypothesis** (hi PAH thuh sus) is a testable explanation of a situation. When enough data from many investigations support a hypothesis, the scientific community accepts the explanation as valid. If the data do not support a hypothesis, the hypothesis is revised and investigated further.

Sometimes scientists make unexpected discoveries. Serendipity is the occurrence of accidental or unexpected, but fortunate, results. For example, penicillin was discovered while a scientist was investigating something else.

Collect the Data

Scientists test a hypothesis through experiments. An **experiment** is an investigation done in a controlled setting that tests the hypothesis. ☑

Picture This

2. Sequence After a researcher draws a conclusion, what does he or she do next, whether or not the hypothesis is supported? (Circle your answer.)

a. draw another conclusion
b. conduct another experiment
c. compare results again

Reading Check

3. Explain the purpose of an experiment.

____

____

What is the purpose of a control group?

Experiments have an experimental group and a control group. The **experimental group** is the group exposed to the factor being tested. For example, suppose scientists wanted to test the effects of a vitamin supplement on energy level. The experimental group would receive the vitamin. The **control group** is the group used for comparison. This group would not receive the vitamin.

How do scientists design an experiment?

In a controlled experiment, scientists change only one factor at a time. The factor that is changed in an experiment is called the **independent variable**. It is the tested factor, and it might affect the outcome of the experiment. A **dependent variable** is something that results from or depends on changes to the independent variable. In our example, the vitamin is the independent variable and energy level is the dependent variable. A **constant** in an experiment remains fixed, while the independent and dependent variables change.

What two kinds of data do scientists collect?

Information gained from observations is called **data**. Data in the form of numbers is called quantitative data. Quantitative data might measure time, temperature, length, mass, area, volume, density, or other factors. Qualitative data are descriptions of what the observer senses. Everyone senses things differently. As a result, qualitative data can vary from one observer to another. ☑

What system of measurement is used?

Scientists use the **metric system** which uses units with divisions that are powers of ten. In 1960, a system of unit standards of the metric system was established called the International System of Units, or **SI**. In biology, the SI units you will use most often are meter (length), gram (mass), liter (volume), and second (time).

Analyze the Data

After biologists collect data from experiments, they interpret the data and look for patterns. They compare their results to expected results to see if the data support their hypothesis. If not, they revise the hypothesis and retest. Even when the data support the hypothesis, the experiment must be repeated many more times. Consistent results from repeated trials give strength to the hypothesis as a valid explanation for the tested phenomenon.

Think it Over

4. Apply Suppose a biologist designed an experiment to study the effect of water pollution on the reproductive rate of salmon. What is the dependent variable in this experiment?

____

____

Reading Check

5. Classify "The average high temperature here in March is 22°C." Is this information qualitative data or quantitative data?

____

____

Why do biologists use tables and graphs?

Picture This

6. Predict what the mass of the anole will be at 21 days.

Biologists often display data in tables and graphs to make patterns easier to detect. The data about the mass of an anole, a type of lizard, are listed in the table below. The data are plotted on the graph. Note the regular pattern in the graph. The mass increases over a three-day period and then levels off for three days. Then it increases again.

Change in Mass of Anole	
Date	**Mass (g)**
April 11	2.4
April 14	2.5
April 17	2.5
April 20	2.6
April 23	2.6
April 26	2.7
April 29	2.7

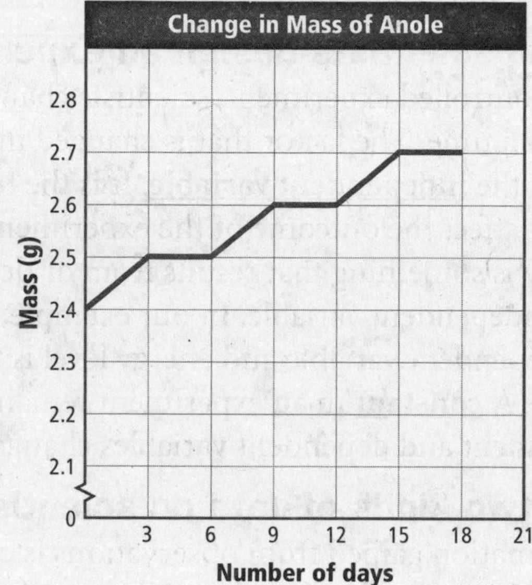

Report Conclusions

Scientists write a report of their experiments for peer review. Other scientists in the same field examine the methods, analysis, and conclusions in the report. If the reviewers agree that the report has value, then the report is published in a scientific journal.

Student Scientific Inquiry

As you study biology, you might have opportunities to do your own investigations. If so, develop a research plan based on the scientific methods described in this chapter. Ask meaningful questions. Form hypotheses. Collect data by conducting careful experiments. Analyze the data. Draw conclusions and report them.

7. Explain the purpose of a safety symbol.

During biology labs, warning statements and safety symbols will alert you to possible hazards. A safety symbol is a logo designed to alert you about a specific danger. Refer to the safety symbols chart at the front of the textbook before beginning any field or lab activity. Learn where safety equipment is located in the classroom. You are responsible for performing your investigations safely at all times. ☑

Principles of Ecology

section 1 Organisms and Their Relationships

Before You Read

On the lines below, list the organisms that you have encountered today. You share the same environment with these organisms. In this section you will learn how many organisms exist in the same environment.

MAIN Idea

Biotic and abiotic factors work together in an ecosystem.

What You'll Learn

- the differences between biotic and abiotic factors
- the levels of biological organization
- the difference between an organism's habitat and its niche

Read to Learn

Ecology

Each living organism depends on nonliving factors for survival in its environment. Each living organism also depends on other living organisms in its environment. Green plants are a food source and can be a place where other organisms live. The animals that eat plants provide food for other organisms. Organisms depend on each other in all types of environments—deserts, tropical rain forests, and grassy meadows. **Ecology** is the study of the interactions between organisms and their environments.

What do ecologists do?

Scientists who study ecology are called ecologists. The German biologist Ernst Haeckel introduced the word *ecology* in 1866. Eventually, it became a separate field of study.

Ecologists use various tools and methods to observe, experiment, and create models. Ecologists conduct tests to learn why and how organisms survive. For example, tests might help explain how some organisms survive in cold water. ☑

Ecologists also learn about the interactions between organisms by observing them in their environments. Sometimes observations are made over long periods of time. This process is called longitudinal analysis.

Study Coach

Make an Outline Create an outline of this section. Use the headings to organize your outline. List details from what you have read to complete your outline.

Reading Check

1. List three ways ecologists study interactions between organisms.

Why do ecologists use models?

Studying organisms in their environments is not always possible. Ecologists use models to represent a process or system in the environment. By using models, ecologists can control the number of variables. Scientists can measure the effect of each variable one at a time, on the model.

The Biosphere

The **biosphere** (BI uh sfihr) is the portion of Earth that supports life. Ecologists study what takes place in the biosphere. The biosphere includes the air, water, and land where organisms can live, both above and below the ground. ☑

Reading Check

2. Define the biosphere.

The biosphere supports a wide variety of organisms in a wide range of conditions. Climates, soils, plants, and animals differ in different parts of the world. Frozen polar regions, deserts, and rain forests contain organisms. The organisms are adapted to survive in the conditions of their environments. The factors in all environments can be divided into two groups—living factors and nonliving factors.

What are biotic factors?

Biotic (bi AH tihk) **factors** are the living factors in an organism's environment. For example, the algae, frogs, and microscopic organisms in the stream are biotic factors for salmon in a stream. Other biotic factors live on the land bordering the stream. These include plants, insects, and small animals. Birds that feed on organisms in the stream are also part of the salmon's biotic factors. These factors interact directly or indirectly. The salmon depend on biotic factors for food, shelter, reproduction, and protection, and in turn can provide food for other organisms.

What are abiotic factors?

The nonliving factors in an organism's environment are called **abiotic** (ay bi AH tihk) **factors**. The abiotic factors for the salmon might be the temperature range of the water, the pH of the water, and the salt concentration of the water. For a plant, abiotic factors might include the amount of rainfall, the amount of sunlight, the type of soil, the range of air and soil temperatures, and the nutrients available in the soil.

Think it Over

3. Evaluate Describe the abiotic factors in the environment where you currently live.

Organisms are adapted to the abiotic factors in their natural environment. If an organism moves to a different location with a different set of abiotic factors, the organism must adjust, or it will die.

Levels of Organization

The biosphere is too large to study all the relationships at one time. Scientists use smaller pieces, or levels of organization, for their studies. The numbers and interactions among organisms increase at higher levels of organization. The following are levels of organization from simplest to most complex:

1. organism
2. population
3. biological community
4. ecosystem
5. biome
6. biosphere

The first four of these levels of organization are shown in the figure below.

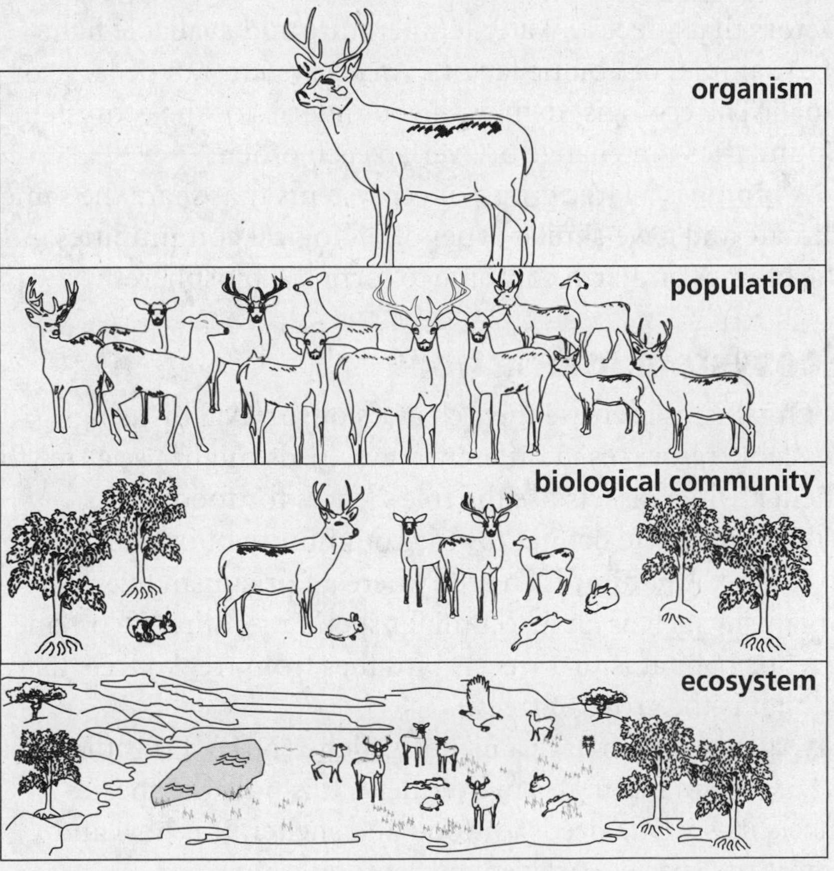

How do available resources affect a population?

The lowest level of complexity that ecologists study is an individual organism. Individual organisms of the same species living in the same geographic location at the same time make up a **population**. A school of fish is a population. Individual organisms in the population must compete to survive. They compete for food, water, mates, and other resources. ☑

Picture This

4. Generalize How do the levels become more complex?

____

____

____

Reading Check

5. Identify Which of the following is a population? (Circle your answer.)

a. all rabbits living on Earth
b. all white-tailed rabbits living in a meadow today
c. all white-tailed rabbits that have ever lived in a meadow

What limits the size of a population?

A population can keep growing as long as resources are available to its members. If a population grows too large, there will not be enough resources for all members of the population. The population will get smaller until it reaches a number that can be supported by the available resources. ☑

Reading Check

6. Draw Conclusions If a population is growing, what can you conclude about the amount of resources available to the organisms?

____

____

____

____

What is a biological community?

A **biological community** is a group of populations that interact in the same geographic area at the same time. Organisms might or might not compete for available resources in a biological community. The plants and animals that live in a park are a biological community.

Who defines the boundaries of an ecosystem?

An **ecosystem** is a biological community and all the abiotic factors that affect it. Water temperature and available light are examples of abiotic factors. An ecosystem can be large or small. The ecologist defines the boundaries of the ecosystem. Boundaries can change or overlap each other.

A **biome** is a large group of ecosystems that share the same climate and have similar types of biological communities. All the biomes on Earth combine to form the biosphere.

Ecosystem Interactions

Organisms increase their chances of survival by using available resources in different ways. Birds might use a tree for shelter, while insects use the tree's leaves for food.

The tree is the habitat for the community of organisms that live there. A **habitat** is an area where an organism lives. An organism such as an insect might spend its entire life on one tree. Its habitat is that tree. A bird flies from tree to tree. Its habitat is the grove of trees.

Organisms also have a niche. A **niche** (NIHCH) is the role an organism has in its environment. It is how the species meets its specific needs for food and shelter. It is how and where the species survives and reproduces.

Think it Over

7. Explain the difference between a habitat and a niche.

____

____

____

____

Community Interactions

Organisms living in biological communities interact constantly. Ecosystems are shaped by these interactions and the abiotic factors. In a biological community, each organism depends on other organisms and competes with other organisms.

When do organisms compete?

Competition occurs when organisms need to use the same resource at the same time. Organisms compete for such resources as food, water, space, and light. When strong organisms compete with weak organisms, the strong organisms usually survive. During a drought, water might be scarce for many organisms. Strong organisms will use the available water. Weak organisms might die or move to another location.

What is predation?

The act of one organism consuming another organism for food is **predation** (prih DAY shun). Most organisms obtain their food by eating other organisms. If you have seen a cat stalk and capture a mouse, you have seen a predator catch its prey. The organism that pursues—the cat—is the predator. The organism that is pursued—the mouse—is the prey. Predators can be plants, animals, or protists.

What is symbiosis?

Some species survive because of relationships with other species. A relationship in which two organisms live together in close association is called **symbiosis** (sihm bee OH sus). The three kinds of symbiosis are mutualism, commensalism, and parasitism.

Mutualism A relationship between two species that live together and benefit from each other is called **mutualism** (MYEW chuh wuh lih zum). A lichen (LI kun) is a mutualistic relationship between algae and fungi. The algae provide food for the fungi. The fungi provide a habitat for the algae. Food and shelter are the benefits of this relationship.

Commensalism A relationship in which one organism is helped and the other organism is not harmed or helped is called **commensalism** (kuh MEN suh lih zum). For example, mosses sometimes grow on tree branches. This does not harm or help the tree, but the mosses benefit from a good habitat.

Parasitism A relationship in which one organism benefits and another organism is harmed is called **parasitism** (PAYR us suh tih zum). When a tick lives on a dog, it is good for the tick but bad for the dog. The tick gets food and shelter, but the dog might get sick. The tick is the parasite and is helped by the relationship. The dog is the host. Usually the parasite does not kill the host, but it might harm or weaken it. If the host dies, the parasite will also die, unless it can find another host.

Think it Over

8. Classify List two more examples of predation that you have seen or of which you have learned.

____

____

____

Think it Over

9. Apply Clown fish live among sea anemones. The anemones provide protection for the clown fish. The clown fish eats food missed by the sea anemones. What term best describes this relationship?

____

chapter 2

Principles of Ecology

section 2 Flow of Energy in an Ecosystem

MAIN Idea

Autotrophs capture energy, making it available for all members of a food web.

What You'll Learn

- the flow of energy through an ecosystem
- food chains, food webs, and pyramid models

Before You Read

If a pet had to survive without your care, how would its diet change? Write your ideas on the lines below. Read about how organisms get food and energy in their environment.

____

____

____

Study Coach

Make Flash Cards Make a flash card for each question heading in this section. On the back of the flash card, write the answer to the question. Use the flash cards to review what you have learned.

Read to Learn

Energy in an Ecosystem

One way to study the interactions within an ecosystem is to trace how energy flows through the system. All organisms are classified by the way they obtain energy.

How do autotrophs obtain energy?

All green plants and other organisms that produce their own food are the primary producers of food in an ecosystem. They are called autotrophs. An **autotroph** (AW tuh trohf) is an organism that captures energy from sunlight or inorganic substances to produce food. Autotrophs make energy available for all other organisms in the ecosystem.

How do heterotrophs differ from autotrophs?

A **heterotroph** (HE tuh roh trohf), also called a consumer, is an organism that obtains energy by consuming other organisms. A heterotroph that consumes only plants is an **herbivore** (HUR buh vor). Cows, rabbits, and grasshoppers are herbivores. ☑

Heterotrophs that prey on other heterotrophs are known as **carnivores** (KAR nuh vorz). Wolves and lions are carnivores. **Omnivores** (AHM nih vorz) eat both plants and animals. Bears, humans, and mockingbirds are examples of omnivores.

Reading Check

1. Define What type of heterotroph consumes only plants for energy?

____

How do detritivores help an ecosystem?

Detritivores (duh TRYD uh vorz) decompose organic materials in an ecosystem and return the nutrients to the soil, air, and water. The nutrients then become available for use by other organisms. Hyenas and vultures are detritivores. They feed on animals that have died. Fungi and bacteria are also detritivores.

Detritivores play an important role in the biosphere. Without them, the biosphere would be littered with dead organisms. The nutrients in these dead organisms would not be available to other organisms. Detritivores make these nutrients available for use by other organisms. ☑

Models of Energy Flow

Ecologists study feeding relationships to learn how energy flows in an ecosystem. Ecologists use food chains and food webs to describe the flow of energy. Each step in a food chain or food web is called a **trophic** (TROH fihk) **level**. Autotrophs are the first trophic level in all ecosystems. Heterotrophs make up the remaining levels.

Organisms at the first trophic level produce their own food. Organisms at all other levels get energy from the trophic level before it.

What is a food chain?

A **food chain** is a simple model that shows how energy flows through an ecosystem. A typical grassland food chain is shown in the figure below. Each organism gets energy from the organism it eats. The flow of energy is always one way—into the consumer. An organism uses part of the energy to build new cells and tissues. The remaining energy is released into the environment and is no longer available to these organisms.

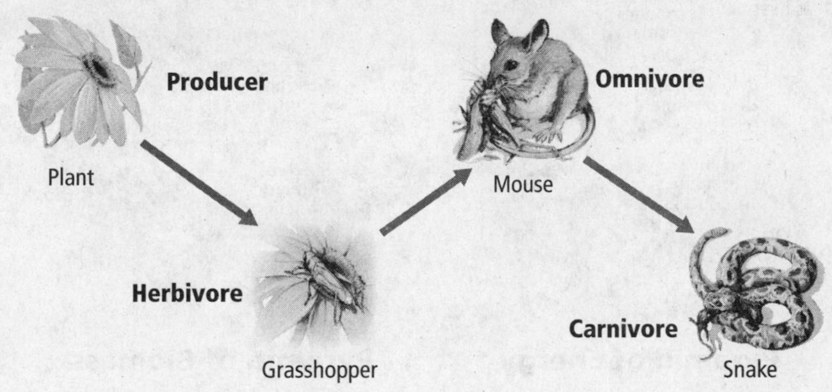

Reading Check

2. Explain How do organisms in an ecosystem depend on detritivores?

______

______

______

Picture This

3. Label Draw a circle around the autotroph. Draw a box around the heterotrophs.

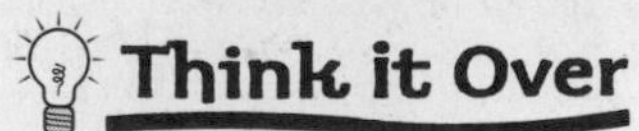

4. Synthesize Why might an ecologist use a food chain for one study and a food web for another study?

What does a food web show?

Feeding relationships are usually more complex than a single food chain model can show. Most organisms feed on more than one species. A **food web** is a model that shows all the possible feeding relationships in an ecosystem. Food webs give a more accurate picture of how energy flows in an ecosystem than food chains.

What do ecologists model with an ecological pyramid?

Ecologists also use ecological pyramids to model how energy flows through ecosystems. A pyramid model can be used to show energy flow in three different ways. Each level of the pyramid represents a trophic level.

A pyramid of energy indicates the amount of energy available to each trophic level. In the energy pyramid below, notice that about 90 percent of the available energy is used by the organisms at each level. Some of the energy is used to build cells and tissues. Some is released into the environment as heat. Only about 10 percent is available to the next level of the pyramid.

The **biomass**, or total mass of living matter at each trophic level, can also be modeled by an ecological pyramid. In a pyramid of biomass, each level shows the amount of biomass consumed by the level above it.

A pyramid of numbers shows the number of organisms consumed at each trophic level in an ecosystem. The number decreases at each level because less energy is available to support organisms.

Picture This

5. Explain How is mass measured on the pyramid of biomass?

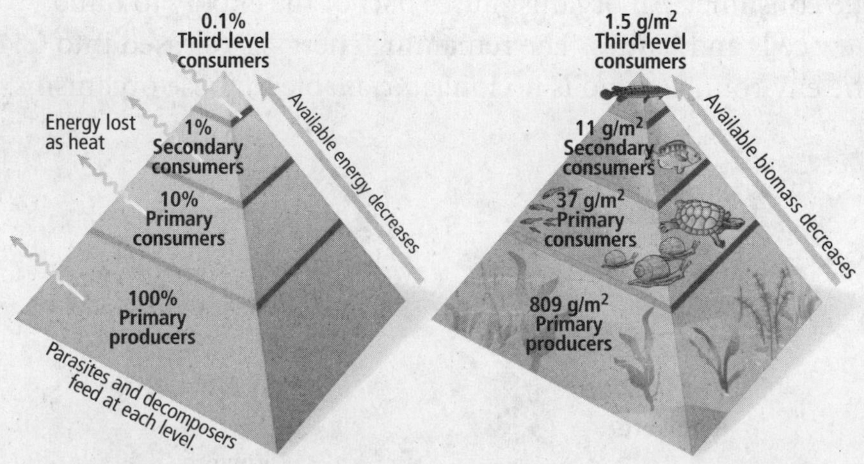

chapter 2 Principles of Ecology

section 3 Cycling of Matter

Before You Read

By looking at calendars, you can observe cycles, such as the cycle of the school year and summer vacation. On the lines below, write about cycles in your life. Read about the cycles in nature.

____

____

MAIN Idea

Nutrients move through the biotic and abiotic parts of an ecosystem.

What You'll Learn

- the importance of nutrients to living organisms
- the biogeochemical cycles of nutrients

Read to Learn

Cycles in the Biosphere

The law of the conservation of mass states that matter is not created or destroyed. Instead, matter is cycled through the biosphere. **Matter** is anything that takes up space and has mass.

Matter provides the nutrients needed for organisms to function. A **nutrient** is a chemical substance that an organism needs to perform life processes. An organism obtains nutrients from its environment. The bodies of all organisms are built from water and nutrients. Common nutrients include carbon, nitrogen, and phosphorus.

Mark the Text

Identify Main Ideas Circle the names of the cycles described in this section. Underline the text that summarizes the steps in each cycle.

How do nutrients cycle through the biosphere?

Nutrients cycle through the biosphere through organisms. Producers begin the cycle. In most ecosystems, plants obtain nutrients from air, water, and soil. Plants convert the nutrients into organic compounds that they use. Most capture energy from the Sun and convert it into carbohydrates. When a consumer eats a producer, the nutrients in the producer pass to the consumer. For example, the nutrients in green grass pass to the cow that eats the grass. The cycle continues until the last consumer dies. Detritivores return the nutrients to the cycle, and the process begins again.

FOLDABLES™

Take Notes Make a five-tab Foldable, as shown below. As you read, take notes and organize what you learn about five cycles in the biosphere.

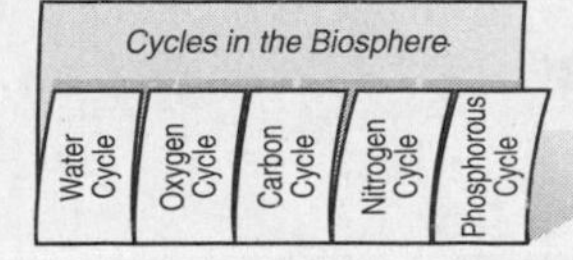

What is the biogeochemical cycle?

Both biological processes and chemical processes are needed to cycle matter in living organisms. The cycle also requires geological processes such as weathering. Weathering breaks down large rocks into small pieces. Plants and other organisms obtain nutrients from these pieces. Scientists use the name **biogeochemical cycle** to describe the combination of processes that exchange matter through the biosphere. ☑

Reading Check

1. Synthesize What three processes form the biogeochemical cycle?

____

____

____

How does water cycle?

Evaporation occurs when liquid water changes into water vapor—a gas—and enters the atmosphere. Water evaporates from bodies of water, from water in the soil, and from the surfaces of plants.

As water vapor rises, it begins to cool in the atmosphere. Clouds form when water vapor condenses into droplets around dust particles in the atmosphere. When the droplets become large and heavy, they fall from the clouds as precipitation. Precipitation can be in the form of rain, hail, sleet, or snow. Most falls directly back into the ocean. The figure below shows the water cycle. It is a model that describes how water moves from the surface of Earth to the atmosphere and back to the surface again.

Picture This

2. Identify Complete the figure by labeling the missing steps in the water cycle.

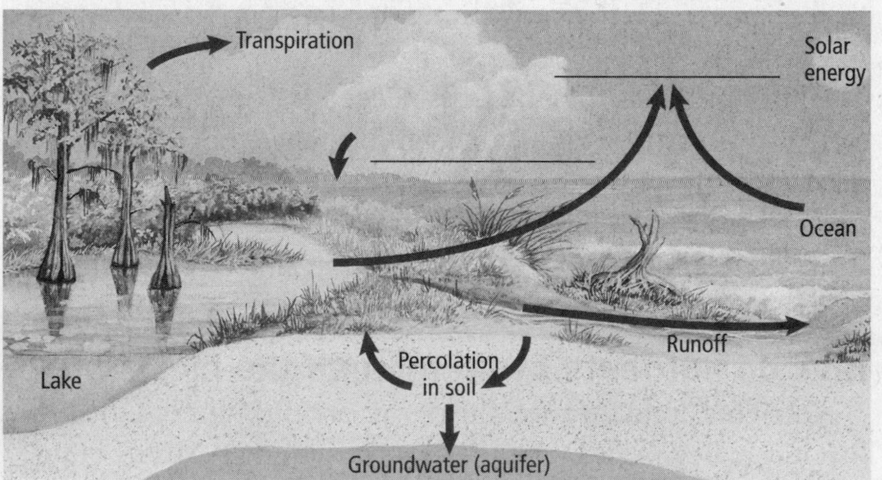

Why are carbon and oxygen important to organisms?

Living organisms are composed of molecules that contain carbon. Living things also need oxygen for many life processes. Carbon and oxygen make up molecules needed for life, including carbon dioxide and sugar.

What are the carbon and oxygen cycles?

During photosynthesis (foh toh SIHN thuh sus), producers change carbon dioxide into carbohydrates and release oxygen into the air. The carbohydrates are a source of energy for all organisms in a food web. Autotrophs and heterotrophs release carbon dioxide into the air during cellular respiration. Carbon and oxygen cycle quickly through living organisms.

Carbon is also part of a cycle that takes much longer. During a process that could take millions of years, carbon is converted into fossil fuels such as gas, peat, or coal. Carbon is released into the atmosphere in the form of carbon dioxide when fossil fuels are burned.

What is the nitrogen cycle?

Organisms need nitrogen to produce proteins. The atmosphere is 78 percent nitrogen. However, most organisms cannot use nitrogen directly from the air. Nitrogen gas is captured from the air by a species of bacteria, as shown in the figure below. These bacteria live in water, the soil, or grow on the roots of some plants. **Nitrogen fixation** is the process of capturing and changing nitrogen into a form that plants can use. Humans add nitrogen to the soil when they apply chemical fertilizers to a lawn or to crops.

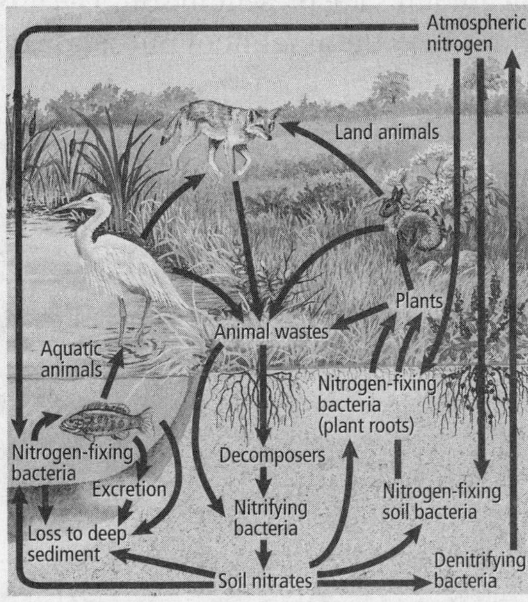

How does nitrogen enter food webs?

Nitrogen enters the food web through plants. Consumers get nitrogen by eating producers or other animals that contain nitrogen. At each step in the food web, organisms reuse nitrogen to make proteins. The amount of nitrogen is limited, and will often determine the growth of producers.

Think it Over

3. Summarize How do photosynthesis and cellular respiration differ?

Picture This

4. Determine What captures the atmospheric nitrogen?

What is denitrification?

Nitrogen returns to the soil when animals urinate and when organisms die and decay. When organisms die, decomposers break down matter in the organisms into a nitrogen compound called ammonia. Ammonia is changed by organisms in the soil into nitrogen compounds that can be used by plants. Some bacteria in the soil change nitrogen compounds into nitrogen gas in a process called **denitrification**. This process releases nitrogen into the atmosphere. ☑

Reading Check

5. **Name** the nitrogen compound that is changed by organisms in the soil.

____

What is the phosphorus cycle?

Organisms must have phosphorus to grow and develop. Large amounts are used to build bones and teeth. There are two phosphorus cycles—a short-term cycle and a long-term cycle. In the short-term cycle, phosphorus is cycled from the soil to producers to consumers. Phosphorus returns to the soil when organisms die or produce waste products, as shown in the figure below.

In the long-term cycle, phosphorus is added to soil from weathering or erosion of rocks that contain phosphorus. Weathering and erosion are long processes. They slowly add phosphorus to the soil. Phosphorus does not dissolve in water, and only small amounts are present in soil. The growth of producers is limited by the amount of phosphorus available to them.

Picture This

6. **Explain** how phosphates are added to water.

____

____

____

____

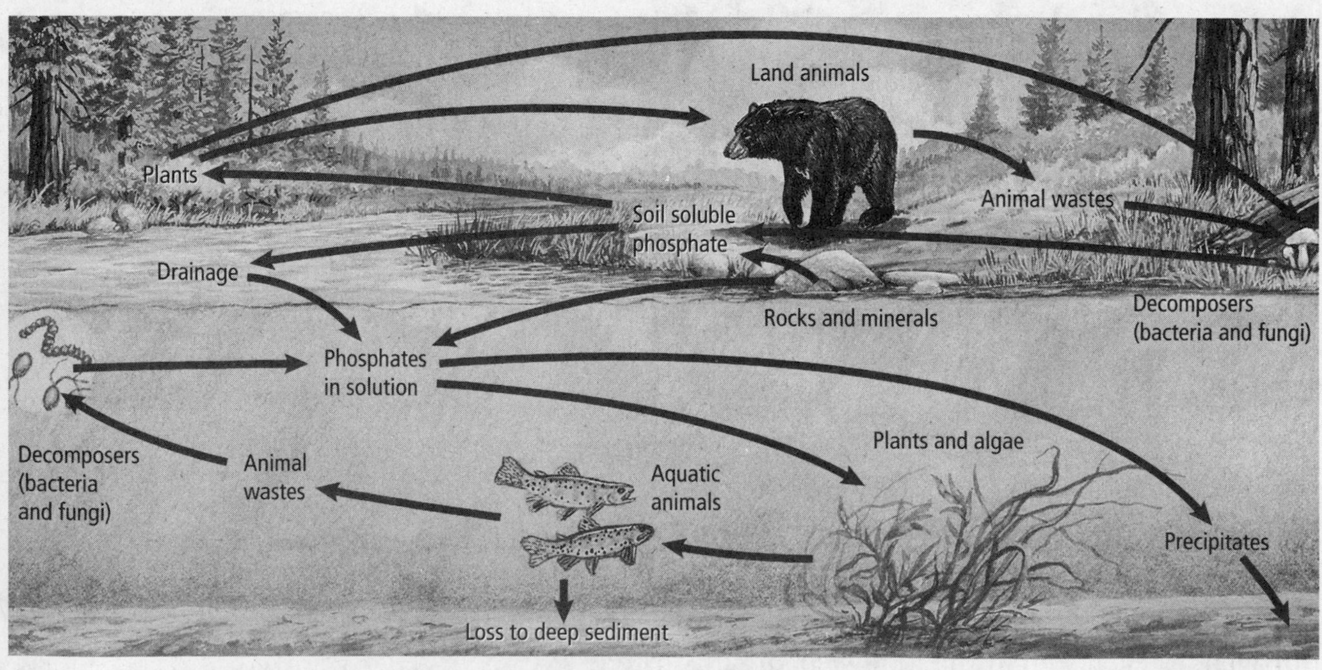

Communities, Biomes, and Ecosystems

section 1 Community Ecology

Before You Read

On the lines below, list several plants and animals that live in your community. Then name one organism that would have trouble surviving where you live. Read the section to learn why some species can live in an area while others cannot.

____

____

____

MAIN Idea

All living organisms are limited by factors in the environment.

What You'll Learn

- how ranges of tolerance affect the distribution of organisms
- the stages of primary and secondary succession

Read to Learn

Communities

Your biological community includes more than just the people around you. It also includes the plants, other animals, bacteria, and fungi in your area. A biological **community** is a group of interacting populations that occupy the same area at the same time. Organisms that live in a desert community are different from organisms that live in a polar community. Organisms that live in a city differ from organisms that live in the country.

You have read that abiotic factors affect individual organisms. Abiotic factors also affect communities. For example, soil is an abiotic factor. If soil becomes too acidic, some species might die. This might affect the food sources of other organisms. As a result, the community would change.

Organisms are adapted to the conditions where they live. A wolf's fur coat enables it to survive in cold winter climates. Depending on which factors are present and in what quantities, organisms can survive in some ecosystems but not in others.

Study Coach

Make an Outline Make an outline of the information you learn in this section. Start with the headings. Include the underlined terms.

Think it Over

1. **Predict** Suppose a fungus killed a species of tree in a forest community. What might happen to the woodpecker species that nests in that kind of tree?

____

____

____

What factors limit populations in communities?

Any abiotic factor or biotic factor that restricts the numbers, reproduction, or distribution of organisms is called a **limiting factor**. Abiotic limiting factors include sunlight, climate, water, fire, and space. Biotic limiting factors include other plant and animal species. Factors that limit one species might enable another to thrive. For example, water is a limiting factor. Organisms that need less water can survive in a desert community. ☑

Reading Check

2. Identify The average winter temperature in the Arctic is about −30°C. What type of limiting factor is this? (Circle your answer.)

a. abiotic

b. biotic

How does range of tolerance affect species?

For any environmental factor, there is an upper limit and a lower limit that defines the conditions in which an organism can live. **Tolerance** is the ability of any organism to survive when exposed to abiotic or biotic factors. The figure below shows a range of tolerance for steelhead trout. The limiting factor in this case is water temperature. Trout can tolerate water temperatures between 9°C and 25°C. Most trout live in the optimum zone, which is the temperature range that is best for trout survival. The zone of physiological stress lies between the optimum zone and the tolerance limits. Fewer trout live in this zone. Trout that do live in this zone experience physiological stress, such as the inability to grow.

Picture This

3. Explain what the curved graph line represents.

____

____

____

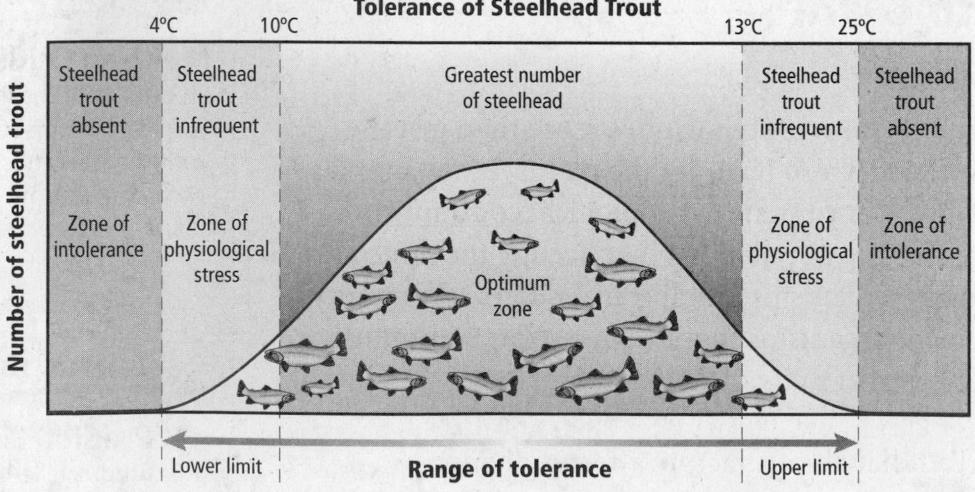

Ecological Succession

Ecosystems constantly change. A tree falling in a forest affects the forest ecosystem. A fire might alter the forest habitat so much that some species cannot survive while others can thrive. The process of one community replacing another as a result of changing abiotic and biotic factors is called **ecological succession**.

How does soil form in primary succession?

There are two types of ecological succession—primary succession and secondary succession. **Primary succession** is the establishment of a community in an area of bare rock that does not have topsoil. For example, suppose a lava flow alters an ecosystem. The lava hardens to form bare rock. Usually, lichens begin to grow on the rock first. Because lichens and some mosses are among the first organisms to appear, they are called pioneer species.

Pioneer species physically and chemically break down rocks. As pioneer species die, their decaying organic materials mix with small pieces of rock. This is the first stage of soil development. Small weedy plants begin to grow in the soil. These organisms die, adding to the soil. Seeds brought by animals, water, and wind begin to grow. Eventually, enough soil forms to support trees and shrubs. ☑

It might take hundreds of years for the ecosystem to become balanced and achieve equilibrium. When an ecosystem is in equilibrium, there is no net change in the number of species. New species come into the community at about the same rate that others leave the community. This is a **climax community**—a stable, mature community in which there is little change in the number of species.

How does secondary succession occur?

Disturbances such as fire or flood can disrupt a community. After a disturbance, new species of plants and animals might occupy the habitat. Over time, the species belonging to the climax community are likely to return. **Secondary succession** is the orderly and predictable change that takes place after a community of organisms has been removed but the soil remains. Pioneer species begin the process of restoring a habitat after a disruption. The figure below shows how the community changes after a forest fire, leading again to a mature climax community.

FOLDABLES™

Record Information Make a two-tab Foldable to record what you learn about the two types of ecological succession.

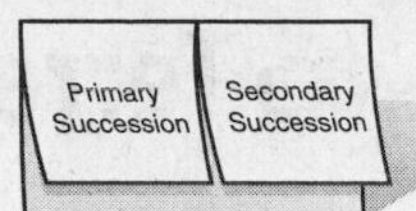

Reading Check

4. Summarize the importance of pioneer species in primary succession.

____

____

____

Picture This

5. Draw Conclusions How many years after a forest fire would trees begin to grow?

____

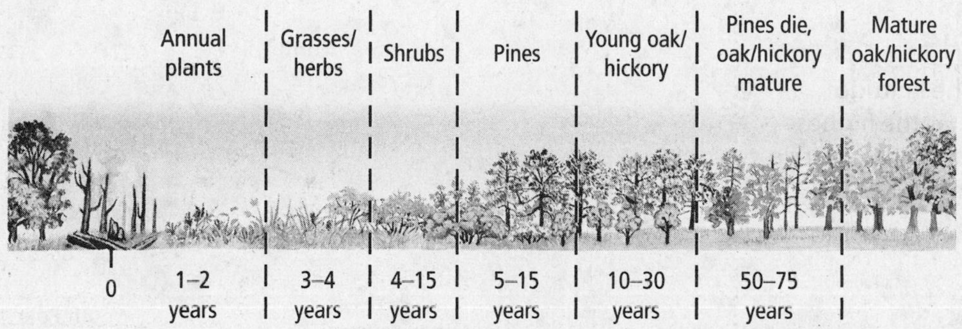

Communities, Biomes, and Ecosystems

section 2 Terrestrial Biomes

MAIN Idea

Ecosystems are grouped into biomes based on the plant communities within them.

What You'll Learn

- the three major climate zones
- the major abiotic factors
- how climate and biotic factors differ among land biomes

Before You Read

On the lines below, describe the climate in your area. Include seasonal differences in temperatures and precipitation. Then read the section to learn how climate influences the location of biomes.

____

____

Mark the Text

Identify Biomes Highlight each term that introduces a biome. Then use a different color to highlight important facts about each biome.

Read to Learn

Effects of Latitude and Climate

Weather is the condition of the atmosphere at a specific place and time. **Climate** is the average weather conditions in an area, including temperature and precipitation. An area's latitude has a large effect on its climate. **Latitude** is the distance of any point on the surface of Earth north or south of the equator. The equator is 0° latitude. The poles, which are the farthest points from the equator, are 90° latitude.

As shown in the figure below, sunlight strikes Earth more directly at the equator than at the poles. As a result, different areas of Earth's surface are heated differently. Ecologists call these areas polar, temperate, and tropical zones.

Picture This

1. **Identify** the latitude below that has the highest average temperature. (Circle your answer.)
   - a. 30°S
   - b. 60°N
   - c. 90°S

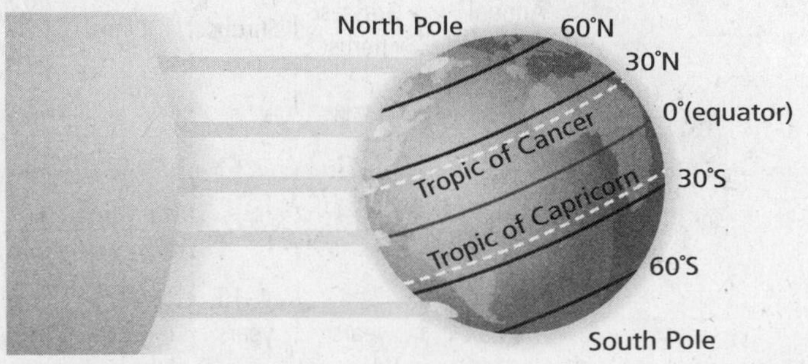

What other factors affect biome location?

Recall that a biome is a large group of ecosystems that share the same climate and have similar types of communities. The plant and animal communities are adapted to the area's climate. Factors that determine climate include elevation, landmass features, winds, and ocean currents.

Major Land Biomes

Biomes are classified mainly by the characteristics of their plants. This section describes each major land biome.

What are the characteristics of tundra?

South of the Arctic ice cap, a band of tundra runs across the regions of northern North America, Europe, and Asia, as shown in the figure below. **Tundra** is a treeless biome with a layer of permanently frozen soil called permafrost beneath the surface. Mosses, lichens, and short grasses can grow in the tundra. ☑

Reading Check

2. Define permafrost.

Where is the boreal forest?

As shown in the figure below, the taiga extends across North America, Europe, and Asia, south of the tundra. The **boreal forest**, or taiga, is a broad band of dense evergreen forest. It is also known as the northern coniferous forest. The boreal forest does not have permafrost because temperatures are a bit warmer and summers are longer than in the tundra. Spruce, fir, and pine trees and low-growing shrubs and bushes grow in the boreal forest.

Picture This

3. Identify Does any tundra lie within the United States? If so, where?

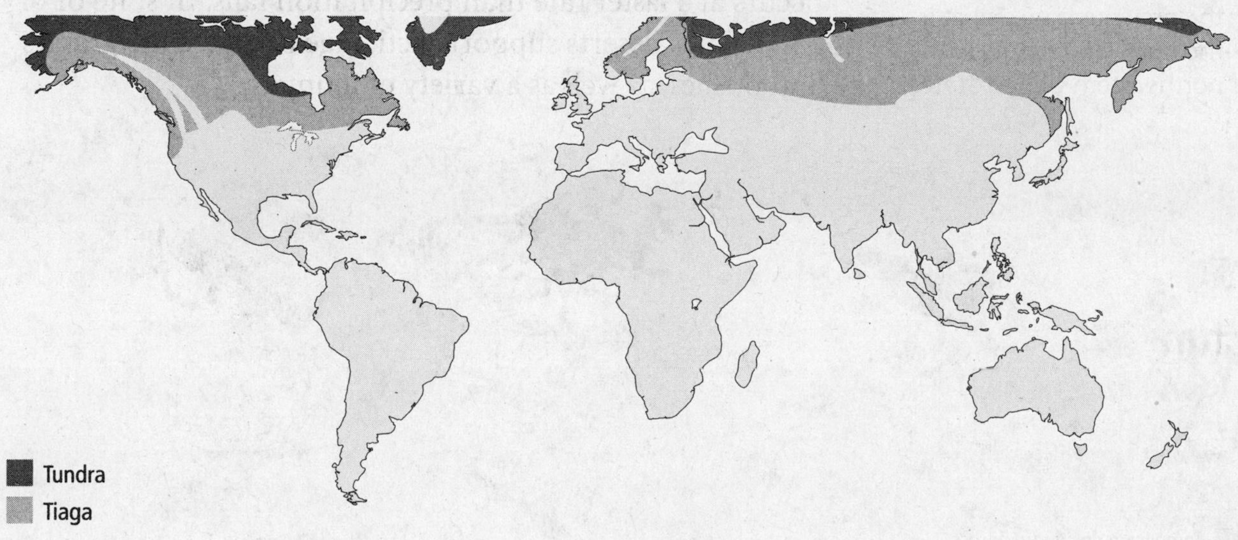

What trees thrive in the temperate forest?

Temperate forests are located south of the boreal forests. **Temperate forests** are composed mostly of broad-leaved deciduous (dih SIH juh wus) trees. These are trees that shed their leaves over a short period. Areas of temperate forest have four seasons, with hot summers and cold winters. As shown in the figure below, temperate forests cover much of southeastern Canada, eastern North America, most of Europe, and parts of Asia and Australia.

Where do woodlands and shrublands occur?

Woodlands occur in areas surrounding the Mediterranean Sea, along the western coasts of North and South America, in South Africa, and in Australia. Open **woodlands** and shrub communities receive less rainfall than temperate forests. Areas with mostly shrubs are called chaparral. Summers are hot and dry. Winters are cool and wet. ☑

Reading Check

4. Contrast How do woodlands differ from temperate forests?

What keeps grasslands from becoming forests?

Temperate grasslands are found on many continents. Grasslands are called steppes in Asia; prairies in North America; pampas and llanos in South America; savannahs and velds in Africa; and rangelands in Australia. The fertile soils of **grasslands** support a thick cover of grasses. Grazing animals and fires keep grasslands from becoming forests. Grasslands have hot summers and cold winters.

Do any plants and animals live in a desert?

As shown in the figure below, deserts exist on every continent except Europe. A **desert** is any area where evaporation occurs at a faster rate than precipitation falls. In spite of dry conditions, deserts support cacti, sage brush, some grasses and bushes, as well as a variety of animals.

Picture This

5. Classify Based on the map below and the map on the previous page, what is the main land biome in the northeastern United States?

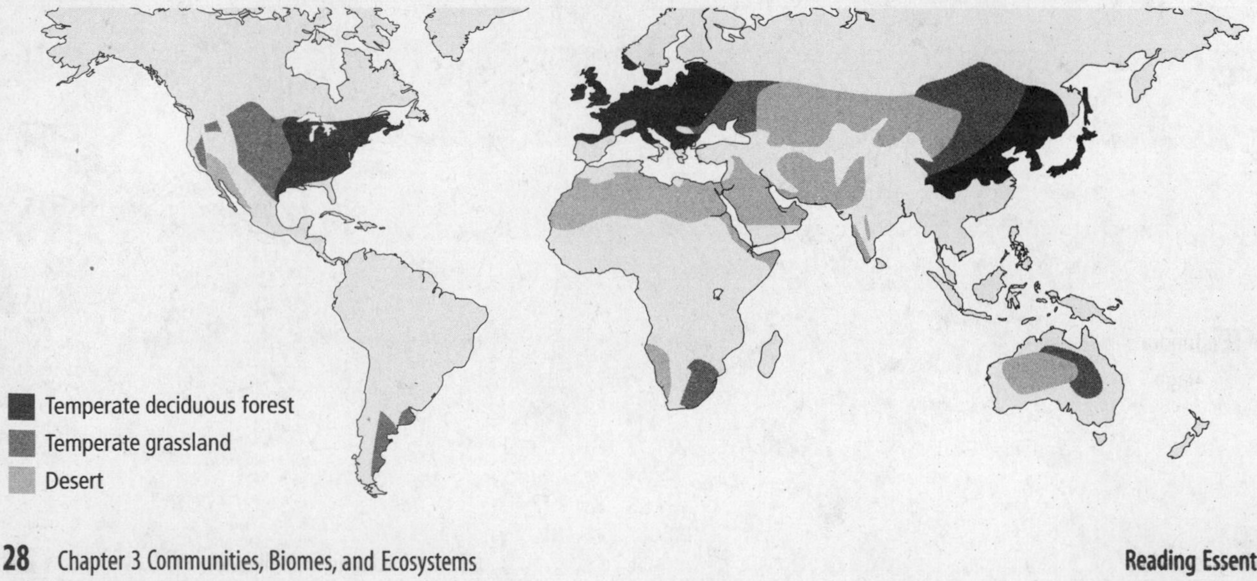

Temperate deciduous forest
Temperate grassland
Desert

What are features of a tropical savanna?

Tropical savannas occur in Africa, South America, and Australia. A **tropical savanna** is characterized by grasses and scattered trees in climates with less precipitation than other tropical areas. Summers are hot and rainy. Winters are cool and dry.

Where are tropical seasonal forests?

Tropical seasonal forests occur in Africa, Asia, Australia, and South and Central America. **Tropical seasonal forests**, also called tropical dry forests, have a wet season and a dry season. During the dry season, almost all of the trees drop their leaves to conserve water.

What biome supports the most diversity?

Tropical rain forests are found in the equatorial regions of Central and South America, Asia, western Africa, and Australia. A **tropical rain forest** has warm temperatures and lots of rainfall throughout the year. The tropical rain forest is the most diverse of all biomes. Tall trees covered with mosses, ferns, and orchids form the canopy, or upper layer. Shorter trees, shrubs, and creeping plants make up the understory, or lower layer.

Other Terrestrial Areas

Mountains do not fit the definition of a biome because their climate and plant and animal diversity are determined by elevation. Polar regions are also not true biomes because they are covered by ice masses and lack land areas with soil.

How do conditions change with elevation?

If you climb a mountain, you might notice that temperatures fall as you climb higher. Also, precipitation varies as you climb. As a result, many communities are able to exist on a mountain. Grasslands are at the bottom, pine trees grow farther up, and the cold elevations at the top support communities similar to the tundra.

Do polar regions support life?

A thick layer of ice covers the polar regions. In spite of year-round cold, polar regions support life. Polar bears and arctic foxes live in the arctic polar region in the north. Antarctica, in the south, supports colonies of penguins. Whales and seals prey on penguins, fish, and shrimplike krill in the coastal waters of Antarctica.

Think it Over

6. Contrast How does rainfall differ between tropical rain forests and tropical seasonal forests?

____
____
____
____
____

Think it Over

7. Draw Conclusions Name one adaptation of polar bears that helps them survive in their polar home.

____
____
____

Communities, Biomes, and Ecosystems

section 3 Aquatic Ecosystems

MAIN Idea

Aquatic ecosystems are grouped by abiotic factors.

What You'll Learn

- how depth and water flow affect freshwater ecosystems
- how to identify transitional aquatic ecosystems
- the zones of marine ecosystems

Before You Read

On the lines below, list some characteristics of a body of water near you. How deep is it? Is the water salty? Is it calm or fast flowing? Then read the section to learn the characteristics of different water ecosystems.

____

____

Study Coach

Make Flash Cards Think of a quiz question for each paragraph. Write the question on one side of the flash card. Write the answer on the other side. Use the flash cards to quiz yourself until you know all the answers.

Read to Learn

The Water on Earth

Most of Earth is covered with water. Aquatic ecosystems include freshwater, transitional, and marine ecosystems.

Freshwater Ecosystems

Ponds, lakes, streams, rivers, and wetlands are freshwater ecosystems. The graph on the left below shows that only about 2.5 percent of Earth's water is freshwater. The graph on the right shows that 68.9 percent of the freshwater is contained in glaciers, 30.8 percent is groundwater, and 0.3 percent is found in lakes, ponds, rivers, streams, and wetlands. Almost all freshwater species live in the 0.3 percent.

Picture This

1. Calculate the percentage of freshwater that is not ice. Show your work.

____

____

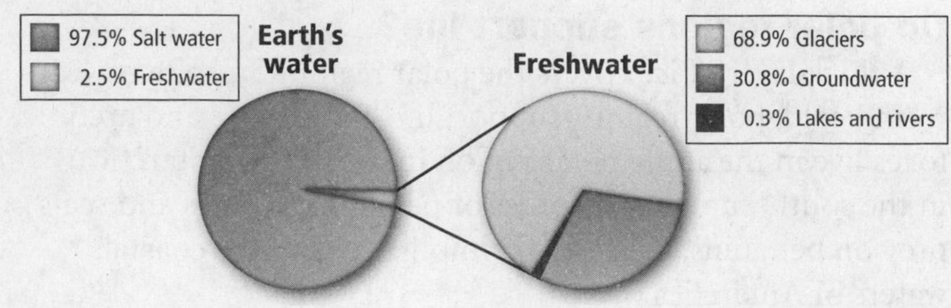

What affects water flow in rivers and streams?

The water in rivers and streams flows in one direction. As illustrated in the figure below, the water flow begins at a source called a headwater. The water flows to the mouth, where it empties into a larger body of water. Rivers and streams also might start from underground springs or from melting snow.

The slope of the land determines direction and speed of the water flow. Water flows quickly down a steep slope. Fast-flowing water picks up a lot of sediment. **Sediment** is material left by water, wind, or glaciers. As the slope levels, the fast-flowing water slows. This causes the sediment to be deposited in the form of sand, silt, and clay.

Rivers and streams change during their journey from source to mouth. Wind stirs up the water's surface and adds oxygen to the water. Water erodes the land, changing the path of the river or stream.

Currents of fast-moving rivers and streams prevent organic materials and sediments from building up. As a result, fewer species live in rapid waters. Organisms living in rivers and streams must be able to withstand the water current. Plants take root in streambeds where rocks and sand bars slow the water flow. In slow-moving water, insect eggs and larvae are the main food source for many fish. Calm water also provides a home for crabs, tadpoles, and frogs.

Picture This

2. Label the area of the river where most sediment will be deposited. On the lines below, explain why.

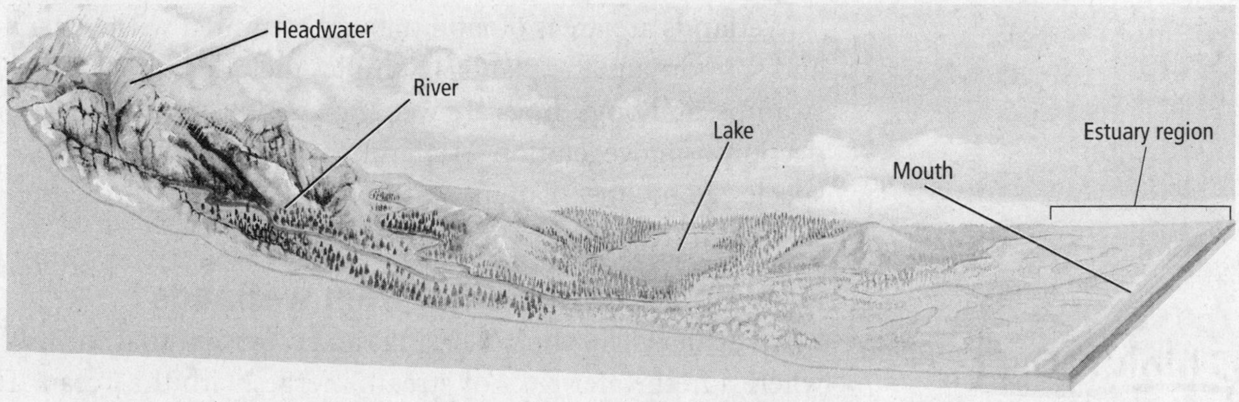

How does altitude affect life in lakes and ponds?

Some lakes and ponds last only a couple of weeks every year. Other lakes might exist for thousands of years. Nutrient-poor lakes, called oligotrophic (uh lih goh TROH fihk) lakes, are found high in the mountains. Few plant and animal species live in these lakes. Many plant and animal species live in nutrient-rich lakes, called eutrophic (yoo TROH fihk) lakes, at lower elevations. ☑

3. Identify the limiting factor in oligotrophic lakes.

What distinguishes zones in lakes and ponds?

Lakes and ponds are divided into three zones that are determined by depth and distance from the shoreline. The area closest to shore is the **littoral** (LIH tuh rul) **zone**. Species in this zone includes algae, rooted and floating plants, snails, clams, insects, fishes, and amphibians. Some insect species lay eggs in the littoral zone and the larvae develop there.

The **limnetic** (lihm NEH tihk) **zone** is the open water area. It is well lit and full of plankton. **Plankton** are free-floating photosynthetic autotrophs that live in freshwater or marine ecosystems. Many species of freshwater fish live in the limnetic zone because food is plentiful there.

The deepest area of a lake is the **profundal** (pruh FUN dul) **zone**. It is much colder and has less oxygen than the other two zones. Less light reaches the profundal zone, which limits the species that are able to live there.

Think it Over

4. Explain why you would expect to find few plankton in the profundal zone.

____

____

____

Transitional Aquatic Ecosystems

Transitional aquatic ecosystems are a combination of two or more different environments. Transitional aquatic ecosystems can be areas where land and water mingle. They can also be areas where salt water and freshwater mix. Examples of transitional aquatic ecosystems are wetlands and estuaries.

What kinds of life thrive in wetlands?

Wetlands are areas of land that are saturated with water and support aquatic plants. Examples include marshes, swamps, and bogs. Bogs are wet and spongy areas of decomposing vegetation. Wetlands support a diversity of species. Pond lilies, cattails, amphibians, reptiles, birds, and mammals live in wetlands.

How do estuaries differ from wetlands?

An **estuary** (ES chuh wer ee) is an ecosystem that forms where a freshwater river or stream merges with the ocean. The mixing of waters with different salt concentrations creates a unique ecosystem. Algae, seaweed, and marsh grasses thrive in estuaries. Animals such as worms, oysters, and crabs feed on tiny organic matter called detritus (dih TRY tus). Many species of fishes, shrimp, ducks, and geese use estuaries as nurseries for their young.

Salt marshes are transitional ecosystems similar to estuaries. Salt-tolerant grasses live along the shoreline. Animals such as shrimp and shellfish live in salt marshes.

Think it Over

5. Contrast How do estuaries differ from wetlands?

____

____

____

____

Marine Ecosystems

Marine ecosystems have a major impact on the planet. For example, marine algae consume large amounts of carbon dioxide from the atmosphere. In the process, they supply much of the oxygen in the atmosphere. Also, water that evaporates from the oceans eventually provides most of Earth's precipitation—rain and snow. Oceans are separated into zones, as shown in the figure below. ☑

How do the tides affect the intertidal zone?

The **intertidal** (ihn tur TY dul) **zone** is a narrow band where the ocean meets land. As tides and waves move in, the intertidal zone is submerged. As tides and waves move out, the intertidal zone is exposed. Only a few species of algae and mollusks live where the highest tides reach. A diversity of species, including algae and small animals such as snails, crabs, sea stars, and fishes, live in areas that are submerged during high tide. The bottom of the intertidal zone is exposed only during the lowest tides. Many species of invertebrates, fishes, and seaweed live here. On sandy coasts, waves constantly shift the sand. The constant shifting makes it hard for algae and plants to grow on sandy beaches. Animals that live on beaches include worms, clams, predatory crustaceans, crabs, and shorebirds.

Reading Check

6. Describe two important ways that marine ecosystems impact the planet.

____

____

____

Picture This

7. Identify the zone in the figure that the tide does not submerge.

____

____

How do layers of the pelagic zone differ?

The open ocean is divided into the pelagic (puh LAY jihk) zone, abyssal (uh BIH sul) zone, and benthic zone. The **photic zone** is the area in the pelagic zone from the surface of the water down to about 200 m. The photic zone is shallow enough for sunlight to penetrate. As depth increases, light decreases. The photic zone supports seaweed, plankton, fishes, turtles, jellyfish, whales, and dolphins.

Below the photic zone lies the **aphotic zone** where sunlight cannot penetrate. This region of the pelagic zone remains in constant darkness. Organisms that depend on sunlight for energy cannot live in the aphotic zone. The water in the aphotic zone is generally cold.

Where are the benthic and abyssal zones?

The **benthic zone** is the area along the ocean floor. It consists of sand, silt, and dead organisms. In shallow areas, sunlight can penetrate to the ocean floor. As depth increases, less sunlight can penetrate and temperatures decrease. As a result, species diversity also decreases as depth increases. Many species of bacteria, fungi, sponges, sea anemones, and fishes live in shallower parts of the benthic zone. ☑

Reading Check

8. Name two limiting factors as depth increases in the benthic zone.

____

____

The **abyssal zone** is the deepest region of the ocean. The water is very cold. Most organisms depend on pieces of food that drift down from the zones above. Hydrothermal vents on the seafloor release hot water, hydrogen sulfide, and other minerals. Communities of bacteria live around these vents. These bacteria can use the sulfide molecules for energy.

What organisms do coral reefs support?

A coral reef is an ecosystem that exists in warm, shallow marine waters. The hard, stony structure of the reef is formed by secretions of tiny animals—coral polyps. Most coral polyps have a symbiotic relationship with algae. The algae provide corals with food. In turn, the corals provide algae with protection and access to light. Corals also feed by extending tentacles to catch plankton. Sea slugs, octopuses, sea urchins, sea stars, and fishes are part of the great diversity of the coral reef. ☑

Reading Check

9. Define What is a coral? (Circle your answer.)

a. a plant
b. an animal
c. a colorful shell

Like all ecosystems, a coral reef is sensitive to changes in the environment. A natural event such as a tsunami as well as human activity such as land development can damage or kill a coral reef. Ecologists monitor coral reef environments to help protect them from harm.

Population Ecology

section 1 Population Dynamics

Before You Read

On the lines below, explain why animals that live in your area might be found in greater or smaller numbers than in other places on Earth. Then read the section to learn about factors that limit the growth of any population.

____

____

MAIN Idea

Populations of species are described by density, spatial distribution, and growth rate.

What You'll Learn

- concepts of carrying capacity and limiting factors
- ways in which populations are distributed

Read to Learn

Population Characteristics

Every organism belongs to a population. A population is a group of organisms of the same species that live in a specific area. Populations of organisms include plants, animals, and bacteria. All populations have certain characteristics, such as population density, spatial distribution, and growth rate.

Mark the Text

Identify Concepts Highlight each question heading in this section. Then use a different color to highlight the answers to the questions.

What are common patterns of dispersion?

Population density is the number of organisms per unit area. For example, there was an average of four American bison per square kilometer in Northern Yellowstone in 2000.

Dispersion is the pattern of spacing of a population within an area. The figure below shows three main types of dispersion—uniform, clumped, and random. Black bears are dispersed in a uniform, or even, arrangement. American bison are dispersed in clumped groups or herds. White-tailed deer are dispersed in random groups.

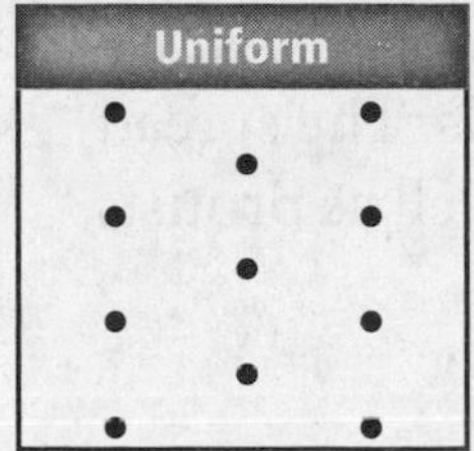

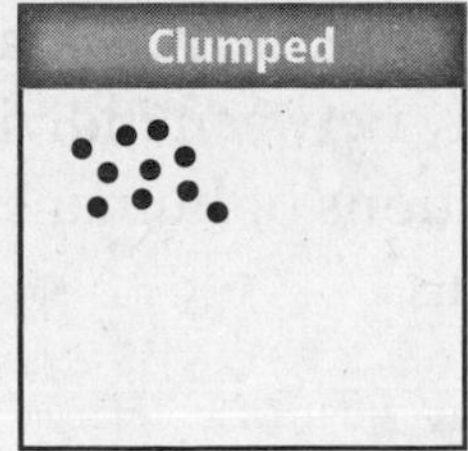

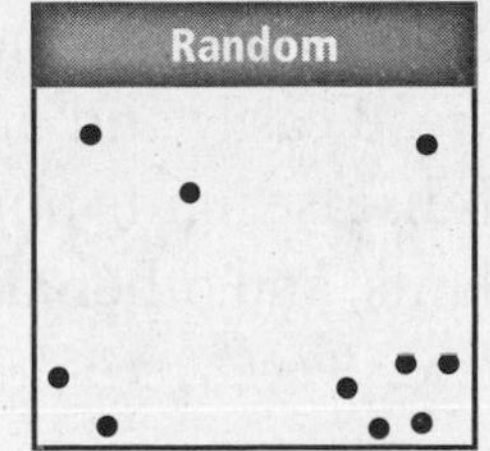

Picture This

1. **Apply** Wolverines spread across their range, with each individual patrolling a territory of about 320 km^2. What type of dispersion do wolverines represent?

____

What limits spatial distribution?

No population lives in all habitats of the biosphere. A species might not be able to expand its spatial distribution because it cannot survive the conditions in the new area. Abiotic factors, such as temperature, humidity, and rainfall, could make the new area unlivable for a species. Biotic factors, such as predators and competitors, also might prevent a species from surviving in the new area.

FOLDABLES™

Record Information
Make a layered Foldable from two sheets of paper to record what you learn about how populations of species are described.

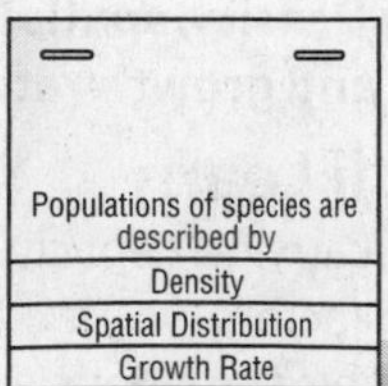

Population-Limiting Factors

All species have limiting factors. Limiting factors keep a population from growing indefinitely. For example, the food supply is a limiting factor. The number of individuals in a population cannot increase beyond the amount of food available to support that number.

There are two categories of limiting factors. They are density-independent factors and density-dependent factors.

What limiting factors are density independent?

Recall that population density is the number of members of a population per unit area. A **density-independent factor** is any factor in the environment that does not depend on population density. Usually these factors are abiotic. For example, populations are limited by weather events such as drought, floods, and hurricanes.

Human activities can also be density-independent limiting factors. For example, dam building alters the water flow of rivers, limiting some species. Pollution resulting from human activities reduces the available resources by making air, water, and land toxic in some areas. ☑

✔ Reading Check

2. Explain why water pollution is a density-independent factor.

____

____

____

What limiting factors are density dependent?

A **density-dependent factor** is any factor in the environment that depends on population density. Often these are biotic factors, such as disease, competition, parasites, and predators.

Disease Outbreaks of disease tend to occur when a population has increased and population density is high. When population density is high, individuals come into contact more frequently. Frequent contact enables disease to spread easily and quickly between individuals. The spread of disease limits populations of humans as well as protists, plants, and other animals.

Think it Over

3. Apply Which is an example of a density-dependent factor? (Circle your answer.)

a. frost that destroys tomato plants

b. fungus that spreads from plant to plant

Competition High population density increases competition among individuals for resources. When a population grows to a size that food and space become limited, individuals must compete for the available resources. Competition occurs within a species or between different species that use the same resources. As a result of competition, some individuals might die of starvation. Others move to different areas in search of resources. As population density decreases, competition decreases.

Parasites When population density is high, parasites spread in a way similar to the way disease spreads. The spread of parasites limits population growth.

Predators The figure below illustrates how the interaction of predators and prey limits the populations of both groups. Before the winter of 1947, there were no wolves on Isle Royale, located in Lake Superior. That winter, a pair of wolves crossed the ice on Lake Superior and reached the island. With plenty of moose available as prey, the wolf population increased. Follow the events in the figure below to see how the populations of wolves and moose depend on one another. As the population density of one decreases, the population density of the other increases.

Picture This

4. Predict how the cycle might change for the moose population if the wolves were removed from Isle Royale.

____

____

____

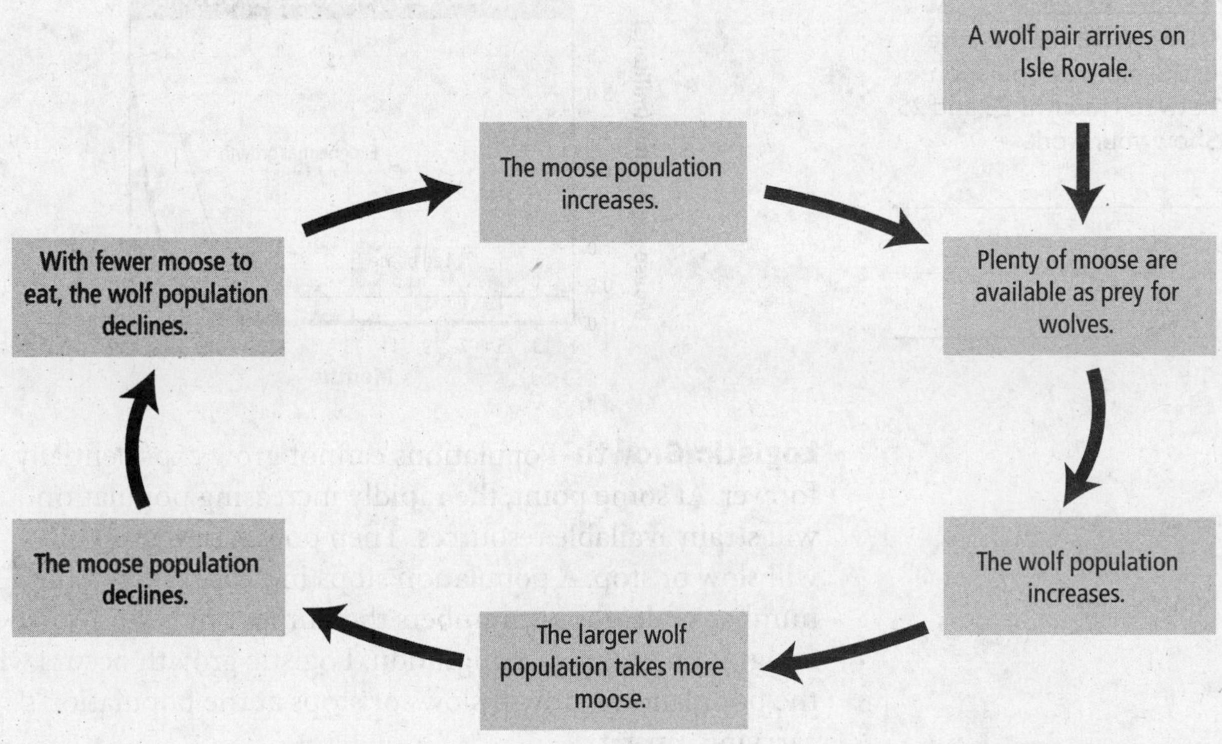

What factors affect a population's growth rate?

The **population growth rate** is a measure of how fast a given population grows. Two factors that influence a population's growth rate are birthrate and death rate. Birthrate, or natality, is the number of individuals that are born in a given time period. Death rate, or mortality, is the number of individuals that die in a given time period.

Emigration and immigration also affect the rate of population growth. **Emigration** (em uh GRAY shun) is the number of individuals moving away from a population. **Immigration** (ih muh GRAY shun) is the number of individuals moving into a population. ☑

Reading Check

5. Name four factors that influence a population's growth rate.

______

______

______

Exponential Growth The graph below shows how a population of mice would grow if there were no environmental limiting factors. The graph starts with a population of two adult mice. They breed and have a litter of young. At first, the population increases slowly. This slow period is the lag phase in the graph. Without limiting factors, all of the young survive and breed. The population increases rapidly, or grows exponentially. All populations grow exponentially until some limiting factor slows the growth. Notice that exponential growth gives the graph a J shape.

Picture This

6. Calculate the increase in the mouse population between months 23 and 25. Show your work.

______

______

______

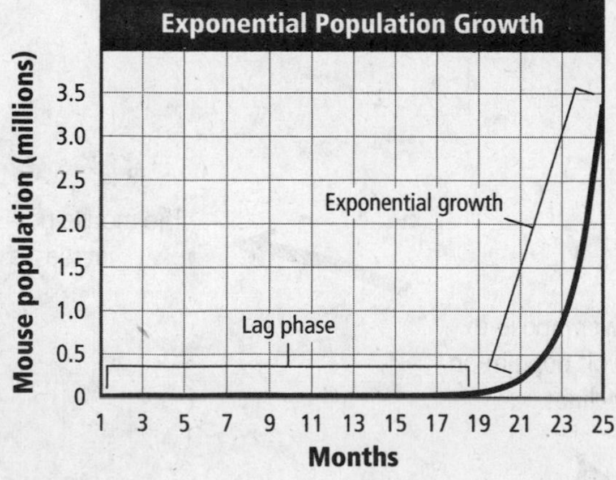

Logistic Growth Populations cannot grow exponentially forever. At some point, the rapidly increasing population will strain available resources. Then population growth will slow or stop. A population stops increasing when the number of deaths outnumbers the number of births or when emigration exceeds immigration. Logistic growth occurs when the population's growth slows or stops at the population's carrying capacity.

Carrying Capacity The graph below shows logistic growth. Notice that the graph begins in a J-shaped pattern of exponential growth, as in the previous graph. Then limiting factors slow population growth, causing the graph to bend into an S-shape. This S-pattern is typical of logistic growth. The population stops growing at the carrying capacity, as shown on the graph. The **carrying capacity** is the maximum number of individuals in a species that an environment can support for the long term. Carrying capacity is limited by resources such as water, oxygen, and nutrients.

When populations develop in an area with plenty of resources, there are more births than deaths. The population reaches the carrying capacity, and resources become limited. If a population becomes larger than the carrying capacity, there will be more deaths than births because there are not enough resources to support the population. The population falls below the carrying capacity as individuals die. Populations tend to stabilize near their carrying capacity.

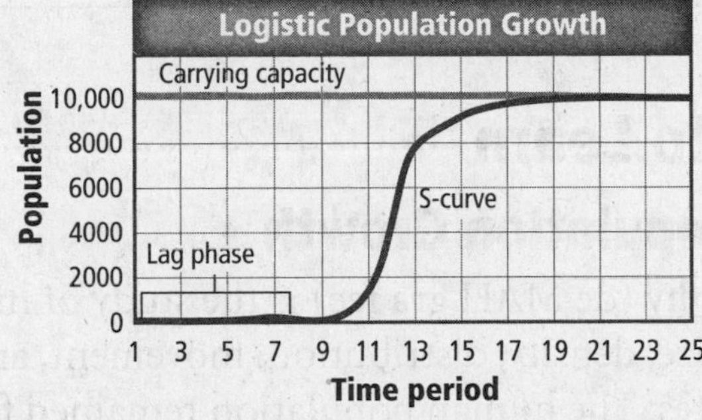

How do reproductive strategies differ?

Species vary in their reproductive factors, such as the number of offspring born during each reproductive cycle, the age that reproduction begins, and the life span of the organism. Both plants and animals are placed into reproductive strategies based on their reproductive factors.

Rate strategists, or *r*-strategists, are small organisms. They usually have short lives. They produce as many offspring as possible and do not nurture them. Typically, the population of *r*-strategists is controlled by density-independent factors and does not stay near the carrying capacity.

Carrying-capacity strategists, or *k*-strategists, are large organisms. They usually have long lives. They produce few offspring and nurture them. Typically, the population of *k*-strategists is controlled by density-dependent factors and stays near the carrying capacity. ☑

Picture This

7. Identify For the population represented on the graph, what is the maximum number of individuals that the environment can support over a long time period?

____

Reading Check

8. Summarize the reproductive strategies of *r*-strategists and *k*-strategists for ensuring continuation of their species.

____

____

____

____

chapter 4

Population Ecology

section 2 Human Population

MAIN Idea

Human population growth changes over time.

What You'll Learn

- the age structure of nongrowing, slowly growing, and rapidly growing countries
- consequences of continued population growth

Before You Read

Think about the characteristics of populations that you read about in Section 1. On the lines below, explain how these characteristics might apply to human populations. Then read the section to learn about human populations.

Study Coach

Create a Quiz As you read this section, write quiz questions based on what you have learned. After you write the questions, answer them.

Read to Learn

Human Population Growth

Demography (de MAH gra fee) is the study of human population size, density, distribution, movement, and birth and death rates. The human population remained fairly stable for thousands of years, but it has recently increased.

How has technology affected growth?

Humans have learned to change their environment in ways that increase the carrying capacity. Agriculture and domestication of animals have increased the human food supply. Technological advances, such as in medicine and shelter construction, have reduced the death rate. The current growth rate is just over 80 million people per year. By 2050, the population is expected to be nine billion.

Although the human population is growing, the rate of growth has slowed. Human population growth peaked at over 2.2 percent in 1962. By 2003, the rate of growth had dropped to almost 1.2 percent. The decline in growth is due primarily to diseases such as AIDS and voluntary population control.

Think it Over

1. **Describe** a technology, not yet invented, that could increase the human carrying capacity.

Trends in Human Population Growth

Events such as disease and war can change population trends. Human population growth is not the same in all countries. However, countries with similar economies tend to have similar population growth trends. ☑

For example, one trend is a change in the population growth rate in industrially developed countries, such as the United States. An industrially developed country has advanced industry and technology and a high standard of living. Early in its history, the United States had a high birthrate and a high death rate. Many children died before reaching adulthood. Typically, individuals died by their early forties. In recent years, the birthrate in the United States has decreased a lot. The average lifespan is now more than 70 years. A change from high birth and death rates to low birth and death rates in a population is called a **demographic transition**.

How is population growth rate calculated?

As an example, we will calculate and compare the 2008 population growth rates for the United States and Honduras, a small country in Central America. The calculation for PGR is [birthrate − death rate + migration rate] = PGR (%)

In our example, we'll have to divide the final answer by 10 to get a percentage because the rates are calculated per 1000. The United States has birthrate 14.1 (per 1000), death rate 8.3 (per 1000), and migration rate 2.9 (per 1000). This gives a PGR of 0.87 percent for the United States.

Honduras has birthrate 26.9 (per 1000), death rate 5.4 (per 1000), and migration rate −1.3 (per 1000). This gives a PGR of 2.02 percent for Honduras.

What is zero population growth?

Zero population growth (ZPG) occurs when the birthrate equals the death rate. According to one estimate, the world will reach zero population growth between 2020 and 2029. Although the population will have stopped growing, births and deaths will continue at the same rate. At zero population growth, the number of people in different age groups should be nearly equal.

Reading Check

2. List two events that can change human population trends.

____

____

Applying Math

3. Apply Canada has birthrate 10.2 (per 1000), death rate 7.7 (per 1000), and migration rate 5.6 (per 1000). Calculate the PGR for Canada.

____

____

____

How does age structure predict growth?

A population's **age structure** is the number of males and females in each of three age groups. The groups are pre-reproductive (up to age 20), reproductive (ages 20 through 44), and post-reproductive (after age 44).

The figure below shows the age structure for three countries. Compare the shapes of the three diagrams. When the largest portion of the population is in the pre-reproductive stage, as in Kenya, the population is growing rapidly. When the smallest portion is pre-reproductive, as in Germany, the population is decreasing. When the reproductive and pre-reproductive groups are roughly equal, as in the United States, the population is growing slowly.

Picture This

4. Label the pre-reproductive stage that represents the largest and the one that represents the smallest portion of its country's population.

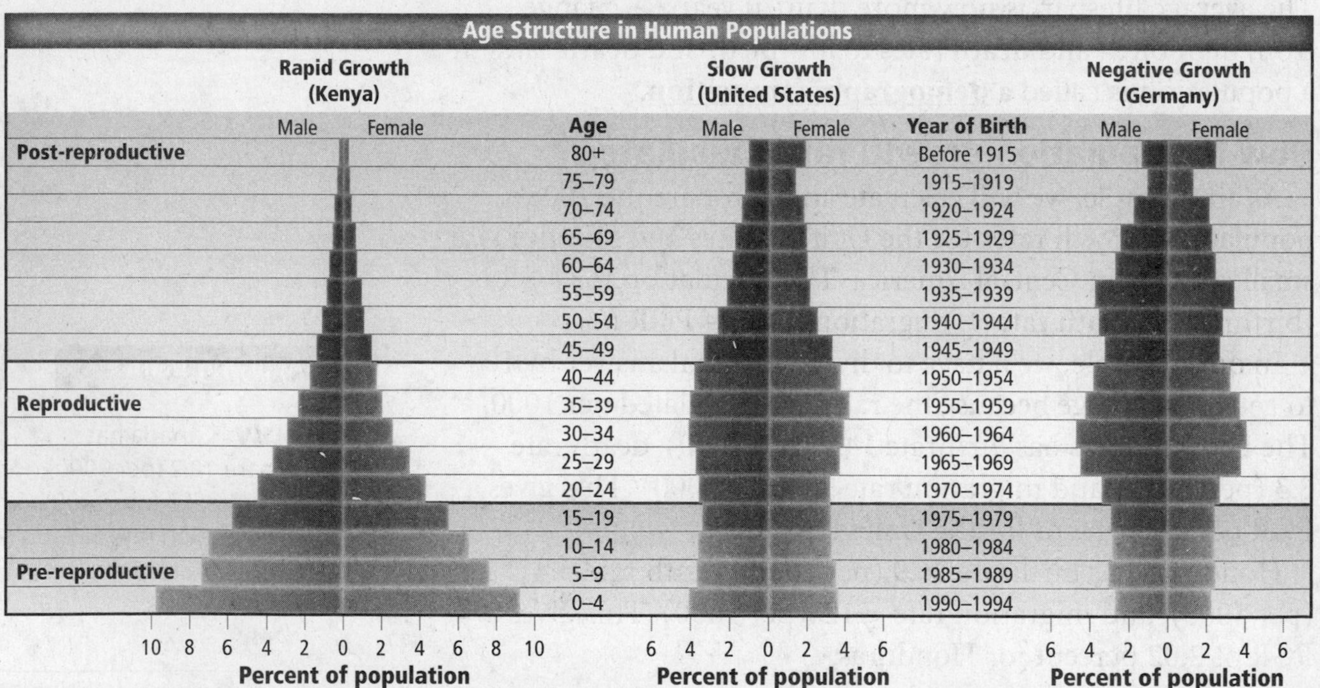

Why is human population growth a concern?

All populations have carrying capacities, including human populations. Scientists are concerned that the human population might exceed Earth's ability to support it. Like populations of other organisms, human overcrowding will lead to disease and starvation. Family planning in many countries is being used to slow the growth rate. ☑

Currently, individuals in industrially developed countries use far more resources than individuals in developing countries. Populations in developing countries are increasing rapidly. As these countries industrialize, resource use will also increase rapidly.

Reading Check

5. Summarize the possible consequences of rapid human population growth.

____

____

____

____

Biodiversity and Conservation

section 1 Biodiversity

Before You Read

Think about the different organisms that live in your area. On the lines below, list as many of them as you can. Then read about the importance of biological diversity.

____

____

____

MAIN Idea

Biodiversity maintains a healthy biosphere.

What You'll Learn

- the three types of diversity
- why biodiversity is important
- the direct and indirect value of biodiversity

Read to Learn

What is biodiversity?

Biodiversity is the variety of life in one area that is determined by the number of different species in that area. The variety of species in the biosphere decreases as species become extinct. **Extinction** occurs when the last member of a species dies.

Biodiversity increases the health and stability of an ecosystem. Three important types of biodiversity are genetic diversity, species diversity, and ecosystem diversity.

Why is genetic diversity important?

Two individuals of the same species will show differences. For instance, two ladybird beetles might differ in color, their ability to resist disease, or their ability to obtain nutrients from a new food source should the old food source disappear. These differences come from differences in the beetles' genes.

Genetic diversity is the variety of genes present in a population. Some populations of a species have a lot of genetic diversity. Other populations have little. A population with more genetic diversity is more likely to survive during environmental changes, an outbreak of disease, or the disappearance of a food source.

Mark the Text

Restate Main Ideas Underline or highlight the main ideas in each paragraph. Stop after every paragraph and state what you just read in your own words.

FOLDABLES™

Compare Make a folded table Foldable to compare the three types of biodiversity. Include a description and explain the importance of each type.

Biodiversity	Genetic Diversity	Species Diversity	Ecosystem Diversity
Description			
Importance			

How does species diversity contribute to biodiversity?

<u>Species diversity</u> is the number of different species and the abundance of each species in a biological community. Areas with many species have a high level of species diversity. Species diversity is higher in tropical regions near the equator and lower in polar regions. This can be seen in the figure below.

Picture This

1. Identify Circle the areas with the lowest and highest amounts of species diversity.

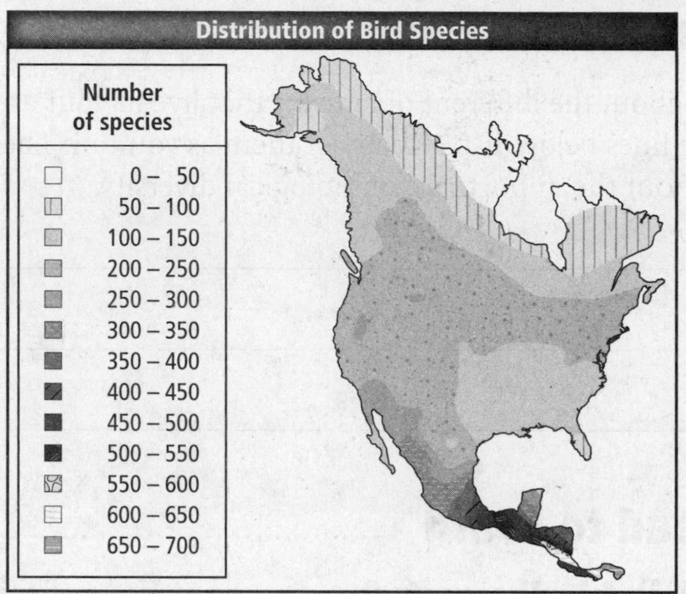

What is ecosystem diversity?

<u>Ecosystem diversity</u> is the variety of ecosystems that are present in the biosphere. Recall that an ecosystem includes all populations that interact and the abiotic, or non-living, factors that support them. The interactions among organisms are important to developing stable ecosystems. Different locations have different abiotic factors that support different types of life.

The Importance of Biodiversity

Many people work to preserve biodiversity for economic and scientific reasons. Other people work to preserve species that are beautiful.

2. Name three things people need that come from ecosystems.

Why is biodiversity valuable to humans?

People depend on other living things for food, clothing, energy, medicine, and shelter. Preserving the genetic diversity of species that people use directly is important. It is also important to preserve the genetic diversity of species that are not used directly. These species are possible sources of desirable genes that might be needed in the future.

Why might a species be valuable someday?

One reason to preserve biodiversity is that wild species might someday be used to create better crops for growing food. Biologists are beginning to learn how to transfer genes that control inherited characteristics from one species to another. Another reason is that scientists continue to find new medicines in nature. Many medicines were first identified in living things. Aspirin was discovered in willow, and penicillin was discovered in bread mold. In remote regions, many plants and other organisms have not been identified. These unknown species offer the promise of new medicines. ☑

Reading Check

3. Name one medicine that was discovered in nature.

____

____

What are the indirect values of biodiversity?

People, like all living things, benefit from a healthy biosphere. Scientists have begun to team up with economists to understand the dollar value of healthy ecosystems.

In the 1990s, New York City needed to clean up its drinking water. Much of the water for the city came from watersheds. Watersheds are land areas where the water on or underneath them drains to the same place. Two of the city's watersheds were not clean enough to supply drinking water. The city faced a choice: build a water-filtration system, which would cost 6 billion dollars, or clean up the watersheds, which would cost 1.5 billion dollars. The city found that cleaning up the ecosystem was a less expensive solution than using technology.

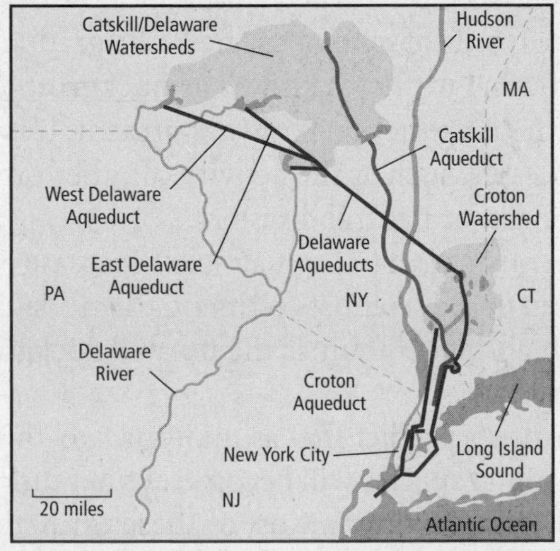

Picture This

4. Explain Highlight the watersheds that supply New York's drinking water.

Are there other values to biodiversity?

Many people work to preserve ecosystems for scientific reasons and also because ecosystems are beautiful. These factors are important and worthwhile, although it is difficult to attach a dollar value to them.

chapter 5 Biodiversity and Conservation

section 2 Threats to Biodiversity

MAIN Idea

Some human activities destroy biodiversity in ecosystems.

What You'll Learn

- factors that threaten biodiversity
- how the decline of one species can impact an ecosystem

Before You Read

You have probably read or heard about environmental issues in the news. On the lines below, list some environmental problems. Then read to learn about the possible consequences of human activities on the environment.

____

____

____

Mark the Text

Identify Threats to Biodiversity Highlight or underline the threats to biodiversity that you read about in this section.

Read to Learn

Extinction Rates

Many species have become extinct during Earth's long history. Scientists have learned a lot about life on Earth by studying the fossils of extinct species. The gradual process of species becoming extinct is known as **background extinction**. This low level of extinction is always present. It is caused by natural processes, such as the activity of other organisms, climate changes, or natural disasters.

Many scientists worry about a recent increase in the rate of extinction. Some scientists estimate that today's rate of extinction is about 1000 times the normal background extinction rate.

Some scientists predict that as many as two-thirds of all plant and animal species will become extinct during the second half of this century. Most of these extinctions will occur near the equator.

Some scientists believe we are in a period of a mass extinction. During a **mass extinction** a large percentage of all living species become extinct in a relatively short period of time. The last mass extinction, in which the dinosaurs became extinct, occurred about 65 million years ago. ☑

Reading Check

1. Define *mass extinction.*

____

____

____

____

How many species have become extinct?

The table below shows the high rate of extinctions since the year 1600. Many extinctions have occurred on islands. For example, 73 percent of mammals that have become extinct in the last 500 years were island species.

Species on islands are vulnerable to extinction for several reasons. Many island species evolved without natural predators. As a result, they do not have the ability to protect themselves. When a cat, dog, or other predator is introduced to the population, it can harm populations of native species. Nonnative species also harm native species by bringing diseases. The native population often does not have resistance to the disease and dies.

Picture This

2. Identify What two groups of living things have the highest rate of extinction?

____

____

Estimated Number of Extinctions Since 1600

Group	Mainland	Island	Ocean	Total	Estimated Number of Species	Percent of Group Extinct
Mammals	30	51	4	85	4000	2.1
Birds	21	92	0	113	9000	1.3
Reptiles	1	20	0	21	6300	0.3
Amphibians	2	0	0	2	4200	0.05
Fish	22	1	0	23	19,100	0.1
Invertebrates	49	48	1	98	1,000,000+	0.01
Flowering plants	245	139	0	384	250,000	0.2

Factors that Threaten Biodiversity

The high extinction rate today is due to the activities of a single species—*Homo sapiens.* Humans are changing conditions on Earth faster than new traits can evolve to cope with the new conditions. Evolving species might not have the natural resources they need. **Natural resources** are all materials and organisms found in the biosphere. Natural resources include minerals, fossil fuels, plants, animals, soil, clean water, clean air, and solar energy.

How does overexploitation harm a species?

One factor that is increasing the current rate of extinction is overexploitation. **Overexploitation** is the excessive use of a species that has economic value. For example, at one time, about 50 million bison roamed the central plains of North America. The bison nearly became extinct because of overhunting. By 1889, there were fewer than 1000 bison left. ☑

Reading Check

3. Define What is overexploitation?

____

____

____

How has overexploitation caused extinction?

At one time, passenger pigeons were numerous in North America. Large flocks of the birds would darken the skies. Passenger pigeons were overhunted and forced from their habitats. By the early 1900s, the birds had become extinct. Animals today that suffer from overexploitation include the ocelot and the white rhinoceros. People kill ocelots for their fur and white rhinoceroses for their horns. ☑

Reading Check

4. Identify What caused passenger pigeons to become extinct? (Circle your answer.)

a. pollution
b. overhunting
c. climate change

Why is habitat loss a problem?

Overexploitation was once the main cause of extinction. Today, the main cause is the loss or destruction of habitat. When a habitat is destroyed, the native species might have to move or they will die.

An example of habitat destruction occurs in tropical rain forests. Clearing of tropical rain forests is a serious threat to biodiversity. Remember that tropical areas have high levels of biodiversity. More than half of the world's plants and animals live in tropical rain forests. Removal of these forests would cause high numbers of extinction.

How can habitat disruption affect biodiversity?

Changing one thing in a habitat can also lead to loss of biodiversity. The figure below shows an example of how the decline of one species can affect an entire ecosystem. This chain of events occurred off the coast of Alaska in the 1970s when plankton-eating whales began to disappear. This caused the number of plankton to increase and began a chain reaction that affected many species, disrupting their habitat.

Picture This

5. Explain What caused the decline in the number of sea otters?

Can biodiversity be preserved in small areas?

Another source of habitat disruption is **habitat fragmentation**, the separation of an ecosystem into small areas. Species stay within the small areas because they are unable or unwilling to cross the human-made barrier. This causes several problems for the survival of species.

First, small areas of land cannot support large numbers of species. Second, individuals in one area cannot reproduce with individuals in another area, causing genetic diversity to decrease. Less genetically diverse populations are less able to resist disease and adjust to environmental changes.

Third, several small areas have more edges than one large area. Environmental conditions along the boundaries of an ecosystem are different, a factor known as the **edge effect**. Temperature, humidity, and wind are often different along the edge of a habitat than they are at its center. Some species are better adapted to living in the edge environment, but other species might find it difficult to survive there.

Think it Over

6. Apply How might edge effect impact a group of organisms in a state park?

____

____

____

____

____

How does pollution impact biodiversity?

Pollution damages ecosystems and decreases biodiversity by releasing harmful substances into the environment. Pesticides and industrial chemicals are examples of pollutants that are in food webs. Organisms ingest these substances in their food or water.

Some pollutants accumulate in the tissues of these organisms. Animals that eat other animals are most affected by the buildup of pollutants. **Biological magnification** happens when pollutants build up to high levels in bodily tissues of carnivores. The amount of pollutants might be relatively low when it enters the food web, but it increases as it spreads to a higher trophic level.

Picture This

7. Identify Circle the animal below that would be most affected by biological magnification.

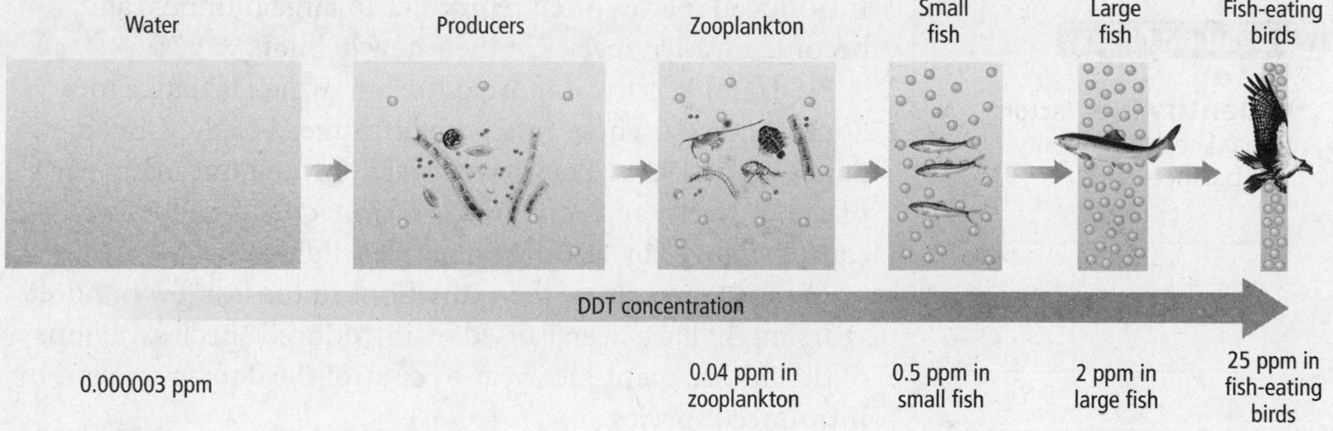

What effects did DDT have on some species of birds?

Some pollutants might disrupt normal bodily functions. The pesticide DDT causes eggshells of birds to be fragile and thin, leading to the death of developing birds. From the 1940s to the 1970s, DDT use caused populations of the American bald eagle and the peregrine falcon to become nearly extinct. DDT is now banned in some parts of the world.

How does acid precipitation affect ecosystems?

Acid precipitation is another pollutant. When fossil fuels are burned, compounds that form sulfuric acid and nitric acid are released into the environment. These acids fall back to Earth in rain, sleet, snow, or fog.

Acid precipitation removes nutrients from the soil. It damages plants and slows their growth. It pollutes lakes, rivers, and streams, killing fish and other organisms. ☑

Reading Check

8. Name two effects of acid precipitation.

____

____

____

____

What is eutrophication?

Water pollution can destroy habitats for fish and other species. **Eutrophication** (yoo troh fih KAY shun) occurs when fertilizers, animal waste, and sewage flow into waterways. These substances are rich in nitrogen and phosphorus, and they cause algae to grow. The algae use up the oxygen, causing other organisms in the water to suffocate. Sometimes the algae release toxins that poison the water.

How do nonnative species change ecosystems?

Organisms that have been moved to a new habitat are known as **introduced species**. In their native habitat, these organisms are kept in balance by predators, parasites, and competition with other species. When they are introduced into a new area, these controlling factors are not in place. Introduced species often reproduce in large numbers and become invasive species in their new habitat. ☑

Reading Check

9. Identify three factors that keep biodiversity in balance.

____

____

____

An example is the imported fire ant, which is native to South America. These ants were introduced to the United States in the 1920s. They have spread throughout many parts of the southern and southwestern United States and have caused damage by feeding on native wildlife.

About 40 percent of the extinctions in the last few hundred years might have been caused by introduced species. Billions of dollars are spent each year to control the damage caused by introduced species.

Biodiversity and Conservation

section 3 Conserving Biodiversity

Before You Read

On the lines below, list some activities that you could do in your home or school to use fewer natural resources. Then read to learn about ways people are preserving biodiversity.

____

____

MAIN Idea

People are working to preserve biodiversity.

What You'll Learn

- two classes of natural resources
- how biodiversity can be conserved
- two methods used to restore biodiversity

Read to Learn

Mark the Text

Locate Information Underline every heading in the reading that asks a question. Then highlight or underline the answers to those questions as you find them.

Natural Resources

There are more than six billion people living in the world today, and the number keeps growing. As the human population grows, the need for natural resources also grows.

The figure below shows the natural resources used by people in different parts of the world. Notice that people in some countries, like the United States and Canada, use more resources, while people in other countries use fewer resources. As countries become industrialized, people living there consume more resources.

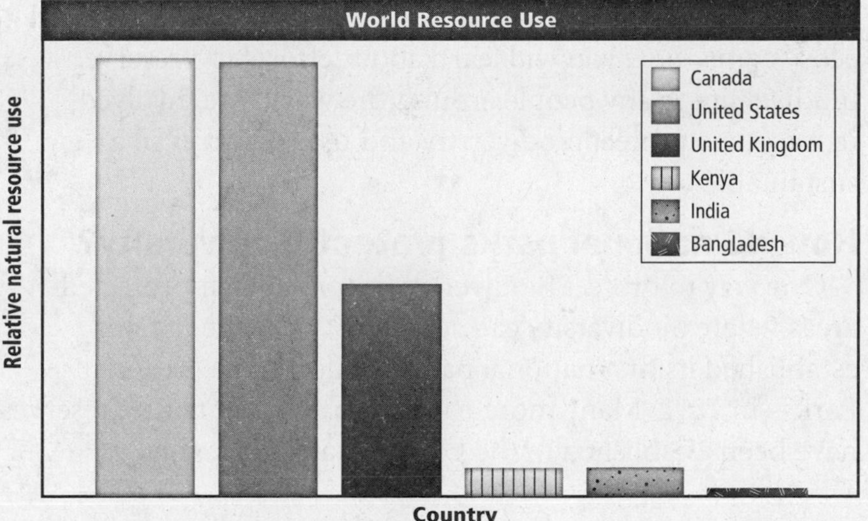

Picture This

1. Name two countries with high consumption and two countries with low consumption.

____

____

____

What are examples of renewable resources?

The two types of natural resources are renewable resources and nonrenewable resources. **Renewable resources** are resources that are replaced by natural processes faster than they are consumed. Solar energy is a renewable resource. Other renewable resources include plants used for food, animals, clean water, and clean air. It is important to remember that the supply of these resources is not endless. These resources might run out if we do not manage them carefully.

Why are some resources nonrenewable?

Think it Over

2. Identify Which of these energy resources is a nonrenewable resource? (Circle your answer.)

a. wind
b. gasoline
c. solar

Nonrenewable resources are resources that are replaced by natural processes slower than they are consumed. Fossil fuels and minerals are nonrenewable resources. Species are considered to be nonrenewable resources because an extinct species cannot be replaced.

A small group of trees in a large forest ecosystem is renewable because replacement trees can be grown from seeds in the soil. Enough of the forest is still intact to be the habitat for the organisms that live there. When an entire forest is cleared of many of its trees, it is not a renewable resource. The organisms that lived in the forest have lost their habitat and might die.

How can natural resources be managed?

Sustainable use means using resources at a rate in which they can be replaced or recycled while preserving the long-term environmental health of the biosphere. Sustainable use includes reducing the amount of resources that are used, recycling, and using resources responsibly. ☑

3. Identify two ways people can sustainably use resources.

____

____

____

Protecting Biodiversity

In Section 2, you learned how human activity has affected ecosystems. Now you will learn about efforts to preserve biodiversity. Many people around the world are involved in efforts to protect biodiversity and use resources in a sustainable way.

How do national parks protect biodiversity?

One way to protect biodiversity is to establish protected areas where biodiversity can succeed. The United States established its first national park—Yellowstone National Park—in 1872. Many more national parks and nature reserves have been established in the United States since then.

How much of the world's land is protected?

Many countries have established their own natural parks and nature reserves. Today about 7 percent of the world's land is protected for biodiversity.

Biodiversity in these areas can be threatened by the activity of people. Many of the protected areas are small and surrounded by areas of human activity. The human activity could damage the ecosystem in the protected areas.

Costa Rica has established megareserves in which one or more zones are surrounded by buffer zones. In a buffer zone, sustainable use of natural resources is permitted. ☑

Reading Check

4. Explain What happens in a buffer zone?

What is a biodiversity hot spot?

Scientists have identified locations around the world that are characterized by many **endemic** species—species that are only found in that one location. These areas are called hot spots. To be called a hot spot, there must be at least 1500 species of vascular plants that are endemic and the area must have lost at least 70 percent of its original habitat.

About one-third of all plant and animal species are found in hot spots. These hot spots originally covered 15.7 percent of Earth's surface. Currently, about a tenth of that habitat remains. ☑

Reading Check

5. State What portion of the world's plant and animal species live in biodiversity hot spots?

Biologists do not always agree about how to preserve biodiversity. Some biologists believe we should focus most of our efforts on hot spots in order to preserve the greatest number of species. Other biologists believe that while focusing on hot spots, other problems might be neglected.

How can corridors between habitats work?

One way biologists hope to improve biodiversity is by providing pathways, or corridors, between habitat fragments. Protected corridors connect small areas of land and give animals a way to move safely from one fragment of habitat to the next. One problem with this approach is that small fragments connected in this way are subject to edge effects.

Restoring Ecosystems

Sometimes, biodiversity in an ecosystem is destroyed. The ecosystem no longer has all the needed biotic and abiotic factors to maintain its health. When this happens, the ecosystem no longer functions properly. People have devised ways of restoring ecosystems.

What kinds of ecosystems need to be restored?

Natural causes, such as volcanic eruptions or floods, can destroy biodiversity. People can destroy biodiversity when they do not use resources sustainably. Damaged ecosystems might take a long time to recover.

Think of what happens when a tropical rain forest is cleared for farmland. After a few years, people abandon the farmland because the soil is in poor condition. The ecosystem might take many years to recover. Another example of a damaged ecosystem that needs restoration occurs when an accidental oil spill or toxic chemical spill pollutes the area and kills native species.

Given time, ecosystems can recover. Typically, larger areas take longer to recover. Some types of disturbances recover more quickly than others. Ecologists use two methods to speed recovery—bioremediation and biological augmentation.

Think it Over

6. Draw Conclusions Why do larger disturbed areas take longer to recover?

______

______

______

How does bioremediation clean up pollution?

Bioremediation is the use of living organisms to remove toxins from a polluted area. Bioremediation relies on bacteria, fungi, or plants to clean up pollutants in the soil.

Some species of plants can be used to remove toxic metals such as zinc, lead, and nickel. The plants are planted in contaminated soils. The plants grow and store the toxic metals in their tissues. People then harvest the plants, removing the metals from the ecosystem.

What is biological augmentation?

Biological augmentation involves adding essential items to a degraded ecosystem. For example, ladybugs are predators that eat other insects. Ladybugs can be introduced to help control insect populations.

Legally Protecting Biodiversity

During the 1970s, people's awareness of environmental problems grew. In 1973 the Endangered Species Act was passed in the United States. It gives legal protection to species that are in danger of becoming extinct. In 1975 an international treaty was signed that outlawed the trade of endangered animals and animal parts, such as elephant tusks and rhinoceros horns. Since then, many more laws and treaties have been enacted with the purpose of preserving biodiversity and the health of the biosphere. ☑

Reading Check

7. Explain the purpose of the Endangered Species Act.

______

______

______

Chemistry in Biology

section 1 Atoms, Elements, and Compounds

Before You Read

On the lines below, describe how you think chemistry relates to living things. Then read the section to learn about the chemical building blocks of matter.

____

____

MAIN Idea

Matter is composed of tiny particles called atoms.

What You'll Learn

- the particles that make up atoms
- the difference between covalent bonds and ionic bonds
- about van der Waals forces

Read to Learn

Mark the Text

Read for Understanding
As you read this section, highlight any sentences that you do not understand. After you finish the section, reread the highlighted sentences.

Atoms

Chemistry is the study of matter. Matter is anything that has mass and takes up space. All organisms are made of matter. **Atoms** are the building blocks of matter.

Atoms are made up of neutrons, protons, and electrons, as shown in the figure below. The **nucleus** is the center of the atom where the neutrons and protons are located. **Protons** are positively charged particles (p^+). **Neutrons** are particles that have no charge (n^0). **Electrons** are negatively charged particles (e^-) that are located outside the nucleus.

Electrons move around the nucleus in energy levels. The atom's structure is the result of the attraction between protons and electrons. Atoms contain an equal number of protons and electrons. As a result, the overall charge of an atom is zero.

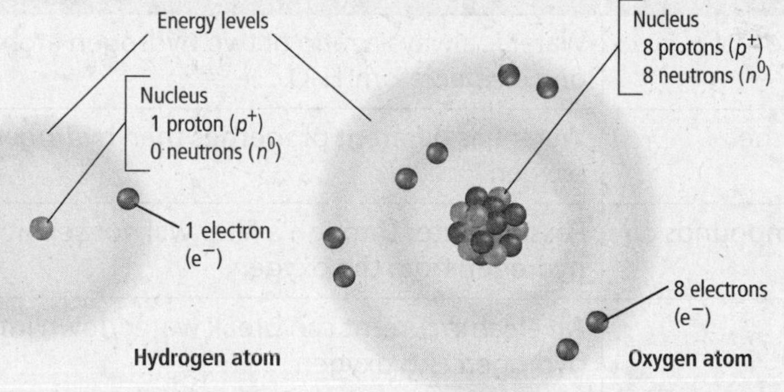

Picture This

1. **Identify** the number of electrons in the outermost energy level of the oxygen atom.

____

Elements

An **element** is a pure substance that cannot be broken down into other substances. The periodic table of elements organizes information about elements in rows, called periods, and columns, called groups. A periodic table is located inside the back cover of this workbook. Each block includes the element's name, number, symbol, and mass. Living things are composed mainly of three elements—carbon, hydrogen, and oxygen.

How are isotopes identified?

Atoms of the same element have the same number of protons and electrons but sometimes different numbers of neutrons. Atoms of the same element with different numbers of neutrons are called **isotopes**. Isotopes are identified by adding the number of protons and neutrons. Carbon-12 has six protons and six neutrons. Carbon-14 has six protons and eight neutrons.

Think it Over

2. Apply Another carbon isotope has six protons and seven neutrons in its nucleus. What do you think this carbon isotope is called?

____

What makes an isotope radioactive?

Changing the number of neutrons in an atom can cause the nucleus to decay, or break apart. When a nucleus breaks apart, it gives off radiation. Isotopes that give off radiation are called radioactive isotopes. All living things contain the radioactive isotope carbon-14. Scientists know the half-life of carbon-14, or the amount of time it takes for half of carbon-14 to decay. By finding how much carbon-14 remains in an object, scientists can calculate the object's age.

Picture This

3. Draw Conclusions Table salt is a compound made of sodium and chlorine. Could you separate the sodium from the chlorine by crushing the salt crystals? Explain.

____

____

____

Compounds

When two or more elements combine, they form a **compound**. Each compound has a chemical formula made up of the chemical symbols from the periodic table. For example, water is made of hydrogen (H) and oxygen (O). Its formula is H_2O. The table below lists characteristics of compounds.

Characteristics of Compounds	Example
Always formed from a specific combination of elements in a fixed ratio	Water is always a ratio of two hydrogen atoms and one oxygen atom: H_2O.
Chemically and physically different than the elements that comprise them	Water has different properties than hydrogen and oxygen.
Cannot be broken down into simpler compounds or elements by physical means	Passing water through a filter will not separate the hydrogen from the oxygen.
Can be broken down by chemical means	An electric current can break water down into hydrogen and oxygen.

Chemical Bonds

The force that holds substances together is called a chemical bond. Chemical bonding involves electrons. Electrons travel around the nucleus of an atom in energy levels. Each energy level can hold only a certain number of electrons. The first energy level, which is closest to the nucleus, can hold up to two electrons. The second level can hold up to eight electrons.

A partially-filled energy level is not as stable as a full or an empty energy level. Atoms become more stable by losing electrons or attracting electrons from other atoms. This electron activity forms chemical bonds between atoms. The forming of chemical bonds stores energy. The breaking of chemical bonds releases energy for an organism's life processes—growth, development, and reproduction. The two main types of chemical bonds are covalent and ionic.

How do covalent bonds form?

A **covalent bond** forms when atoms share electrons. The figure below shows the covalent bonds between oxygen and hydrogen to form water. Each hydrogen (H) atom has one electron in its outer energy level, and the oxygen (O) atom has six. The outer energy level of oxygen is the second level, so it can hold up to eight electrons. Oxygen has a strong tendency to fill the energy level by sharing electrons from the two nearby hydrogen atoms. Hydrogen also has a strong tendency to share electrons with oxygen to fill its outer energy level. Two covalent bonds form a water molecule.

Most compounds in living things are molecules. A **molecule** is a compound in which the atoms are held together by covalent bonds. Covalent bonds can be single, double, or triple. A single bond shares one pair of electrons. A double bond shares two pairs of electrons. A triple bond shares three pairs of electrons.

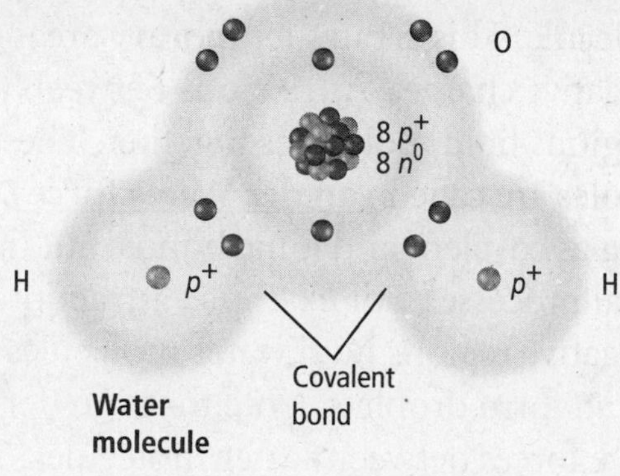

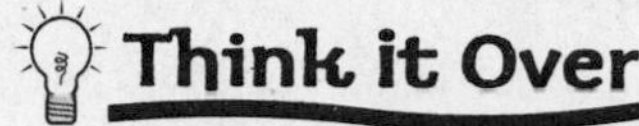

4. Apply Study the oxygen atom illustrated below. Is the second energy level of the oxygen atom full? Explain.

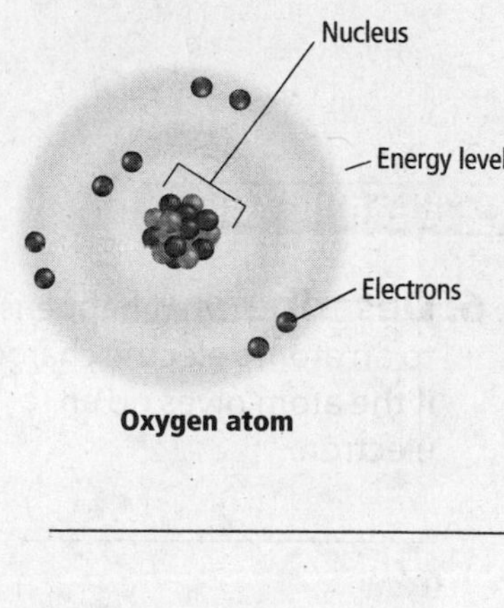

____

____

____

Picture This

5. Label the first energy level and second energy level in the oxygen atom. Include in each label the number of electrons required to fill the level.

How do ionic bonds form?

Recall that atoms do not have an electric charge. Also recall that an atom is most stable when its outer energy level is either empty or full. To become more stable, an atom might give up electrons to empty its outer energy level. Or, the atom might accept electrons to fill the outer energy level. An atom that has given up or gained one or more electrons becomes an **ion** and carries an electric charge.

For example, the outer energy level of sodium (Na) has one electron. Sodium can become more stable if it gives up this electron to empty the energy level. When it gives up this electron, the neutral sodium atom becomes a positively charged sodium ion (Na^+). Chlorine (Cl) needs just one electron to fill its outer energy level. When it accepts an electron from another atom, chlorine becomes a negatively charged ion (Cl^-). ☑

Reading Check

6. Describe what happens to an atom's electric charge if the atom gives up an electron.

____

____

____

An **ionic bond** is an electrical attraction between two oppositely charged ions. When sodium gives its electron to chlorine, the positively charged sodium ion (Na^+) is attracted to the negatively charged chlorine ion (Cl^-). The ionic bond between them forms the ionic compound sodium chloride (NaCl), or table salt.

Ions in living things help maintain homeostasis as they travel in and out of cells. Ions also help transmit signals that enable you to see, taste, hear, feel, and smell.

Some atoms give up or accept electrons more easily than other atoms. The elements identified as metals in the periodic table tend to give up electrons. The elements identified as nonmetals tend to accept electrons.

van der Waals Forces

Electrons travel around the nucleus randomly. The random movement can cause an unequal distribution of electrons around the molecule. This creates temporary areas of slightly positive and negative charges. Attractions between these positive and negative regions hold molecules together. These attractions between molecules are called **van der Waals forces**. These forces are not as strong as covalent and ionic bonds, but they play a key role in biological processes. For example, attractions between positive and negative regions hold water molecules together. As a result, water can form droplets. Note that van der Waals forces are the attractive forces between water molecules. They are not the forces between the atoms that make up water. ☑

Reading Check

7. Identify the substances that are held together by van der Waals forces. (Circle your answer.)

a. atoms
b. molecules

Chemistry in Biology

section 2 Chemical Reactions

Before You Read

On the lines below, explain why you think rust forms on metal. Then read the section to learn the role of chemical reactions in living things.

____

____

MAIN Idea

Chemical reactions allow living things to grow, develop, reproduce, and adapt.

What You'll Learn

- the parts of a chemical reaction
- how energy changes relate to chemical reactions
- the importance of enzymes in organisms

Read to Learn

Reactants and Products

Chemical reactions occur inside your body all the time. You digest food. Your muscles grow. Your cuts heal. These functions and many others result from chemical reactions.

A <u>**chemical reaction**</u> is the process by which atoms or groups of atoms in substances are reorganized into different substances. Chemical bonds are broken and formed during chemical reactions. For example, rust is a compound called iron oxide. It forms when oxygen in the air reacts with iron.

What was once silver and shiny becomes dull and orange-brown. Other clues that a chemical reaction has taken place include the production of heat or light, and formation of gas, liquid, or solid.

Study Coach

Create a Quiz After you read this section, create a five-question quiz from what you have learned. Then, exchange quizzes with another student. After taking the quizzes, review your answers together.

How are chemical equations written?

Scientists express chemical reactions as equations. On the left side of the equation are the starting substances, or <u>**reactants**</u>. On the right side of the equation are the substances formed during the reaction, or the <u>**products**</u>. An arrow is between these two parts of the equation. You can read the arrow as "yield" or "react to form." The general form of a chemical equation is shown below.

$$\text{Reactants} \longrightarrow \text{Products}$$

Picture This

1. **Describe** how this general chemical equation would be expressed in words.

____

____

____

Why must chemical equations balance?

The following chemical equation describes the reaction between hydrogen (H) and oxygen (O) to form water (H_2O).

$$2H_2 + O_2 \longrightarrow 2H_2O$$

Picture This

2. Label the subscripts and the coefficients in this equation after you read the discussion on this page.

Matter cannot be created or destroyed in chemical reactions. This is the principle of conservation of mass. Therefore, mass must balance in all chemical equations. This means that the number of atoms of each element on the reactant side must equal the number of atoms of the same element on the product side. In our example, the number of H atoms on the left side must equal the number of H atoms on the right side. The same must be true of O atoms.

The larger 2 to the left of the element H is called a coefficient. Coefficients are used to balance chemical equations. If no coefficient or subscript appears with an element, both are assumed to be 1.

To see that the above equation is balanced, multiply the coefficient by the subscript for each element. Then add up the total number of atoms of each element. Follow along in the equation above as you read the analysis below.

Reactant side:
2 (coefficient of H) × 2 (subscript of H) = 4 H atoms
1 (coefficient of O) × 2 (subscript of O) = 2 O atoms

Product side:
2 (coefficient of H) × 2 (subscript of H) = 4 H atoms
2 (coefficient of O) × 1 (subscript of O) = 2 O atoms

The equation has the same number of H atoms on both sides. It also has the same number of O atoms on both sides. No mass has been gained or lost. The equation balances.

FOLDABLES™

Take Notes Make a three-tab Foldable from a sheet of notebook paper. As you read, record what you learn about reactants, products, and the energy required to start a chemical reaction.

Energy of Reactions

Energy is required to start a chemical reaction. The minimum amount of energy needed for reactants to form products in a chemical reaction is called the **activation energy**. For example, a candle will not burn until you light the wick. The flame from a match provides the activation energy for the candle wick to react with oxygen in the air. Some reactions need higher activation energy than others.

How does energy change in chemical reactions?

Chemical reactions can be exothermic or endothermic. In exothermic reactions, energy is released in the form of heat or light. As a result, the energy of the product is lower than the energy of the reactants. In endothermic reactions, energy is absorbed. As a result, the energy of the product is higher than the energy of the reactants. ☑

Reading Check

3. Explain why the energy of the product might be lower than the energy of the reactants.

____

____

____

____

Enzymes

Some chemical reactions occur slowly in a laboratory because the activation energy is high. To speed up the chemical reaction, scientists use catalysts. A **catalyst** is a substance that lowers the activation energy needed to start a chemical reaction. A catalyst does not increase how much product is made, and it does not get used up in the reaction.

In living things, special proteins called **enzymes** are biological catalysts. Enzymes speed up the rate of chemical reactions in the body. Like all catalysts, enzymes are not used up by the chemical reaction. They can be used again. Also, most enzymes act in just one type of reaction. For example, the enzyme amylase is found in saliva. Amylase helps begin the process of food digestion in the mouth.

The figure below shows how an enzyme works. The reactants that bind to the enzyme are called **substrates**. The specific location where a substrate binds on an enzyme is called the **active site**. The substrate and active site are shaped to fit together exactly. Only substrates shaped to fit the active site will bind to the enzyme.

The bond between the enzyme and substrates creates the enzyme-substrate complex. This complex helps to break bonds in the reactants and form new bonds, changing the substrates into products. The enzyme then releases the products.

Enzymes are the chemical workers in cells. The actions of enzymes enable cell processes that supply energy. Factors such as pH and temperature affect enzyme activity.

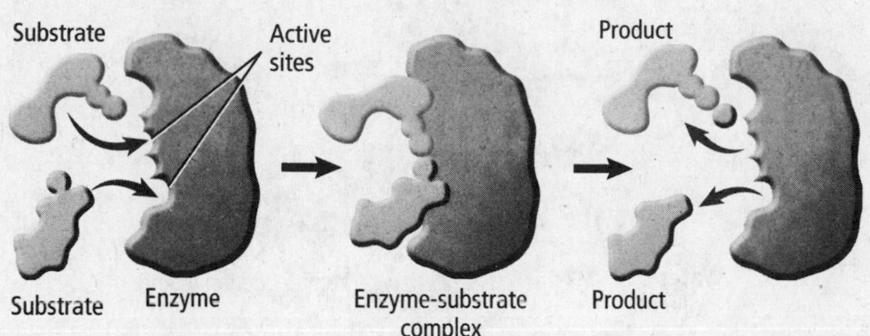

Picture This

4. Label each of the three parts of this process with a brief description of what the part shows.

chapter 6

Chemistry in Biology

section 3 Water and Solutions

MAIN Idea

The properties of water make it well-suited to help maintain homeostasis in an organism.

What You'll Learn

- why water is a good solvent
- the differences between suspensions and solutions
- how acids differ from bases

Before You Read

Have you ever stirred a spoonful of a powdered drink into water? On the lines below, describe what happened to the powder. Then read the section to learn the properties of different types of mixtures.

____

____

Mark the Text

Restate the Main Point
Highlight the main point in each paragraph. Then restate each main point in your own words.

Read to Learn

Water's Polarity

Earlier you learned that water molecules are formed by covalent bonds that link two hydrogen (H) atoms to one oxygen (O) atom. The electrons in a water molecule are attracted more strongly to an oxygen atom's nucleus. As a result, the electrons in the covalent bond are not shared equally. The electrons spend more time near the oxygen nucleus than near the hydrogen nuclei, as shown in the figure below.

Note that the water molecule has a bent shape. This shape and the unequal distribution of electrons result in oppositely charged regions. The oxygen end has a slightly negative charge. The hydrogen end has a slightly positive charge.

Picture This

1. Label the H and O atoms. Then label each electron with the symbol for a negative charge (−). Most negative charges are close to the nucleus of which atom?

____

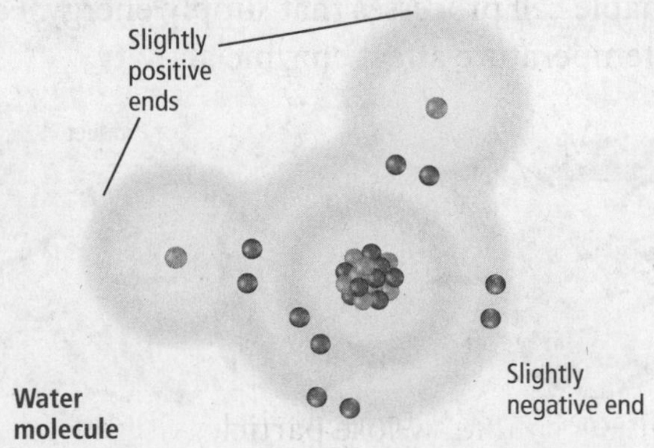

Why is polarity important?

Molecules that have an unequal distribution of charges are called **polar molecules**. Polarity means having two opposite poles, or ends. A magnet has polarity. When the opposite poles of a magnet are close to each other, they attract. In the same way, when oppositely charged regions of polar molecules are close together, they attract each other. In water, the attraction is called a hydrogen bond. A **hydrogen bond** is a weak interaction involving a hydrogen atom and a fluorine, oxygen, or nitrogen atom. The hydrogen bonds between water molecules are illustrated by dotted lines in the diagram below.

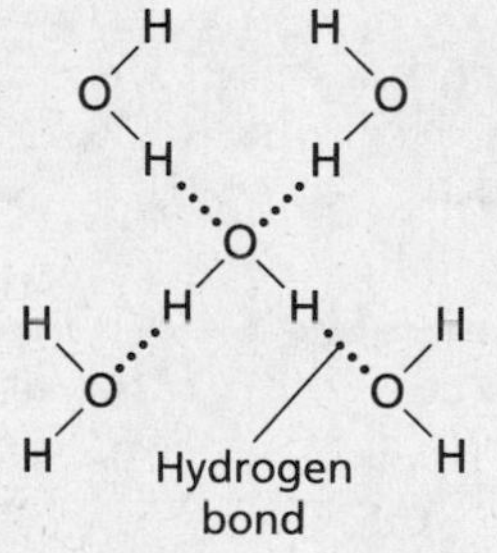

H H
O O
H H
O
H H H H
O O
H Hydrogen bond H

Picture This

2. Circle each water molecule in this diagram. Remember, a water molecule is made up of two H atoms and one O atom. Hydrogen bonds link the water molecules together. How many water molecules are shown in this diagram?

____

Mixtures with Water

When you make a fruit-flavored drink, you dissolve drink powder in water. It does not react with water to form a new product. A mixture has been created. A **mixture** is a combination of two or more substances in which each substance keeps its individual characteristics and properties.

What is a homogeneous mixture?

A homogeneous (hoh muh JEE nee us) mixture has the same composition throughout. A **solution** is another name for a homogeneous mixture. A solution has two parts: a solvent and a solute. A **solvent** is a substance in which another substance is dissolved. A **solute** is the substance that is dissolved in the solvent. In the fruit-flavored drink, water is the solvent and drink powder is the solute.

How does a heterogeneous mixture differ?

In a heterogeneous mixture, the parts remain distinct—that is, you can identify the individual parts. For example, in a salad, you can tell the lettuce from the tomatoes.

Sand mixed with water is a suspension—a type of heterogeneous mixture. Over time, the particles in a suspension will settle to the bottom. Paint is a heterogeneous mixture called a colloid, whose particles do not settle out.

Think it Over

3. Apply Are coins in your pocket a homogeneous mixture or heterogeneous mixture? Explain.

____

____

____

How do acids differ from bases?

Water's polarity enables many solutes to dissolve easily in water. The human body is about 70 percent water and contains many solutions. When a substance containing hydrogen is dissolved in water, the substance might release a hydrogen ion (H^+), as illustrated in the figure below. Substances that release hydrogen ions when dissolved in water are called **acids**. The more hydrogen ions released, the more acidic the solution. ☑

Reading Check

4. Identify the property of water that makes water a good solvent.

____

Substances that release hydroxide ions (OH^-) when dissolved in water are called **bases**. The more hydroxide ions released, the more basic the solution.

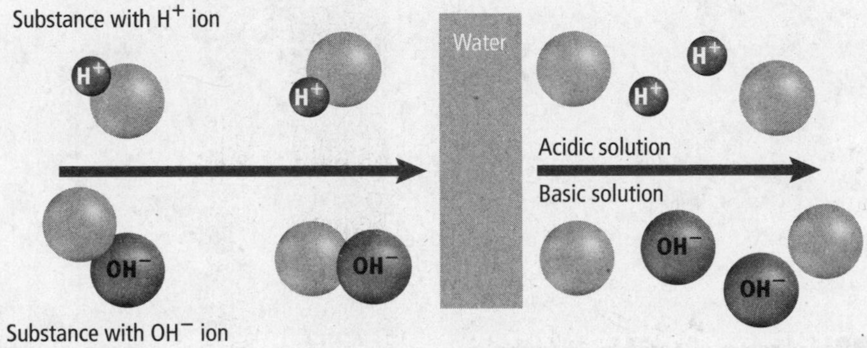

Picture This

5. Apply Suppose you want to make the acid solution in this figure more acidic. Add drawings that would result in a more acidic solution.

How do buffers affect pH?

The measure of concentration of H^+ in a solution is called **pH**. Scientists use a pH scale like the one below to compare the strengths of acids and bases. Water is neutral and has a pH of 7.0. Acidic solutions have more H^+ and have pH values lower than 7. Basic solutions have more OH^- and have pH values higher than 7. To maintain homeostasis, H^+ levels must be controlled. **Buffers** are mixtures that can react with acids or bases to keep the pH within a certain range.

Picture This

6. Determine which solutions are acidic and which are basic. Draw a circle around the buffer.

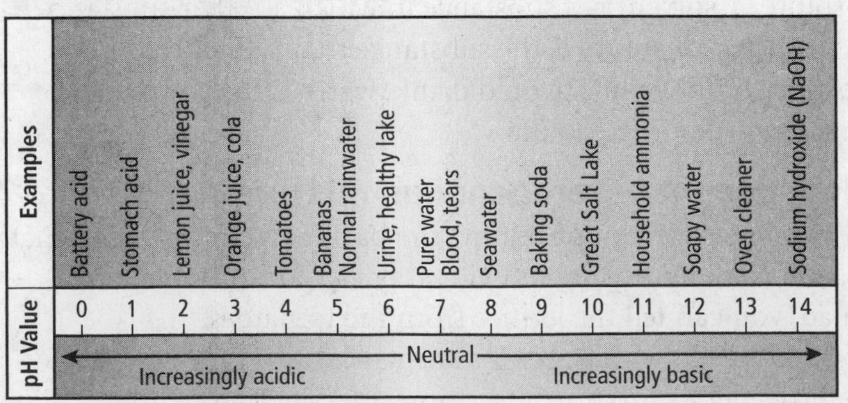

Chemistry in Biology

section 4 The Building Blocks of Life

Before You Read

You have probably heard about DNA— the "genetic code." On the lines below, describe what you think DNA does. Then read the section to learn about DNA and the other compounds that make up all living things.

____

____

MAIN Idea

Organisms are made up of carbon-based molecules.

What You'll Learn

- the four major families of biological macromolecules
- the functions of each group of biological macromolecules

Read to Learn

Organic Chemistry

Almost all biological molecules contain the element carbon. For this reason, all life is considered carbon-based. Organic chemistry is the study of organic compounds—the compounds that contain carbon.

In the figure below, notice that carbon has four electrons in its outer energy level. Recall that the second energy level can hold eight electrons. Therefore, a carbon atom can form four covalent bonds with other atoms. Carbon atoms can bond with each other, forming a variety of organic compounds. These organic compounds can take the form of straight chains, branched chains, and rings, as illustrated in the figure below. Carbon compounds are responsible for the diversity of life on Earth.

Study Coach

Make an Outline Make an outline of the information you learn in this section. Start with the headings. Include the boldface terms.

Picture This

1. **Calculate** What percentage of the carbon atom's second energy level is filled?

____

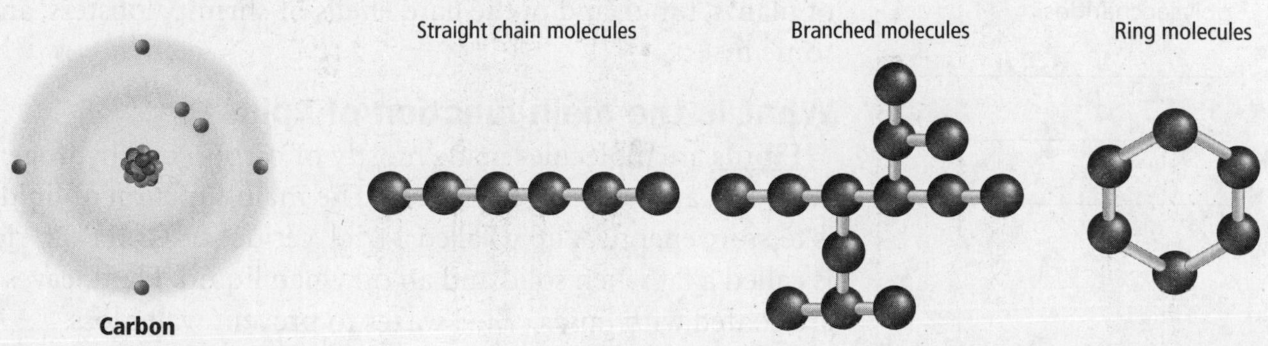

Macromolecules

Macromolecules are large molecules that are formed by joining smaller molecules together. Macromolecules are also called polymers. **Polymers** are made from repeating units of identical or nearly identical compounds called monomers. The monomers are linked together by a series of covalent bonds.

The four major groups of biological macromolecules are carbohydrates, lipids, proteins, and nucleic acids. The table below summarizes the functions of each group.

Picture This

2. Draw Conclusions To what group of macromolecules do you think DNA belongs?

Biological Macromolecules	
Group	**Function**
Carbohydrates	• stores energy • provides structural support
Lipids	• stores energy • provides steroids • waterproofs coatings
Proteins	• transports substances • speeds reactions • provides structural support • provides hormones
Nucleic acids	• stores and communicates genetic information

What roles do carbohydrates play in biology?

Carbohydrates are composed of carbon, hydrogen, and oxygen with a ratio of one oxygen and two hydrogen atoms for each carbon atom: CH_2O. Short chains of carbohydrates are monosaccharides (mah nuh SA kuh ridz), or simple sugars. A disaccharide (di SA kuh rid) is two monosaccharides linked together. Longer carbohydrate chains are called polysaccharides.

Carbohydrates serve as energy sources for organisms. Also, carbohydrates provide structural support in the cell walls of plants, fungi, and in the hard shells of shrimp, lobsters, and some insects.

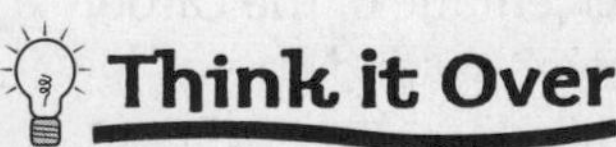

3. Contrast monosaccharides, disaccharides, and polysaccharides.

What is the main function of lipids?

Lipids are molecules made mostly of carbon and hydrogen. Fats, oils, and waxes are all lipids. The main function of lipids is to store energy. A lipid called a triglyceride (tri GLIH suh rid) is called a fat when solid and an oil when liquid. Plant leaves are coated with lipids called waxes to prevent water loss.

Saturated and Unsaturated Fats When the carbon atoms in a fat cannot bond with any more hydrogen atoms, the fat is a saturated fat. The carbon atoms of unsaturated fats can bond with more hydrogen atoms.

Phospholipids A lipid called a phospholipid is responsible for the structure and function of the cell membrane. Lipids do not dissolve in water. This characteristic enables lipids to serve as barriers in biological membranes. ☑

Steroids Cholesterol and hormones are types of steroids, another group of lipids. In spite of its bad reputation, cholesterol provides the starting point for other important lipids, such as the hormones estrogen and testosterone.

What compounds make up proteins?

A **protein** is made of small carbon compounds called amino acids. **Amino acids** are made of carbon, hydrogen, oxygen, nitrogen, and sometimes sulfur.

Amino Acids There are 20 different amino acids. Proteins are made of different combinations of all 20 amino acids. Covalent bonds called peptide bonds join amino acids together to form proteins.

Protein Structure A protein's amino acid chain folds into a three-dimensional shape. The figure below shows two basic protein shapes—the helix and the pleat. A protein might contain many helices, pleats, and folds. Hydrogen bonds help the protein hold its shape.

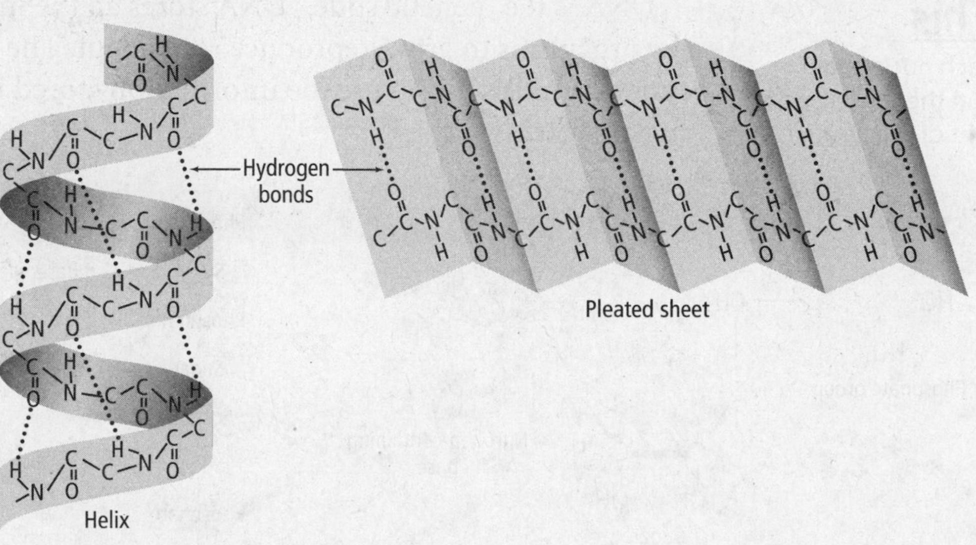

Reading Check

4. Identify the key characteristic of lipids that enables them to keep unwanted substances from penetrating cell membranes.

____

____

Picture This

5. Label Add these descriptive labels to the appropriate protein shape in the figure: *folded paper* and *spiral*.

Think it Over

6. Draw Conclusions What protein function listed here suggests that most enzymes are proteins?

____

____

____

Protein Function Proteins are involved in nearly every function of your body. Your muscles, skin, and hair are made of proteins. Your cells contain about 10,000 different proteins that serve many functions. They

- provide structural support;
- transport substances inside the cell and between cells;
- communicate signals within the cell and between cells;
- speed up chemical reactions;
- control cell growth.

What roles do nucleic acids play in organisms?

Nucleic acids are the fourth group of biological macromolecules. **Nucleic acids** are complex macromolecules that store and transmit genetic information. Repeating subunits, called **nucleotides,** make up nucleic acids.

Nucleotides are composed of carbon, hydrogen, oxygen, nitrogen, and phosphorus. All nucleotides have the three units shown in the figure below—a phosphate, a nitrogenous base, and a sugar.

To form a nucleic acid, the sugar of one nucleotide bonds to the phosphate of another nucleotide, as illustrated in the figure on the right. The nitrogenous base sticks out from the chain. It is available to bond with bases in other nucleic acids.

Two types of nucleic acids are found in living things. One is deoxyribonucleic (dee AHK sih rib oh noo klay ihk) acid, or DNA. The other is ribonucleic (rib oh noo KLAY ihk) acid, or RNA.

DNA is the "genetic code." DNA stores all the instructions for organisms to grow, reproduce, and adapt. The main function of RNA is to use the information stored in DNA to make proteins.

Picture This

7. Circle each nucleotide grouping in the nucleic acid on the right of the figure.

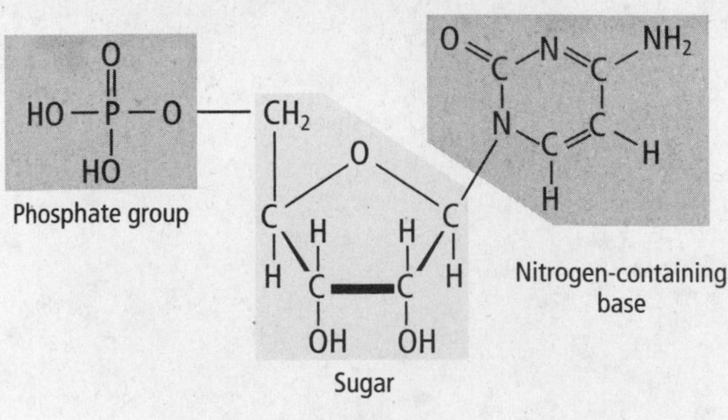

Nucleotide

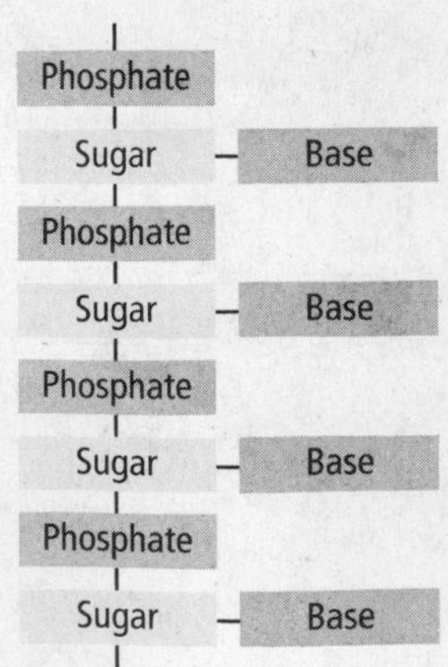

Nucleic acid

Cellular Structure and Function

section 1 Cell Discovery and Theory

Before You Read

Have you ever looked at anything through a magnifying glass or a microscope? Describe on the lines below how the magnifying glass or microscope changed the object. In this section you will learn about some important discoveries made using microscopes.

MAIN Idea

The microscope led to the discovery of cells.

What You'll Learn

- the principles of the cell theory
- how compound light microscopes differ from electron microscopes
- how prokaryotic and eukaryotic cells differ

Read to Learn

Mark the Text

Focus As you read, underline or highlight the main ideas in each paragraph.

History of the Cell Theory

A cell is the basic structural and functional unit of all living things. The human body consists of trillions and trillions of cells. But cells are too small to see with the human eye. The invention of the microscope allowed scientists to discover that cells existed.

In 1665, an English scientist named Robert Hooke made a simple microscope. He used the microscope to look at a piece of cork, which is the dead cells of oak bark. Hooke saw small, box-shaped structures in the cork, which he called *cellulae.* Today, we call them cells.

In the late 1600s, Anton van Leeuwenhoek (LAY vun hook), a Dutch scientist, made another microscope. He examined pond water, milk, and other substances. He was surprised to find living organisms in these substances.

What discoveries led to the cell theory?

In 1838, German scientist Matthias Schleiden studied plants under microscopes. He concluded that all plants are composed of cells. Another German scientist, Theodor Schwann, declared that animal tissues were made up of cells. ☑

Reading Check

1. **Compare** What is one thing that plants and animals have in common?

What is the cell theory?

Scientists continued to learn more about cells. Scientist Rudolf Virchow proposed that cells divide to form new cells. He suggested that every cell came from a cell that already existed. The observations and ideas of the various scientists who studied cells are summarized as the cell theory. The **cell theory** is a fundamental idea of modern biology and includes the principles listed in the table below.

Picture This

2. Highlight the principle in the cell theory that resulted from the discoveries of Matthias Schleiden and Theodor Schwann.

The Cell Theory

Principle	Explanation
1. All living organisms are made up of one or more cells.	An organism can have one or many cells. Most plants and animals have many cells.
2. The cell is the basic unit of organization in living organisms.	Even in complex organisms such as humans, the cell is the basic unit of life.
3. All cells come from living cells. Cells pass copies of their genetic material on to their daughter cells.	Cells contain hereditary information that passes from cell to cell during cell division.

Microscope Technology

The development of the microscope made the discovery of cells possible. Improvements made to early microscopes have helped scientists learn much more about cells.

What is a compound light microscope?

The modern compound light microscope uses a series of glass lenses to magnify, or enlarge, an object. When visible light passes through each lens, it magnifies the image of the previous lens. For example, two lenses that each magnify an image 10× result in a microscope that magnifies the object 100×, as shown in the figure below.

Picture This

3. Calculate If each lens in this example magnified the image 20×, what is the total magnification? (Show your work.)

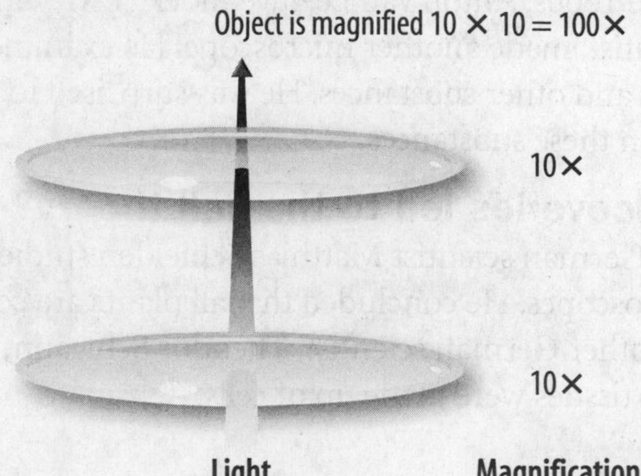

What is an electron microscope?

The best compound light microscopes only magnify an image about 1000×. Scientists needed more powerful microscopes to learn more about cells. The electron microscope was invented in the 1940s. It doesn't use lenses. Instead, the transmission electron microscope (TEM) uses magnets to aim a beam of electrons at the image to be magnified. Some TEMs can magnify an image 500,000×.

The scanning electron microscope (SEM) was a further improvement in technology. It produces a three-dimensional image of the cell. One problem with both the TEM and SEM is that only nonliving cells can be examined. A more recent invention, the scanning tunneling electron microscope (STM), can magnify living cells.

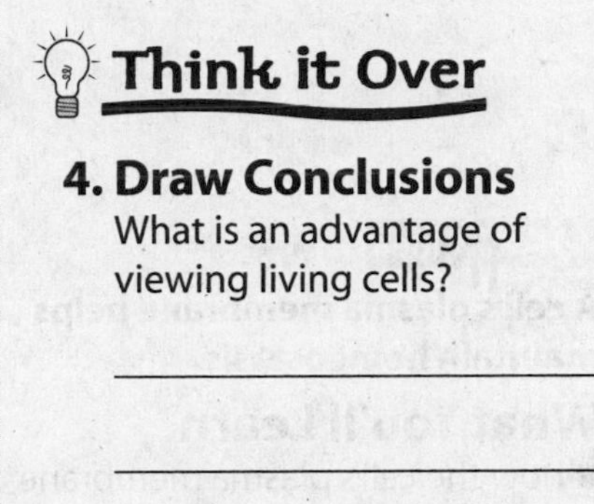

Think it Over

4. Draw Conclusions What is an advantage of viewing living cells?

Basic Cell Types

Cells have different sizes, shapes, and functions, but all cells have a plasma membrane. A **plasma membrane** is a boundary that helps control what enters and leaves the cell.

Some basic functions are common to most cells. For example, most cells have some form of genetic material that provides instructions for making substances that the cell needs. In addition, all cells break down molecules to generate energy.

What are the two categories of cells?

Scientists group cells into two broad categories based on their internal structures. These categories are prokaryotic cells and eukaryotic cells.

Simple cells that have no specialized structures are known as **prokaryotic** (pro kar ee AW tik) **cells.** Cell functions in these simple cells occur in the plasma membrane. Most unicellular organisms, such as bacteria, are prokaryotic cells. Thus, they are called prokaryotes. Prokaryotic cells are believed to be similar to the first cells on Earth.

Eukaryotic (yew kar ee AW tik) cells are the other category of cells. They are usually larger and more complex. **Eukaryotic cells** contain a nucleus and other structures called organelles. **Organelles** are specialized structures that carry out specific functions. The **nucleus** contains the genetic material for the cell. Organisms that are made up of eukaryotic cells are called eukaryotes. Eukaryotes can be unicellular or multicellular.

5. Compare Which cells are more complex? (Circle your answer.)

a. prokaryotic cells
b. eukaryotic cells

chapter 7

Cellular Structure and Function

section 2 The Plasma Membrane

MAIN Idea

A cell's plasma membrane helps maintain homeostasis.

What You'll Learn

- how the cell's plasma membrane functions
- the role of proteins, carbohydrates, and cholesterol in the plasma membrane

Before You Read

A window screen in your home allows air to pass through while keeping insects out. In this section, you will learn about a cell structure that has the same basic function. On the lines below, list some things you think would be allowed to pass into a cell and some things that would be kept out.

____

____

Mark the Text

Make Flash Cards Make a flash card for each question heading in this section. On the back of the flash card, write the answer to the question. Use the flash cards to review what you have learned.

Read to Learn

Function of the Plasma Membrane

A cell's survival depends on maintaining balance, called homeostasis. The plasma membrane is the cell structure primarily responsible for homeostasis. It is the thin, flexible boundary between the cell and its watery environment. Nutrients enter the cell and wastes leave the cell through the plasma membrane.

Selective permeability (pur mee uh BIH luh tee) of the plasma membrane allows some substances to pass through while keeping others out. The figure below shows selective permeability of the cell's plasma membrane. The arrows show common substances that enter and leave the cell. The plasma membrane controls how, when, and how much of these substances enter and leave the cells.

Picture This

1. **Highlight** the items in the figure that enter the cell through the plasma membrane. Circle the items that exit the cell.

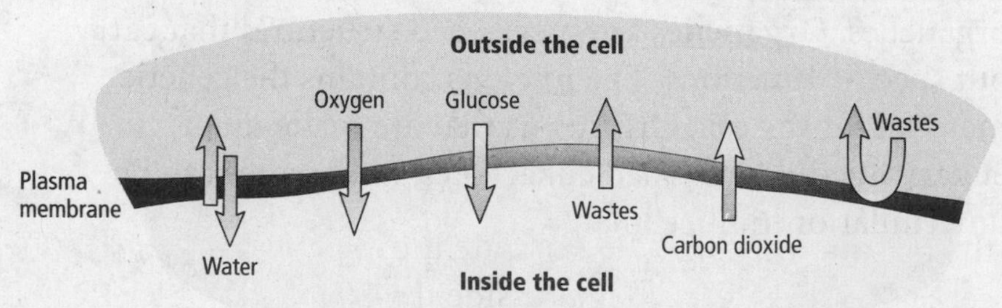

Structure of the Plasma Membrane

You have learned that lipids are large molecules made up of glycerol and three fatty acids. A phospholipid (fahs foh LIH pid) is made up of glycerol, two fatty acids, and a phosphate group. The plasma membrane is made up of two layers of phospholipids arranged tail-to-tail in what is called a **phospholipid bilayer.** The phospholipid bilayer allows the plasma membrane to survive and function in its watery environment. ☑

What is the structure of the phospholipid bilayer?

Each phospholipid has a polar head and two nonpolar tails. The phosphate group in the phospholipid makes it polar. The polar head is attracted to water because water is also polar. The nonpolar tails, made of the fatty acids, are repelled by water.

The phospholipid bilayer is arranged so that the polar heads can be closest to the water that is inside and outside the cell. Likewise, the nonpolar tails are farthest from the water because they are inside the phospholipid bilayer, as shown in the figure below. This bilayer structure is important for the formation and function of the plasma membrane.

Reading Check

2. Explain the purpose of the phospholipid bilayer.

____

____

____

____

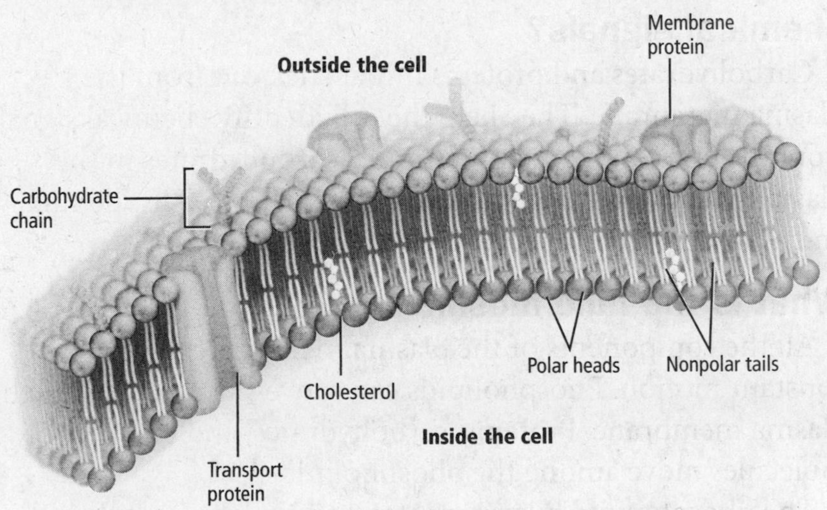

Picture This

3. Identify Circle one phospholipid. Label its head and tails.

How does the phospholipid bilayer function?

The phospholipid bilayer forms a barrier that is polar on the surface and nonpolar in the middle. Substances that can dissolve in water will not pass through the plasma membrane because they are stopped by the nonpolar middle. This allows the plasma membrane to separate the environment inside the cell from the environment outside the cell.

What else is found in the plasma membrane?

Cholesterol, proteins, and carbohydrates move among the phospholipids in the plasma membrane. Proteins are found on both the inner surface and the outer surface of the plasma membrane. Proteins on the outer surface are called receptors because they send signals to the inside of the cell. Proteins on the inner surface anchor the plasma membrane to the cell's internal support structure. These proteins give the cell its shape.

What are transport proteins?

Proteins also create tunnels through the plasma membrane. These proteins, known as **transport proteins,** move needed substances or waste materials through the plasma membrane. Transport proteins contribute to the selective permeability of the plasma membrane. ☑

Reading Check

4. Define the role of transport proteins.

____

____

____

How does cholesterol help cells?

Cholesterol molecules are nonpolar. They move among the tails of the phospholipids. Cholesterol helps prevent the fatty-acid tails from sticking together, keeping the plasma membrane fluid. Cholesterol also helps maintain homeostasis in a cell.

What substances help identify chemical signals?

Carbohydrates and proteins might stick out from the plasma membrane. They help the cell identify chemical signals from the environment. For example, carbohydrates in the plasma membrane might help disease-fighting cells identify and attack a potentially harmful cell.

What is the fluid mosaic model?

All the components of the plasma membrane are in constant motion. Phospholipids can move sideways within the plasma membrane. Proteins, carbohydrates, and cholesterol molecules move among the phospholipids. ☑

Reading Check

5. Name three substances that move among the phospholipids of the plasma membrane.

____

____

____

The phospholipid bilayer creates a sea in which all the other molecules float. As the individual molecules move around, a pattern, or mosaic, is formed on the surface of the plasma membrane. This organization of the plasma membrane is called the **fluid mosaic model.** It is fluid because the molecules are moving and being rearranged. It is called a mosaic because scientists can observe clear patterns on the surface of the plasma membrane.

Cellular Structure and Function

section 3 Structures and Organelles

Before You Read

For cells to function correctly, each part must do its job. Members of families have jobs or chores that help the whole family. On the lines below, list your family members and their jobs.

____

____

MAIN Idea

The eukaryotic cell contains organelles.

What You'll Learn

differences in the structures of plant and animal cells

Read to Learn

Cytoplasm and Cytoskeleton

The environment inside the plasma membrane is a semifluid material called **cytoplasm.** Scientists once thought the organelles of eukaryotic cells floated freely in the cell's cytoplasm. As technology improved, scientists discovered more about cell structures. They discovered a structure within the cytoplasm called the cytoskeleton. The **cytoskeleton** is a network of long, thin protein fibers that provide an anchor for organelles inside the cell. The cell's shape and movement depend on the cytoskeleton.

Two types of protein fibers make up the cytoskeleton. Microtubules are long, hollow protein cylinders that form a firm skeleton for the cell. They assist in moving substances within the cell. Microfilaments are thin protein threads that help give the cell shape and enable the entire cell or parts of the cell to move.

Mark the Text

Identify the Parts Highlight each cell structure as you read about it. Underline the function of each part.

Cell Structures

All chemical processes of a typical eukaryotic cell take place in the organelles, which move around in the cell's cytoplasm. Proteins are produced, food is transformed into energy, and wastes are processed in the organelles. Each organelle has a unique structure and function. ☑

Reading Check

1. **Name** one cell function that takes place in organelles.

____

____

____

____

How are plant and animal cells different?

The figure below shows a typical plant cell and a typical animal cell. Note how many organelles are found in both types of cells. Also, note a few differences, such as the chloroplast that appears only in the plant cell. Observe that the vacuole in the plant cell is much larger than the vacuole in the animal cell.

Picture This

2. Highlight the names of structures found in both plant cells and animal cells. Circle the names of structures that are found only in plant cells. Underline the names of structures that are found only in animal cells.

Plant Cell

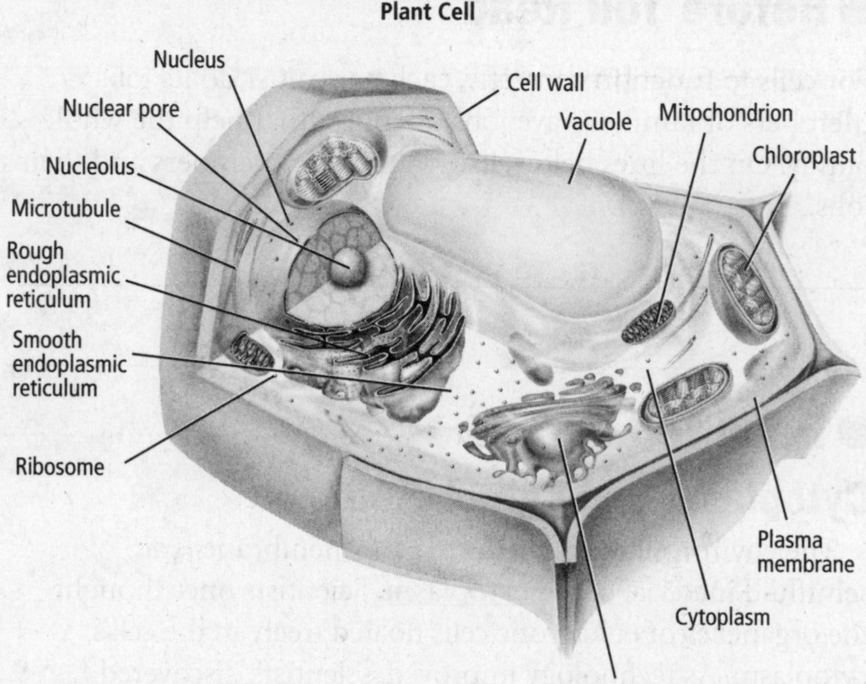

Animal Cell

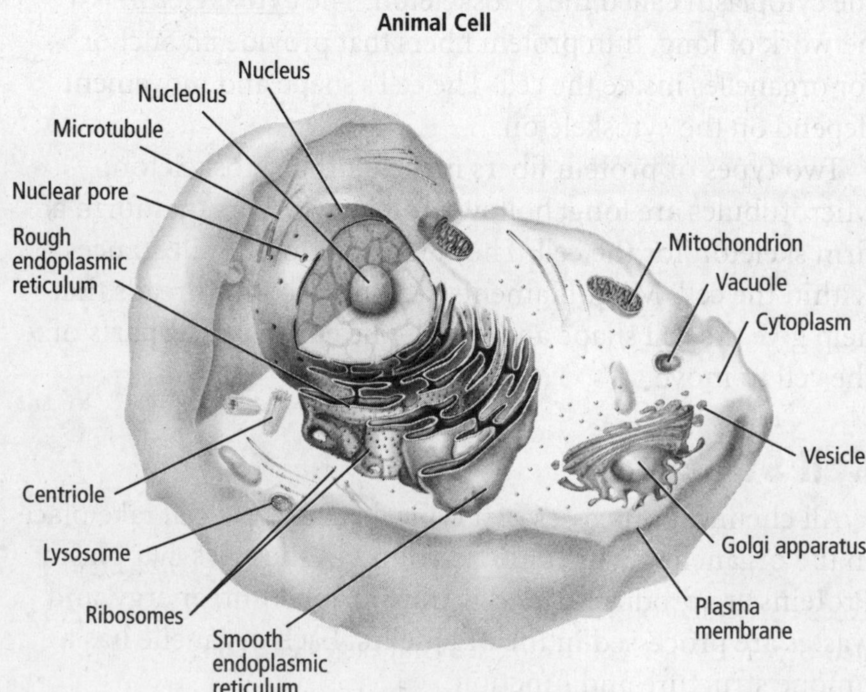

What structure manages cell processes?

The nucleus is the cell's managing structure. Most of the cell's genetic material (DNA) is in the nucleus. DNA defines the cell and controls protein production. A nuclear envelope surrounds the nucleus. Substances pass through the nuclear envelope to move in and out of the nucleus. ☑

Reading Check

3. Explain What is the role of DNA?

____

____

____

Which organelle produces proteins?

Ribosomes produce proteins and are made of two components—RNA and protein. Ribosomes are produced in the **nucleolus,** a structure located inside the nucleus. Some ribosomes float freely in the cytoplasm. They produce proteins that will be used by other cells. Other ribosomes attach to an organelle called the endoplasmic reticulum.

What attaches to rough endoplasmic reticulum?

The **endoplasmic reticulum,** (en duh PLAZ mihk • rih TIHK yuh lum) also called ER, is a membrane system of folded sacs and channels to which ribosomes are attached. There are two types of ER. The first type is called rough endoplasmic reticulum. This is the area where ribosomes attach to the ER's surface. The ribosomes appear to create bumps or rough places on the membrane. The second type, smooth endoplasmic reticulum, has no ribosomes attached. Smooth ER produces complex carbohydrates and lipids.

What is the purpose of the Golgi apparatus?

Once proteins are created, they move to another organelle, the Golgi (GAWL jee) apparatus. The **Golgi apparatus** modifies, sorts, and packs the proteins into sacs called vesicles. The vesicles fuse with the cell's plasma membrane. There the vesicles release the proteins, which move through the plasma membrane to the environment outside the cell.

What is stored in vacuoles?

Cells have vesicles called **vacuoles** that act as temporary storage for materials in the cytoplasm. Vacuoles can store food and other material needed by a cell. They can also store wastes. Plant cells normally have one large vacuole. Animal cells might or might not have a few small vacuoles.

What are lysosomes?

Lysosomes are vesicles that contain substances that digest excess or worn-out organelles and food particles. Lysosomes also digest bacteria and viruses that enter the cell.

What makes up a centriole?

Centrioles are organelles made of microtubules that function during cell division. They usually are found near the nucleus of the cell. Plant cells do not contain centrioles.

Which organelle produces energy?

Cells need energy to survive. The organelles that convert fuel particles such as sugars into usable energy are called **mitochondria** (mi tuh KAHN dree uh). A mitochondrion has an outer membrane and an inner membrane with many folds, as shown in the figure below. The membrane provides a large surface area for breaking the bonds of sugar molecules. Energy is produced when the bonds are broken.

Picture This

4. Explain why the inner membrane has many folds.

Mitochondrion

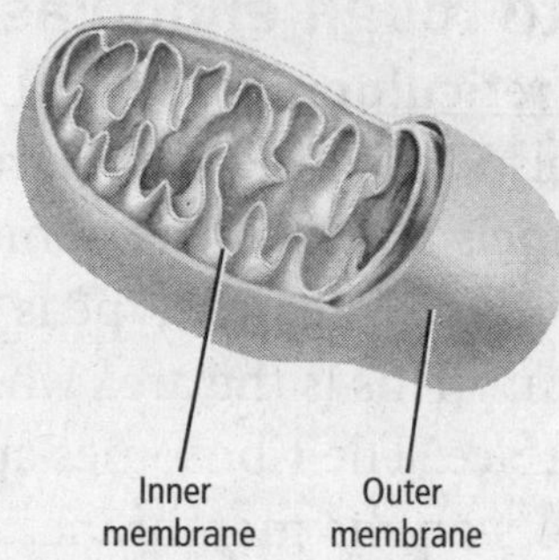

How do plant and animal cells differ?

In addition to mitochondria, plant cells contain chloroplasts. **Chloroplasts** are organelles that capture light energy and convert it to chemical energy through a process called photosynthesis. Plants can use light energy from any light source—usually the Sun. Animal cells do not have chloroplasts and cannot use solar energy as fuel for cell processes.

Plants also have cell walls. The **cell wall** is a mesh of fibers that surrounds the plasma membrane. It protects and supports the cell. Plant cell walls are made of a carbohydrate known as cellulose.

Think it Over

5. Recall What is another function of microtubules?

What are cilia and flagella?

Some animal cell surfaces have cilia or flagella that extend beyond the plasma membrane. **Cilia** are short projections that look like hairs. They move back and forth, similar to the motion of the oars of a rowboat. **Flagella** are longer, whiplike projections that propel cells. Both cilia and flagella are composed of microtubules. They move cells through their watery environments. Cilia also move substances along the surface of the cell.

Comparing Cells

The table below summarizes the structures of eukaryotic plant and animal cells. The function of each structure is described. Note that prokaryotic cells lack most of the organelles found in eukaryotic cells.

Picture This

6. Highlight the organelle to which ribosomes attach.

Cell Structure	Function	Present in Plant/Animal Cells
Cell wall	protects and supports plant cells	plant cells only
Centriole	important in cell division	animal cells only
Chloroplast	site where photosynthesis occurs	plant cells only
Cilia	aids in moving the cell and moving substances along the surface of the cell	some animal cells
Cytoskeleton	a framework for the cell within the cytoplasm	both
Endoplasmic reticulum (ER)	site of protein synthesis; where ribosomes attach	both
Flagellum	aids in moving and feeding the cell	some animal cells
Golgi apparatus	modifies and packages proteins for distribution outside the cell	both
Lysosome	contains digestive enzymes for substance break down	animal cells only
Mitochondrion	supplies energy to the rest of the cell	both
Nucleus	directs the production of proteins and cell division	both
Plasma membrane	controls the movement of substances in and out of the cell	both
Ribosome	produces proteins	both
Vacuole	stores materials temporarily	plant cell—one large; animal cell—a few small

Organelles at Work

The structures in the cell work together to perform cell functions. The synthesis of proteins is a major cell function, which begins in the nucleus. Protein synthesis continues with the ribosomes on the rough ER and the ribosomes that float freely in the cytoplasm. Most proteins made on the rough ER are sent to the Golgi apparatus. There they are packaged in vesicles and sent to other organelles or out of the cell. Like each member of a soccer team, each cell structure has a specific task to do to make the cell function properly. ☑

7. Identify a major cell function.

chapter 7

Cellular Structure and Function

section 4 Cellular Transport

MAIN Idea

Cellular transport moves substances in and out of a cell.

What You'll Learn

- the processes of diffusion, facilitated diffusion, and active transport
- effect of hypotonic, hypertonic, or isotonic solutions on a cell
- how large particles enter and exit a cell

Before You Read

Describe on the lines below how you would move a large box that weighs more than you do. Then read the section to learn how large particles move in and out of cells.

____

____

____

Mark the Text

Create a Quiz As you read this section, write quiz questions based on what you have learned. After you write the questions, answer them.

FOLDABLES™

Record Information Make a three-pocket Foldable from an 11 × 17 sheet of paper. As you read, record information about cellular transport on quarter sheets of notebook paper and store them in the appropriate pocket.

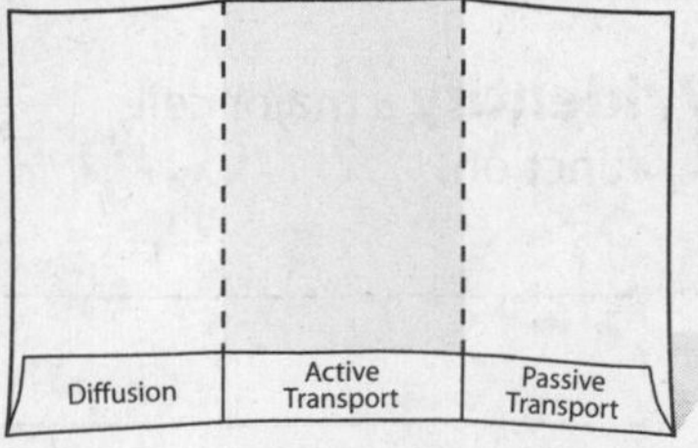

Read to Learn

Diffusion

Substances dissolved in water move constantly and randomly. Imagine you place a drop of red ink on the left side and a drop of blue ink on the right side of a dish of water. The ink moves randomly through the water and turns the water purple as the colors mix. The ink has diffused in the water. **Diffusion** is the net movement of particles from an area where there are more particles of the substance to an area where there are fewer particles. Diffusion does not require additional energy because the particles are already in motion.

Concentration is the amount of a substance in an area. Diffusion continues until the concentrations are the same in all areas of the water. The dish of water has reached **dynamic equilibrium,** in which the particles continue to move randomly, but the overall concentration does not change.

What affects the rate of diffusion?

Concentration, temperature, and pressure affect the rate of diffusion. Diffusion occurs more quickly when the concentration, temperature, or pressure are high because the particles collide more often. The size and charge of a substance also affects the rate of diffusion.

What is facilitated diffusion?

Water can diffuse across the plasma membrane. However, other ions and molecules that cells need to function cannot diffuse across the plasma membrane. Molecules such as sugars and chlorine need help to move from outside the cell's environment to inside the cell. **Facilitated diffusion** uses transport proteins to help move some ions and small molecules across the plasma membrane. One type of facilitated diffusion is shown in the figure below.

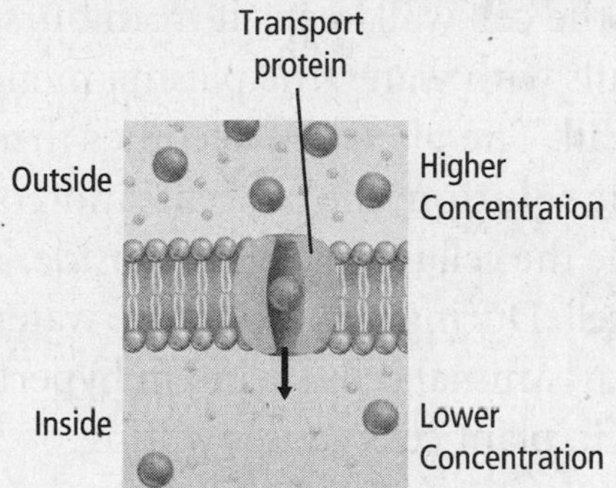

Picture This

1. Explain Use this figure to explain facilitated diffusion to a partner.

Diffusion of water and facilitated diffusion of ions and small molecules occur without additional energy because the particles are already moving. When no energy is added, the transport is referred to as passive transport.

Osmosis: Diffusion of Water

Water passes in and out of the cell through the plasma membrane. The diffusion of water across a selectively permeable membrane is called **osmosis** (ahs MOH sus). Osmosis helps the cell maintain homeostasis.

What is the result of osmosis?

Most cells undergo osmosis because they are surrounded by watery solutions. These solutions have different concentrations than the inside environment of the cell. Before osmosis, the concentration inside and outside the cell have not reached dynamic equilibrium. After osmosis, the concentrations are the same on both sides of the membrane, and dynamic equilibrium has been reached.

What happens to a cell in an isotonic solution?

A cell in an **isotonic solution** has the same concentration in its cytoplasm as its surrounding watery environment. Water continues to move through the plasma membrane, but water enters and leaves the cell at the same rate. The cell is at equilibrium with its surrounding environment. ✔

Reading Check

2. Explain Why is the cell at equilibrium in an isotonic solution?

How do hypotonic solutions and hypertonic solutions differ?

If a cell is placed in a solution that has a lower concentration of dissolved substances, the cell is in a **hypotonic solution.** There is more water outside the cell than inside the cell. Osmosis moves water into the cell. ☑

Reading Check

3. Analyze Why does water move into a cell placed in a hypotonic solution?

____

____

____

As water moves into an animal cell, the plasma membrane swells. If the solution is too hypotonic, pressure builds inside the cell, and it might burst.

In a plant cell, the cell wall keeps it from bursting. As the central vacuole fills with water, the plasma membrane pushes against the cell wall. The plant cell becomes firmer.

In a **hypertonic solution,** the concentration of dissolved substances outside the cell is higher than inside. There is more water inside the cell. During osmosis, more water moves out of the cell than into it. Animal cells shrink in hypertonic solutions. The loss of water in plant cells causes wilting.

Active Transport

Substances might need to move from an area of lower concentration to an area of higher concentration. Transport proteins help move substances across the plasma membrane against the normal flow. This movement against the normal flow requires energy and is called **active transport.**

Transport of Large Particles

Some substances are too large to move by diffusion or active transport. **Endocytosis** is the process by which a cell surrounds a substance in the outside environment with a portion of the plasma membrane, then pinches off the membrane, bringing the substance inside the cell.

Exocytosis is the reverse process by which large substances exit the cell. Both processes, as shown in the figure below, require energy. As with other forms of transport, endocytosis and exocytosis help cells maintain homeostasis.

Picture This

4. Label the cell structure through which substances pass as they leave the cell during exocytosis.

Endocytosis

Inside the cell

Outside the cell

Plasma membrane

Vacuole

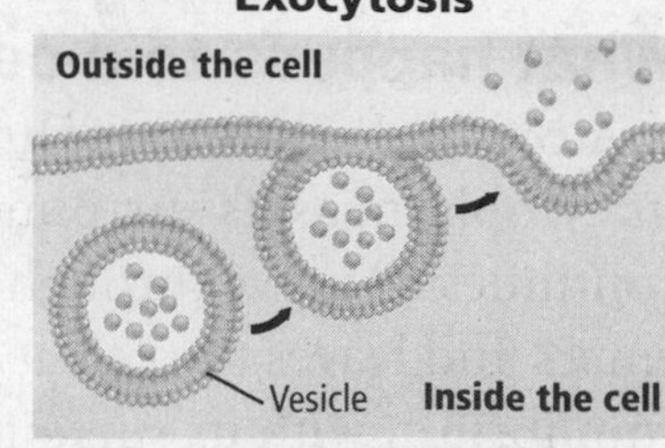

Cellular Energy

section 1 How Organisms Obtain Energy

Before You Read

Think about the objects in your home that use energy. On the lines below, describe the ways that these objects get energy. Then read about how organisms obtain energy.

____

____

MAIN Idea

All living organisms use energy to carry out all biological processes.

What You'll Learn

- the two laws of thermodynamics
- the difference between autotrophs and heterotrophs
- how ATP works in a cell

Read to Learn

Study Coach

Create a Quiz After you read this section, create a quiz based on what you have learned. Then be sure to answer the quiz questions.

Transformation of Energy

Cells need energy. They need energy to move molecules across membranes and to make and break down molecules. **Energy** is the ability to do work. **Thermodynamics** is the study of how energy flows and changes in the universe.

What are the laws of thermodynamics?

Two laws of thermodynamics explain the flow of energy. The first law states that energy can change form, but it cannot be created or destroyed. For example, your body changes the chemical energy in food into a more useable form. Then when you move, your body changes that energy into mechanical energy. ☑

Reading Check

1. State the first law of thermodynamics.

____

____

____

The second law of thermodynamics states that systems change from states of order to states of disorder on their own. This disorder is known as entropy (EN truh pee). Entropy is always increasing. This means that when your body changes forms of energy, some of the energy is lost as heat. The energy is still present, but it can no longer be used.

How do organisms get energy from the Sun?

Nearly all the energy for life on Earth comes from the Sun. Some organisms make their own food. Some autotrophs use inorganic substances as a source of energy. Other autotrophs change light energy from the Sun into chemical energy. Plants and some bacteria are autotrophs.

Heterotrophs get their energy by eating food. Heterotrophs get energy from the Sun indirectly. They do this by eating autotrophs. Animals are heterotrophs. The figure below shows the relationship between autotrophs and heterotrophs.

Picture This

2. Circle the organism that makes its own food.

Metabolism

All of the chemical reactions that go on inside a cell are known as the cell's **metabolism**. A series of reactions in which the product of one reaction becomes the reactant for the next reaction is called a metabolic pathway.

What are the two metabolic pathways?

There are two types of metabolic pathways: catabolic (ka tuh BAH lik) pathways and anabolic (a nuh BAH lik) pathways. In catabolic pathways, energy is released by breaking larger molecules into smaller molecules. In anabolic pathways, the energy released by catabolic pathways is used to build larger molecules from smaller molecules. ☑

Energy flows between the metabolic pathways of organisms in an ecosystem. Photosynthesis is an anabolic pathway. Cellular respiration is a catabolic pathway. These pathways work together to meet the energy needs of cells.

Reading Check

3. Compare the energy usage in anabolic and catabolic pathways.

____

____

____

____

How is energy changed during photosynthesis?

Photosynthesis is a series of reactions that change light energy from the Sun into chemical energy that can be used by the cell. During photosynthesis, light energy, carbon dioxide, and water are changed into organic molecules and oxygen. The energy stored in organic molecules made during photosynthesis can be passed to other organisms. When an animal eats a plant, the plant's stored energy is passed to the animal.

What happens during cellular respiration?

Cellular respiration is a series of reactions that break down organic molecules into carbon dioxide, water, and energy. The energy is used by the cell. The processes of cellular respiration and photosynthesis form a cycle, which is shown in the figure below. The products of photosynthesis are the reactants for cellular respiration, and the products of cellular respiration are the reactants for photosynthesis.

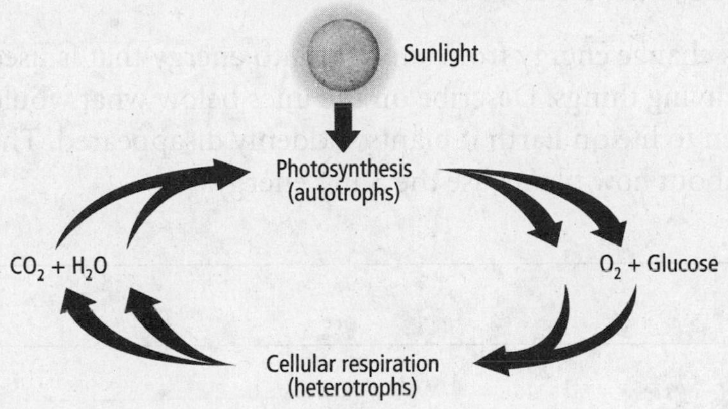

Picture This

4. Identify Draw a circle around the anabolic process and a square around the catabolic process.

ATP: The Unit of Cellular Energy

Cells store chemical energy in biological molecules. The most important biological molecule is **adenosine triphosphate** (uh DEN uh seen • tri FAHS fayt), or **ATP**.

How does ATP store energy?

ATP is the most abundant energy-storing molecule. It is found in all kinds of organisms. The structure of ATP is shown below. It is made of an adenine base, a ribose sugar, and three phosphate groups.

ATP releases energy when the bond between the second and third phosphate groups is broken, forming a molecule called adenosine diphosphate (ADP). ADP can be changed back into ATP by adding a phosphate group.

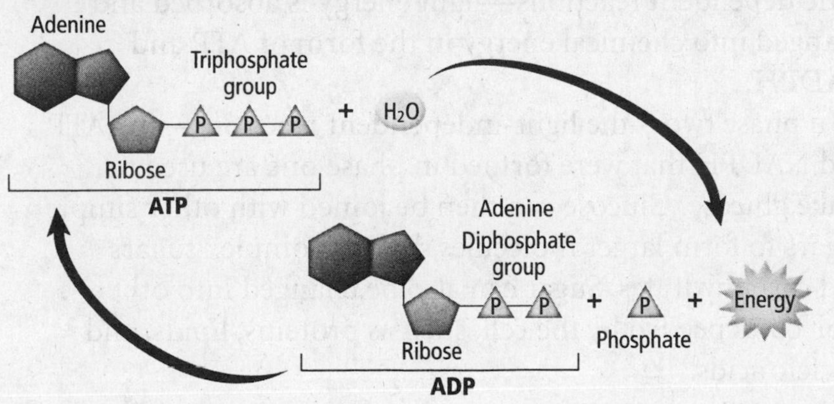

Picture This

5. Identify Circle the high-energy bond that is broken when ATP is converted to ADP.

chapter 8

Cellular Energy

section 2 Photosynthesis

MAIN Idea

Light energy is trapped and converted into chemical energy during photosynthesis.

What You'll Learn

- the two phases of photosynthesis
- how a chloroplast works during light reactions
- how electron transport works

Before You Read

Plants change energy from sunlight into energy that is used by other living things. Describe on the lines below what would happen to life on Earth if plants suddenly disappeared. Then read about how plants use the Sun's energy.

____

____

____

____

Mark the Text

Identify Details As you read, highlight or underline the events of each stage of photosynthesis.

Read to Learn

Overview of Photosynthesis

Photosynthesis is the process in which light energy from the Sun is changed into chemical energy. Nearly all life on Earth depends on photosynthesis. The chemical equation for photosynthesis is shown below.

$$6CO_2 + 6H_2O \xrightarrow{\text{light}} C_6H_{12}O_6 + 6O_2$$

Photosynthesis occurs in two phases. In phase one—the light-dependent reactions—light energy is absorbed and changed into chemical energy in the form of ATP and NADPH.

In phase two—the light-independent reactions—the ATP and NADPH that were formed in phase one are used to make glucose. Glucose can then be joined with other simple sugars to form larger molecules such as complex sugars and carbohydrates. Sugar can also be changed into other molecules needed by the cell, such as proteins, lipids, and nucleic acids. ✓

✓ Reading Check

1. **Identify** one way cells can use glucose.

____

____

____

____

Phase One: Light Reactions

Plants have special organelles called chloroplasts to capture light energy. Photosynthesis begins when sunlight is captured. The captured energy is stored in two energy storage molecules—ATP and NADPH—that will be used in light-independent reactions.

What happens in chloroplasts?

Chloroplasts are large organelles that capture light energy from the Sun. They are found in plants and other photosynthetic organisms. The figure below shows a chloroplast.

A chloroplast is a disc-shaped organelle that contains two compartments. **Thylakoids** (THI la koyds) are flattened saclike membranes. The thylakoids are arranged in stacks called **grana**. The fluid-filled space outside the grana is the **stroma**. Phase one takes place in the thylakoids. Phase two takes place in the stroma.

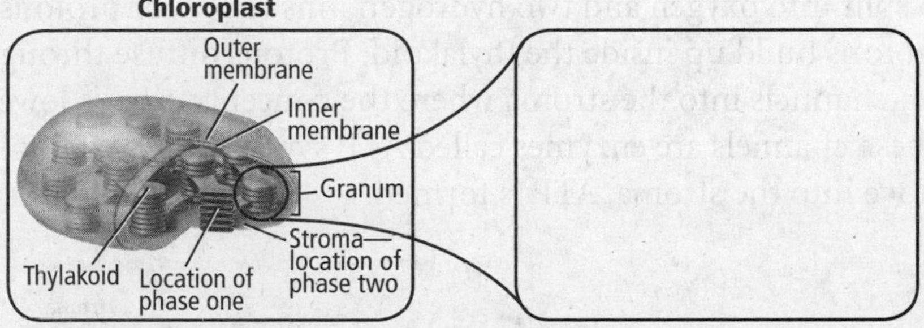

What is the role of pigments in photosynthesis?

Thylakoids contain light-absorbing colored molecules known as **pigments**. Different pigments absorb different wavelengths of light. Chlorophylls are the major light-absorbing pigments in plants. They absorb energy from violet-blue light and reflect green light, giving plants their green color.

Accessory pigments help plants absorb additional light. For instance, carotenoids (kuh ROH tuh noyds) absorb blue and green light and reflect yellow, orange, and red light. Carotenoids give carrots and sweet potatoes their orange color.

Accessory pigments are the reason leaves change colors in autumn. In green leaves, there is so much chlorophyll that it masks the other pigments. In autumn, as trees prepare to lose their leaves, the chlorophyll molecules break down, revealing the colors of other pigments. The colors red, yellow, and orange can be seen. ☑

2. Name Which organism has chloroplasts? (Circle your answer.)

a. mushroom
b. oak tree
c. earthworm

Picture This

3. Illustrate In the box, draw an enlarged picture of a granum.

Reading Check

4. Explain Why do the leaves of some trees change colors in autumn?

____

____

____

____

____

How does electron transport work?

Photosystem I and photosystem II are made of pigments that absorb light and proteins that are important in light reactions. They are in the thylakoid membrane. Follow along in the figure below as you read about their role in photosynthesis. ☑

Photosynthesis begins when light energy causes electrons in photosystem II to go into a high energy state. The light energy also causes a water molecule to split, releasing an electron into the electron transport system, a hydrogen ion into the thylakoid space, and oxygen as a waste product. The excited electrons move from photosystem II and move along a series of electron-carriers to photosystem I. Photosystem I absorbs more light, and the excited electrons move along electron-carriers again. Finally, the electrons are moved to **NADP+**, forming the energy-storage molecule NADPH.

Reading Check

5. Describe Of what are photosystem I and photosystem II made?

______

______

______

How is ATP made during photosynthesis?

ATP is made when light energy causes the water molecule to split into oxygen and two hydrogen ions (H^+), or protons. Protons build up inside the thylakoid. Protons diffuse through ion channels into the stroma where the concentration is lower. These channels are enzymes called ATP synthases. As protons move into the stroma, ATP is formed.

Picture This

6. Identify On the figure, highlight the path that electrons follow. What molecule is the electron's final destination?

______

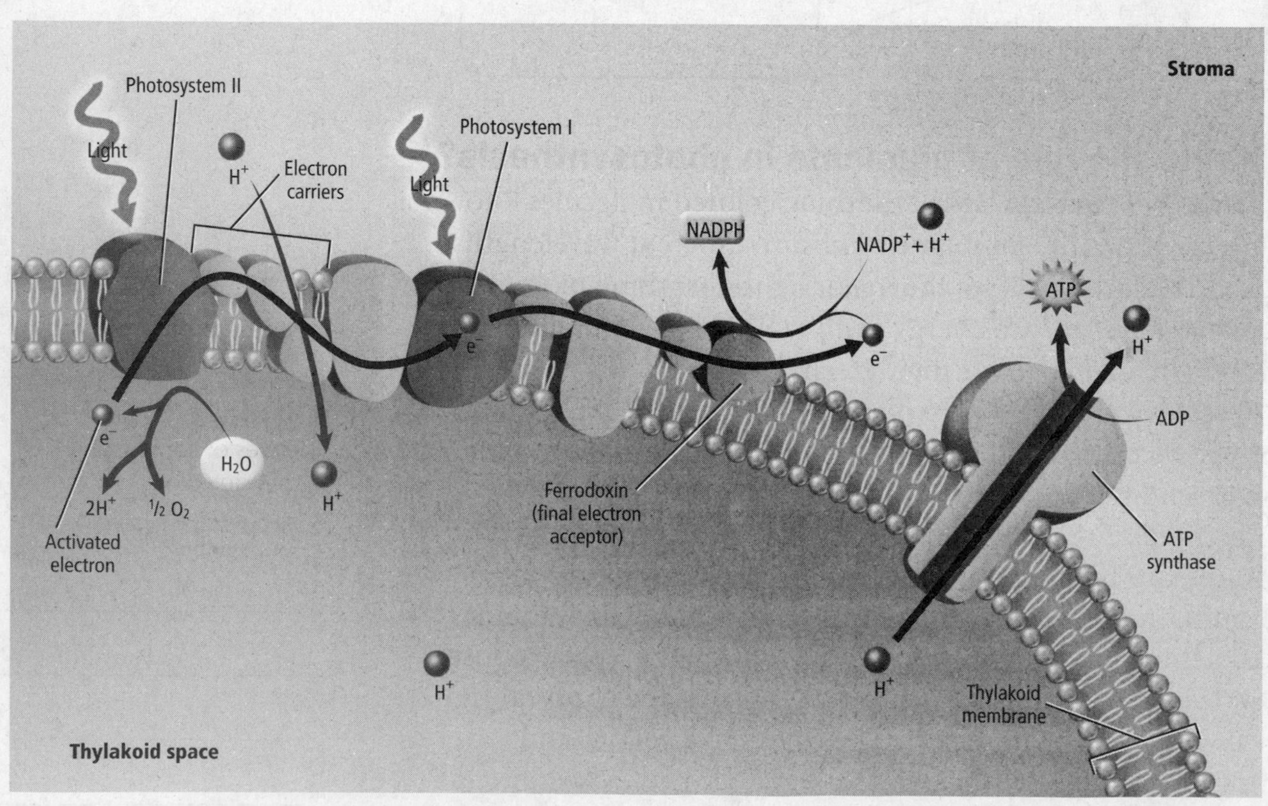

Phase Two: The Calvin Cycle

NADPH and ATP are temporary storage molecules. During phase two, also known as the **Calvin cycle**, the energy in these molecules is stored in organic molecules, such as glucose.

What happens in the Calvin cycle?

The Calvin cycle builds sugars out of carbon dioxide and water using the energy stored in ATP and NADPH. The Calvin cycle's reactions do not require sunlight, which is why they are also referred to as light-independent reactions.

In the Calvin cycle, carbon dioxide molecules combine with six 5-carbon compounds to make twelve 3-carbon molecules. The chemical energy stored in ATP and NADPH is passed to the 3-carbon molecules. Two 3-carbon molecules leave the cycle to be used to make glucose and other organic compounds. The enzyme **rubisco** changes ten 3-carbon molecules into six 5-carbon molecules to continue the cycle. Because rubisco changes carbon dioxide molecules into organic molecules that can be used by the cell, it is considered one of the most important enzymes. Sugar formed in the Calvin cycle can be used as energy and as building blocks for complex carbohydrates, such as starch.

Think it Over

7. Name the main energy-storing products of each phase of photosynthesis.

______

______

______

Alternative Pathways

Photosynthesis might be difficult for plants that grow in hot, dry environments. Many plants in extreme climates have evolved other photosynthesis pathways.

Tropical plants such as sugar cane and corn use the C_4 pathway. Instead of the 3-carbon molecules of the Calvin cycle, C_4 plants fix carbon dioxide into 4-carbon molecules. Less water is lost in the C_4 pathway. These plants keep their stomata closed during hot days to minimize water loss.

What are CAM plants?

Another alternative pathway is called the CAM pathway. CAM plants live in deserts, salt marshes, and other environments where access to water is limited. Cacti and orchids are CAM plants. Carbon dioxide enters the leaves of CAM plants only at night, when the atmosphere is cooler and more humid. The plants also fix carbon dioxide into organic compounds at night. During the day, carbon dioxide is released from organic compounds in the plants. The carbon dioxide enters the Calvin cycle at that point. The CAM pathway minimizes water loss, while allowing for adequate carbon uptake. ☑

Reading Check

8. Name two places where CAM plants live.

______

______

Cellular Energy

section 3 Cellular Respiration

MAIN Idea

Living organisms obtain energy during cellular respiration.

What You'll Learn

- the role of electron carriers in cellular respiration
- the difference between alcoholic fermentation and lactic-acid fermentation

Mark the Text

Identify Main Ideas

As you read, underline or highlight the main ideas in each paragraph.

FOLDABLES™

Compare Make a three-tab Venn diagram Foldable from one sheet of paper to compare and contrast aerobic and anaerobic respiration.

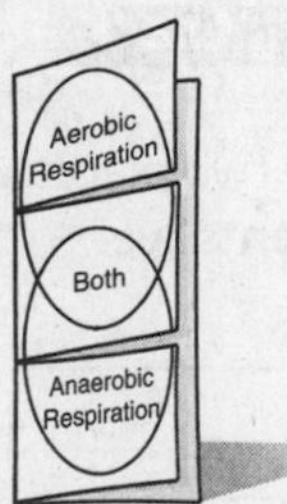

Before You Read

The energy your body uses comes from the Sun. On the lines below, explain how energy from the Sun is passed to you.

____

____

Read to Learn

Overview of Cellular Respiration

Organisms get energy through cellular respiration. Electrons from carbon compounds such as glucose are collected, and the energy is used to make ATP. ATP is used by cells. The equation for respiration, shown below, is the opposite of the equation for photosynthesis.

$$C_6H_{12}O_6 + 6O_2 \rightarrow 6CO_2 + 6H_2O + \text{energy}$$

Cellular respiration begins with glycolysis, a process in which glucose is broken down into pyruvate. Glycolysis is an **anaerobic process**, meaning it does not need oxygen. Glycolysis is followed by **aerobic processes**, which require the presence of oxygen. During **aerobic respiration**, pyruvate is broken down and ATP is made. Aerobic respiration occurs in two parts: the Krebs cycle and electron transport.

Glycolysis

During **glycolysis**, two phosphate groups are joined to glucose, using two molecules of ATP. The 6-carbon molecule is then broken down into two 3-carbon compounds. Two phosphates are added, and electrons and protons combine with two NAD^+ molecules to form two NADH molecules. The two 3-carbon molecules are changed into two molecules of pyruvate. Four molecules of ATP are made.

Krebs Cycle

Next, the pyruvate, made during glycolysis, is transported into the mitochondria. There it is converted into carbon dioxide in a series of reactions called the **Krebs cycle**.

What are the steps of the Krebs cycle?

Before the pyruvate enters the Krebs cycle, it reacts with coenzyme A (CoA) to form a 2-carbon intermediate called acetyl CoA. Carbon dioxide is released, and NAD^+ is changed to NADH. Acetyl CoA then moves to the mitochondria, where it combines with a 4-carbon molecule to form citric acid. Citric acid is then broken down, releasing two molecules of carbon dioxide and making one ATP, three NADH, and one $FADH_2$. Acetyl CoA and citric acid are made, and the cycle continues. Two pyruvate molecules are made during glycolysis, resulting in two turns of the Krebs cycle for each glucose molecule. ☑

1. Identify In the Krebs cycle, what is pyruvate converted to?

Electron Transport

Electron transport, the final stage of cellular respiration, takes place in the mitochondria. The high-energy electrons and protons from NADH and $FADH_2$ are used to change ADP to ATP.

Electrons are passed along a series of proteins. Electrons and protons are released from NADH and $FADH_2$ into the mitochondria. Protons and electrons are transferred to oxygen to make water. Electron transport makes 24 ATP molecules.

Picture This

2. Identify Complete the figure by writing the location of each stage of cellular respiration.

Overview of Cellular Respiration

	Location	Main Activity	High-Energy Molecules Made per Glucose Molecule
Glycolysis		Glucose is converted to pyruvate.	2 ATP, 2 NADH
Krebs cycle		Pyruvate is converted to carbon dioxide.	2 ATP, 8 NADH, 2 $FADH_2$
Electron transport		Electrons and protons combine with oxygen to make water.	28 ATP

Do prokaryotes use cellular respiration?

Some prokaryotes also undergo aerobic respiration. Since they have no mitochondria, electron transport occurs in the cellular membrane instead. Pyruvate does not move to mitochondria, saving the prokaryotic cell two ATP. Prokaryotes make 38 molecules of ATP from one molecule of glucose.

Anaerobic Respiration

Anaerobic respiration takes place when oxygen is low. Some prokaryotes that do not need oxygen use anaerobic respiration all the time. Other cells use anaerobic respiration when oxygen levels are low.

How is ATP made during anaerobic respiration?

Anaerobic respiration, or **fermentation**, follows glycolysis when oxygen is absent. Glycolysis makes two ATP from each glucose molecule. Fermentation makes a small amount of ATP and regenerates the cell's supply of NAD^+ so glycolysis can continue. Two important types of fermentation are lactic-acid fermentation and alcohol fermentation. ☑

Reading Check

3. Define What two processes make up anaerobic respiration?

____

____

____

What are the types of fermentation?

Lactic-acid fermentation changes pyruvate into lactic acid. It takes place in skeletal muscle cells during strenuous exercise, when the body cannot supply enough oxygen. It is also used to make foods like cheese, yogurt, and sour cream.

Yeast and some bacteria undergo a type of fermentation known as alcohol fermentation. These organisms use pyruvate to make ethyl alcohol and carbon dioxide.

Photosynthesis and Cellular Respiration

Photosynthesis and cellular respiration are important ways that cells get and use energy. These processes are related in important ways. The products of photosynthesis—oxygen and glucose—are needed for cellular respiration. The products of respiration—carbon dioxide and water—are needed for photosynthesis. The figure below shows this relationship.

Picture This

4. Classify What type of organisms have cells that carry out all of the processes shown at the right?

____

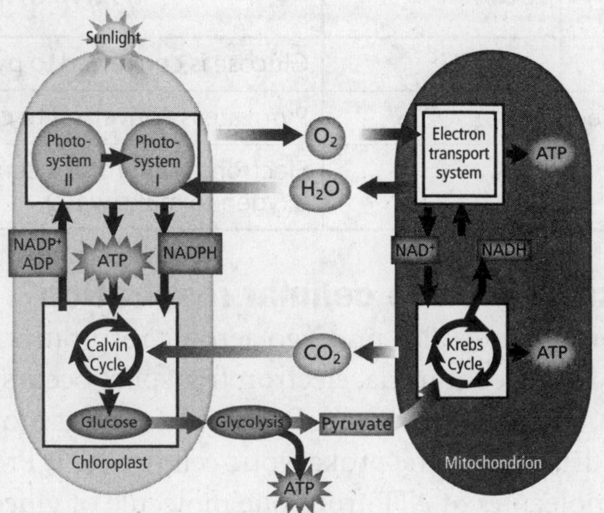

chapter 9

Cellular Reproduction

section 1 Cellular Growth

Before You Read

Think about the life cycle of a human. On the lines below, write some of the stages that occur in the life cycle of a human. In this section, you will learn about the life cycle of a cell.

____

____

MAIN Idea

Cells grow until they reach their size limit, then they either stop growing or divide.

What You'll Learn

- why cells are small
- the stages of the cell cycle
- the stages of interphase

Read to Learn

Cell Size Limitations

Most cells are smaller than the period at the end of this sentence. In this section, you will learn why cells are so small.

What is the surface area to volume ratio?

Recall that all cells are surrounded by a plasma membrane. All substances moving into or out of the cell must cross the plasma membrane. The surface area of the cell is the area covered by the plasma membrane. The volume of a cell is the space taken by the inner contents. Because cells are small, their surface area is high in relation to their volume. This relationship is called the ratio of surface area to volume.

To understand this ratio, imagine a cube-shaped bacterial cell with sides measuring one micrometer (μm) in length. Surface area can be calculated as length times width times the number of sides, or $1\ \mu m \times 1\ \mu m \times 6$ sides $= 6\ \mu m^2$. Volume is length times width times height, or $1\ \mu m \times 1\ \mu m \times 1\ \mu m$, which equals $1\ \mu m^3$. The ratio of surface area to volume of this cell is 6:1.

What happens as the cell grows larger? Imagine a cube-shaped bacterial cell that is 2 μm per side. The surface area is now $24\ \mu m^2$, and the volume is $8\ \mu m^3$. The ratio of surface area to volume is 3:1. Notice that the ratio of the larger cell is lower than that of the smaller cell.

Mark the Text

Identify Main Ideas As you read, underline or highlight the main ideas in each paragraph.

Applying Math

1. **Explain** Show the equations used to calculate $24\ \mu m^2$ for the surface area and $8\ \mu m^3$ for the volume of the larger cell.

____

____

____

____

What are the benefits of a high surface area to volume ratio?

As cells grow in size, the ratio of surface area to volume gets smaller. A low ratio means a cell might have trouble bringing nutrients into and moving wastes out of the cell. ☑

Small cells also have an easier time moving substances around inside the cell. Substances are moved by diffusion or by motor proteins pulling them along the cytoskeleton. Movement of substances over long distances is slow and difficult, so cells remain small to maintain efficiency.

Cells use signaling proteins to communicate. Signaling proteins move around the cell to relay messages. In larger cells, communication becomes slow because signaling proteins have to move over longer distances.

Reading Check

2. Explain What happens to ratio of surface area to volume as cells grow?

____

____

The Cell Cycle

When a cell reaches its size limit, it will either stop growing or it will divide. Cell division keeps cells from getting too large. It is also the way that cells reproduce. A cell's cycle of growing and dividing is called the **cell cycle**.

The cell cycle has three main stages. They are described in the table below. During **interphase**, the cell grows, carries out cellular functions, and copies its DNA. Interphase is followed by **mitosis** (mi TOH sus), the period when the nucleus divides. **Cytokinesis** (si toh kih NEE sis) follows mitosis and is the stage when the cytoplasm divides and two cells are created.

The time it takes a cell to complete the cell cycle varies depending on the type of cell. A typical animal cell takes 12–24 hours to complete the cell cycle. Some cells might complete the cycle in eight minutes. Other cells might take as long as a year to complete one cycle.

Picture This

3. State Complete the table by writing the number of cells present at the end of each stage. The first one has been done for you.

Stage	Description	Number of Cells
Interphase	The cell grows in size, performs normal functions, and copies its DNA.	one cell
Mitosis	The cell nucleus divides, and the chromosomes separate into the two nuclei.	
Cytokinesis	The cytoplasm of the cell divides, forming two daughter cells.	

What happens during interphase?

The stages of the cell cycle are shown in the figure below. Most of the cell cycle is taken up by interphase. During interphase the cell grows, performs normal cell functions, and copies its DNA in preparation for cell division. Interphase is divided into three stages: G_1, S, and G_2, also called Gap 1, Synthesis, and Gap 2.

As soon as a cell divides, it enters the G_1 stage. During G_1, a cell grows, performs normal cell functions, and prepares to copy its DNA. Some cells, such as muscles and nerve cells, exit the cycle at this stage and do not divide again.

During the S stage, the cell copies its DNA. **Chromosomes** (KROH muh sohmz) are the structures in the nucleus that contain DNA, the genetic material that is passed from generation to generation of cells. The DNA in chromosomes is tightly wound. **Chromatin** (KROH muh tin) is the relaxed, or unwound, form of DNA and proteins in the nucleus.

The G_2 stage is the period when the cell prepares for the division of its nucleus. When preparations are complete, the cell enters mitosis.

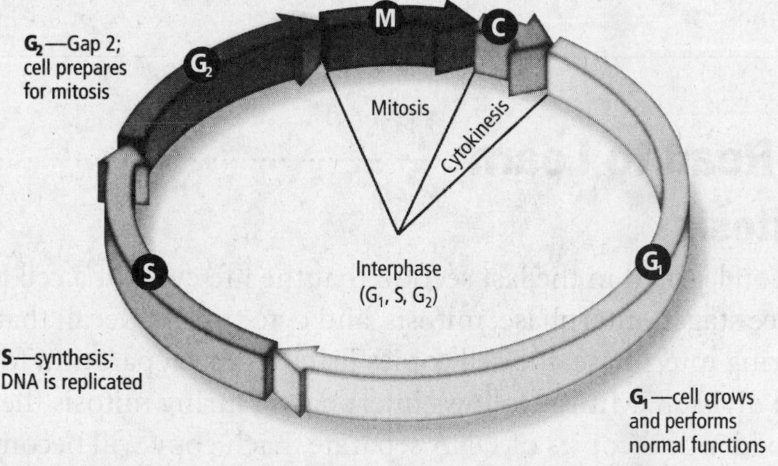

How is the cell cycle completed?

Mitosis and cytokinesis follow interphase. They are described in the next section. At the end of the cell cycle, cell division is complete, and the original cell has become two daughter cells.

How do prokaryotic cells divide?

The cell cycle you have just learned about—interphase, mitosis, and cytokinesis—occurs in eukaryotic cells. As you have learned, prokaryotic cells are simpler cells. They reproduce by a method called binary fission. ✔

Picture This

4. Label Circle the stage of interphase during which DNA is copied.

Reading Check

5. Name What is cell division in prokaryotes called?

____

____

Cellular Reproduction

section 2 Mitosis and Cytokinesis

MAIN Idea

Eukaryotic cells reproduce by mitosis and cytokinesis.

What You'll Learn

- the events of each stage of mitosis
- the process of cytokinesis

Before You Read

Recall the last time you got a cut. Skin heals itself with the help of cell division. Skin cells divide, creating new cells to replace the damaged cells. On the lines below, list some other times when your body might need to create new cells. In this section, you will read about two ways that cells reproduce.

____

____

____

____

Mark the Text

Identify Details As you read, highlight or underline the events of each stage of mitosis.

Read to Learn

Mitosis

You learned in the last section that the life cycle of a cell has three stages: interphase, mitosis, and cytokinesis. Recall that during interphase, the cell copies its DNA in preparation for cell division. Mitosis follows interphase. During mitosis the two identical copies of DNA separate. Each copy will become part of a new cell, called a daughter cell. Daughter cells are genetically identical because they each have the same DNA.

Mitosis increases the number of cells as a young organism grows to its adult size. Mitosis also replaces damaged cells, such as skin cells that are damaged when you get a cut. ☑

Reading Check

1. Name What is one function of mitosis in a multicellular organism?

____

____

____

____

The Stages of Mitosis

Like interphase, mitosis is divided into stages. These four stages are prophase, metaphase, anaphase, and telophase. The figure on the next page shows the four stages of mitosis. Follow the diagram as you read about each stage.

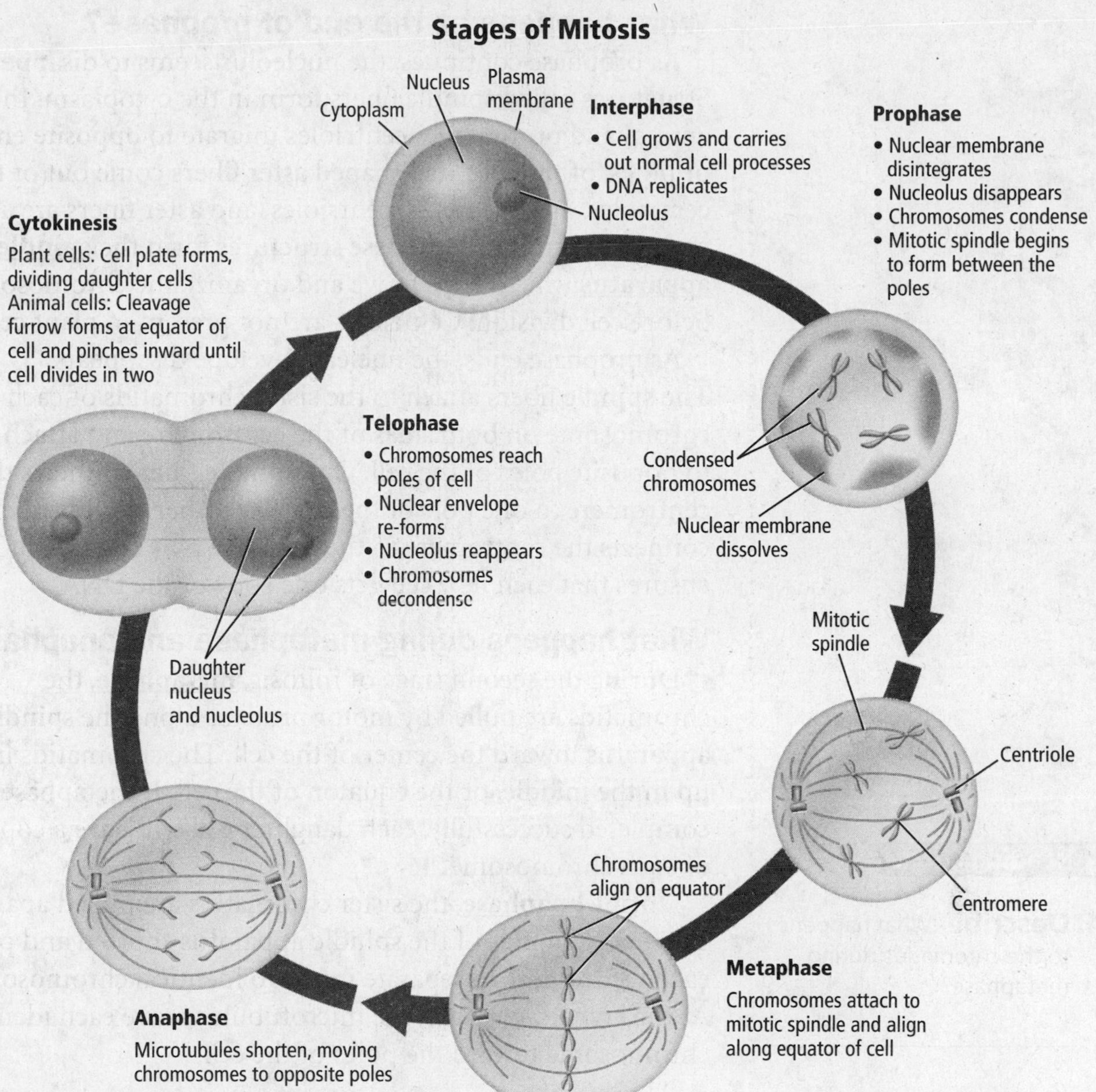

Stages of Mitosis

Cytoplasm

Nucleus

Plasma membrane

Nucleolus

Interphase
- Cell grows and carries out normal cell processes
- DNA replicates

Prophase
- Nuclear membrane disintegrates
- Nucleolus disappears
- Chromosomes condense
- Mitotic spindle begins to form between the poles

Condensed chromosomes

Nuclear membrane dissolves

Mitotic spindle

Centriole

Centromere

Metaphase

Chromosomes attach to mitotic spindle and align along equator of cell

Chromosomes align on equator

Anaphase

Microtubules shorten, moving chromosomes to opposite poles

Telophase
- Chromosomes reach poles of cell
- Nuclear envelope re-forms
- Nucleolus reappears
- Chromosomes decondense

Daughter nucleus and nucleolus

Cytokinesis

Plant cells: Cell plate forms, dividing daughter cells
Animal cells: Cleavage furrow forms at equator of cell and pinches inward until cell divides in two

What happens during prophase?

The first and longest stage of mitosis is called **prophase**. Before prophase, DNA is in a relaxed, or unwound, form known as chromatin. During prophase, chromatin becomes tightly wound, or condenses, into chromosomes.

During prophase, each chromosome is X-shaped. The left half of the X is one copy of DNA. The right half is the other identical copy. Each half of the X is a **sister chromatid** containing identical copies of DNA. The sister chromatids are attached near the center of the chromosome by a structure called the **centromere**. The centromere ensures that a complete copy of the DNA copy becomes part of the daughter cell at the end of the cell cycle. ☑

Picture This

2. Highlight the steps that occur in prophase.

3. Identify How are the sister chromatids attached?

____

What happens at the end of prophase?

As prophase continues, the nucleolus seems to disappear. Structures called spindle fibers form in the cytoplasm. In animal and protist cells, centrioles migrate to opposite ends, or poles, of the cell. Star-shaped aster fibers come out of the centrioles. Spindle fibers, centrioles, and aster fibers are all made of microtubules. These structures form the **spindle apparatus** which helps move and organize the chromosomes before cell division. Centrioles are not present in plant cells.

As prophase ends, the nuclear envelope disappears. The spindle fibers attach to the sister chromatids of each chromosome on both sides of the centromere and attach to opposite poles of the cell. One spindle fiber connects the centromere to one pole of the cell. The other spindle fiber connects the centromere to the opposite pole of the cell. This ensures that each new cell gets one copy of the DNA.

What happens during metaphase and anaphase?

During the second stage of mitosis, **metaphase**, the chromatids are pulled by motor proteins along the spindle apparatus toward the center of the cell. The chromatids line up in the middle, or the equator, of the cell. If metaphase is completed successfully, each daughter cell will have a copy of each chromosome. ☑

Reading Check

4. Describe What happens to the chromatids during metaphase?

______

______

______

During **anaphase**, the sister chromatids are pulled apart. The microtubules of the spindle apparatus shorten and pull at each centromere to separate into two identical chromosomes. At the end of anaphase, the microtubules move each identical chromosome toward the poles of the cell.

What happens during telophase?

Telophase is the final stage of mitosis. During **telophase**, the chromosomes arrive at the poles of the cell and begin to change back into chromatin. Two new nuclear membranes form around each set of chromosomes, the nucleoli reappear, and the spindle apparatus is taken apart, as shown below.

Picture This

5. Identify What does the cell still need to do at the end of telophase?

______

______

Cytokinesis

Near the end of mitosis, cytokinesis begins. During cytokinesis, the cytoplasm divides. The result of cytokinesis is two daughter cells, each with an identical nucleus.

What is the result of cytokinesis in animal cells?

In animal cells, cytokinesis is accomplished by using microtubules to constrict, or pinch, the cytoplasm of the cell in half. The cell splits into two daughter cells. ☑

How is cytokinesis different in plant cells?

Plant cells complete cell division a different way, as shown in the figure below. Recall that plant cells are surrounded by a rigid cell wall. During cytokinesis, plant cells form a new structure, called the cell plate, between the two daughter nuclei. New cell walls then form on either side of the cell plate, dividing the cell into two identical daughter cells.

How is cell division different in prokaryotic cells?

Prokaryotic cells undergo cell division in a different way. Recall that prokaryotic cells divide by a process known as binary fission. When prokaryotic DNA is copied, both identical copies of DNA attach to the plasma membrane. As the plasma membrane grows, the attached DNA molecules are pulled apart. The cell completes fission, producing two new prokaryotic cells.

Reading Check

6. Name the cell structure that pinches the cytoplasm in half.

____

FOLDABLES

Compare and Contrast
Make a Venn diagram Foldable, as shown below. As you read, compare and contrast cytokinesis in plant and animal cells.

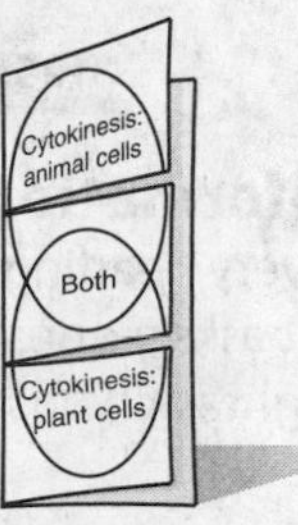

Picture This

7. Identify What features in the figure can you use to identify this cell as a plant cell?

____

____

____

Cellular Reproduction

section 3 Cell Cycle Regulation

MAIN Idea

The normal cell cycle is regulated by cyclin proteins.

What You'll Learn

- how cancer relates to the cell cycle
- the role of apoptosis
- the two types of stem cells and their potential uses

Before You Read

Cancer results when cells lose control of the cell cycle. A healthy lifestyle reduces the risk of cancer. On the lines below, write activities you think might reduce the risk of cancer. Then read to learn about how cancer forms.

____

____

Mark the Text

Locate Information Highlight every heading in the reading that asks a question. Then highlight each answer as you find it.

Read to Learn

Normal Cell Cycle

The timing and rate of cell division are important to the health of an organism. The rate of cell division varies depending on the type of cell. A mechanism involving proteins and enzymes controls the cell cycle.

The cell cycle in eukaryotic cells is controlled by a combination of two substances that signals the cellular reproduction process. Proteins called **cyclin** bind to enzymes called **cyclin-dependent kinases** (CDKs) in the stages of interphase and mitosis to trigger the various activities that take place in the cell cycle, as shown below.

Picture This

1. Name When are some of the important combinations active?

____

____

____

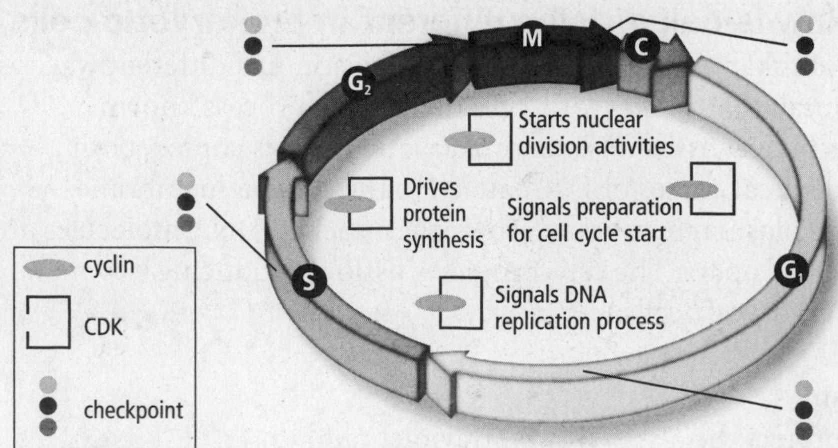

How do cells use cyclin/CDK combinations?

Cells use different combinations of cyclin/CDK to control activities at different stages in the cell cycle. For instance, in the G_1 stage, one cyclin/CDK combination signals the start of the cell cycle. Other cyclin/CDK combinations signal other activities, including DNA replication, protein synthesis, and nuclear division. Cyclin/CDK combinations also signal the end of the cell cycle.

How do quality control checkpoints work?

The cell cycle has built-in quality control checkpoints that monitor the cell cycle and can stop it if something goes wrong. For instance, near the end of the G_1 stage of interphase, the cell monitors its DNA for damage and can stop the cell cycle before entering the S stage if something is wrong. There are other quality control checkpoints during the S stage and after DNA duplication in the G_2 stage before entering mitosis. During mitosis, the cell checks the spindle fibers before it undergoes cytokinesis. If the cell detects a failure, the cell cycle stops.

Abnormal Cell Cycle: Cancer

Sometimes control of the cell cycle fails. When cells do not respond to control mechanisms, cancer results. **Cancer** is the uncontrolled growth and division of cells. ☑

Cancer cells grow and divide as long as they receive nutrients. They crowd normal cells causing tissues and organs to stop working. Cancer can kill an organism.

What causes cancer?

Cancer is caused by mutations, or changes, in segments of DNA that code for production of proteins, including those that regulate the cell cycle. Often, cells can fix mutations in DNA. If the repair system fails, cancer can result.

Environmental factors can increase the risk of cancer. Substances that are known to cause cancer are called **carcinogens** (kar SIH nuh junz). Tobacco, tobacco smoke, alcohol, some viruses, and radiation from the Sun or X-rays are examples of carcinogens.

Avoiding carcinogens can help reduce the risk of cancer. Federal laws protect people from exposure to carcinogens in the workplace and in the food supply. People can reduce their risk of cancer by avoiding all tobacco (including secondhand smoke and smokeless tobacco) and by using sunscreen to protect their skin from ultraviolet radiation from the Sun.

Reading Check

2. Define What is cancer?

____

____

____

Think it Over

3. Identify What is an example of a carcinogen that occurs in nature?

____

____

____

Who can get cancer?

Cancer can occur in people of all ages, but older people have a higher risk. This might be because it takes more than one DNA mutation to change an abnormal cell into a cancer cell. Older cells have had more time to accumulate the mutations that lead to cancer.

Cancer runs in some families. People might inherit one or more DNA mutations from their parents, increasing their risk of developing cancer.

Apoptosis

Some cells in an organism are no longer needed. **Apoptosis** (a pup TOH sus) is a natural process of programmed cell death. Cells that are no longer needed are destroyed by apoptosis.

Apoptosis occurs in the embryo to remove tissue between developing fingers and toes. Apoptosis also occurs in cells that are damaged beyond repair or that could turn into cancer cells. It is also part of the process by which leaves fall from trees in autumn.

Think it Over

4. Identify Which statement about apoptosis is correct? (Circle your answer.)

a. Apoptosis can lead to cancer.

b. Apoptosis can prevent cancer.

Stem Cells

Most cells in a multicellular organism are designed for a special purpose. Some cells might be part of your skin, and other cells might be part of your heart.

Some cells are not specialized. **Stem cells** are unspecialized cells that have the potential to develop into specialized cells. There are two types of stem cells—embryonic stem cells and adult stem cells. ☑

Reading Check

5. Describe What are stem cells?

______

______

______

______

What is the potential for stem cells?

Embryonic stem cells are created after fertilization, when the fertilized egg divides to create 100 to 150 stem cells. Each of these cells has the potential to develop into a wide variety of specialized cells. As the embryo develops, the cells specialize into tissues, organs, and organ systems. Adult stem cells are found in various tissues in the body. They are present in people from birth to adulthood.

Stem-cell research could lead to the treatment of many diseases and conditions. Embryonic stem-cell research is controversial. Ethical issues about the source of embryonic stem cells limit their availability to researchers. Adult stem cell research is less controversial because the cells can be obtained with the consent of the donor.

Sexual Reproduction and Genetics

section 1 Meiosis

Before You Read

Think about the traits that make people unique. Some people are tall, while others are short. People can have brown, blue, or green eyes. On the lines below, list a few traits that make you look different from other people. In this section, you will learn how meiosis rearranges genes.

____

____

____

____

MAIN Idea

Meiosis produces haploid genes.

What You'll Learn

- how chromosome number decreases during meiosis
- the stages of meiosis
- how meiosis provides genetic variation

Read to Learn

Chromosomes and Chromosome Number

All students in your class have characteristics passed on to them by their parents. Each characteristic, such as hair color, eye color, and height, is called a trait.

The instructions for each trait are found on chromosomes. Recall that chromosomes are found in the nuclei of cells. The DNA on the chromosomes is arranged in sections that control the production of proteins. These DNA sections are called <u>**genes**</u>. Each chromosome has about 1500 genes. Each gene has a role in the characteristics of the cell and how the cell works. Living things have thousands of genes.

Human body cells have 46 chromosomes. Chromosomes come in pairs. You have 23 chromosomes from your father and 23 chromosomes from your mother, making 23 pairs of chromosomes.

Study Coach

Create a Quiz After you have read this section, create a quiz based on what you have learned. After you have completed writing the quiz questions, be sure to answer them.

Applying Math

1. Calculate the approximate number of genes humans have. Show your work.

____

____

What are homologous chromosomes?

The chromosomes that make up a pair, one from each parent, are called **homologous** (huh MAH luh gus) **chromosomes**. Homologous chromosomes are the same length and have the centromere in the same place. They also carry genes for the same traits at the same place. Look at the picture below, and see if you can spot the homologous pair.

Homologous chromosomes are similar but not identical. For example, the gene for ear shape will be located at the same place on each homologous chromosome. Although these genes code for ear shape, the gene on one chromosome might code for one ear shape. The gene on the other chromosome might code for a different ear shape.

Picture This

2. Identify Circle a pair of homologous chromosomes.

How is chromosome number maintained in a species?

The number of chromosomes does not change from generation to generation. You have the same number of chromosomes as your parents. **Gametes** (GA meets), or sex cells with half the number of chromosomes, ensure the chromosome number stays the same.

The symbol *n* represents the number of chromosomes. In humans, *n* is equal to 23. A cell with *n* number of chromosomes is called a **haploid** cell. Gametes are haploid cells. ☑

3. Identify How many chromosomes are in a human gamete? (Circle your answer.)

a. 46
b. 23
c. 10

The process in which one haploid gamete joins with another haploid gamete is called **fertilization**. After fertilization, the cell has *2n* chromosomes—*n* chromosomes from the female parent plus *n* chromosomes from the male parent. A cell with *2n* chromosomes is called a **diploid** cell. Notice that *n* also represents the number of chromosome pairs in an organism.

Meiosis I

Recall that most cells are formed by mitosis. During mitosis the chromosome number stays the same. Because sex cells need half the number of chromosomes, a different process of cell division is needed. Gametes are formed during a process called meiosis. **Meiosis** is a kind of cell division that reduces the number of chromosomes by half through the separation of homologous chromosomes. Meiosis takes place in the reproductive organs of plants and animals. During meiosis, there are two cell divisions. They are called meiosis I and meiosis II.

What happens during interphase I?

Just as in mitosis, a cell goes through interphase before undergoing meiosis. A cell in interphase carries out a variety of metabolic functions, copies its DNA, and makes proteins.

What happens during prophase I?

Meiosis I begins with prophase I. During prophase I, replicated chromosomes, consisting of two sister chromatids, condense. When that happens, the chromosomes become visible under a light microscope.

As the homologous chromosomes condense, they begin to form homologous pairs in a process called synapsis (suh NAP sus). The homologous chromosomes are held tightly together along their lengths by a protein that acts like a zipper. Prophase I continues as the chromosomes move to opposite sides of the cell.

What is crossing over?

During synapsis, the chromosomes often swap pieces of DNA. **Crossing over** occurs when a section of one chromosome changes place with a section of its homologous chromosome. This is shown in the figure below. The centrioles move to the opposite poles of the cell. Spindle fibers form and bind to the sister chromatids at the centromere.

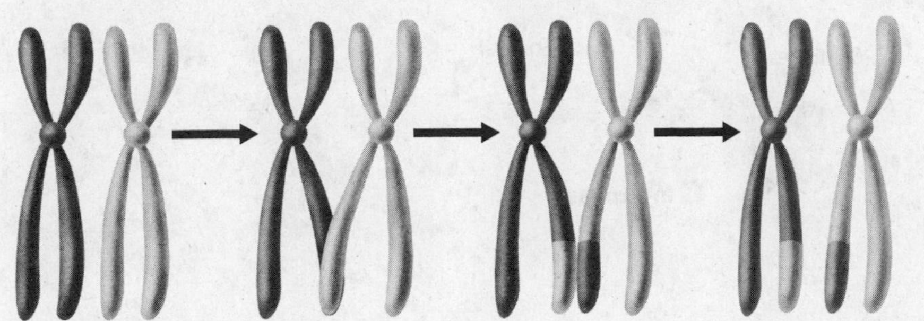

FOLDABLES™

Take Notes Make a folded table Foldable, as shown below. As you read, take notes and organize what you learn about meiosis I and meiosis II.

	Meiosis I	Meiosis II
Prophase		
Metaphase		
Anaphase		
Telophase		
Result		

Picture This

4. Label Circle the part of each chromosome that was swapped during crossing over.

What happens during metaphase I?

The next phase is metaphase I. During metaphase I, the pairs of homologous chromosomes line up in the center of the cell. The spindle fibers attach to the centromere of each homologous chromosome.

What happens during anaphase I?

Next is anaphase I. During anaphase I, each homologous chromosome is guided by the spindle fibers toward opposite poles of the cell. When this happens, the chromosome number is reduced from $2n$ to n. Notice that the sister chromatids do not split during meiosis I. Each homologous chromosome still has two sister chromatids. ☑

Reading Check

5. Identify During what phase is the chromosome number reduced from $2n$ to n?

____

What is the final stage of meiosis I?

The final stage of meiosis I is telophase I. During telophase I, the homologous chromosomes reach opposite poles of the cell. Each pole contains only one member of a pair of homologous chromosomes.

The sister chromatids might not be identical because crossing over might have occurred during synapsis in prophase I. Crossing over is one way that meiosis leads to more genetic diversity. This diversity helps explain how species can change over time.

At the end of telophase I, the cell undergoes cytokinesis, meaning it divides into two cells. The cells then might go into interphase again, but this time, the DNA is not copied during interphase. The events of meiosis I are shown below.

Picture This

6. Label In the space provided, write the chromosome number ($2n$ or n) for each phase.

MEIOSIS I

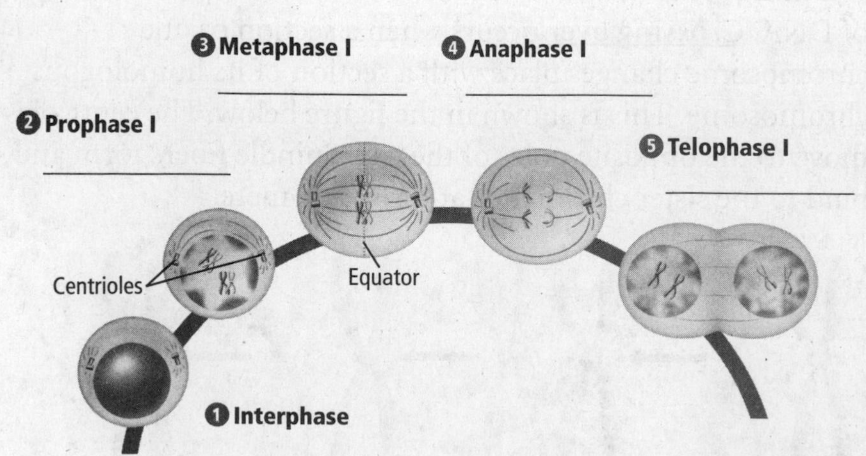

Meiosis II

Meiosis is now half finished. To complete meiosis, the cell must go through meiosis II. Meiosis II is similar to mitosis.

What events occur during meiosis II?

During prophase II, the spindle apparatus forms, and the chromosomes condense. During metaphase II, a haploid number of chromosomes lines up near the center of the cell by the spindle fibers. During anaphase II, the sister chromatids are pulled apart at the centromere by the spindle fibers, and the sister chromatids are pulled to the opposite poles of the cell. In telophase II, the chromosomes reach the poles, and the nuclear membrane and nuclei reform. Cytokinesis, or cell division, occurs. The result is four haploid cells, each with *n* number of chromosomes.

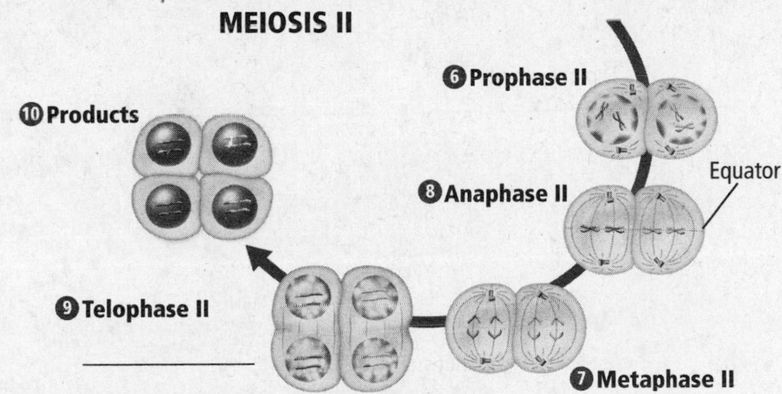

Picture This

7. Label Circle the phase in which the sister chromatids are pulled apart. Count the cells in telophase II, and write the number in the space provided.

The Importance of Meiosis

The figure below shows that meiosis and mitosis have similar steps, but they are different in important ways. An important difference is that mitosis produces two identical diploid daughter cells, while meiosis produces four different haploid daughter cells.

Picture This

8. Compare Fill in the blanks in the table with the number of daughter cells produced.

	Mitosis	Meiosis
Number of cell divisions	one	two
Synapsis of homologous chromosomes	does not occur	occurs during prophase I
Products	________ identical, diploid cells	________ nonidentical, haploid cells
Type of cells produced	body cells	reproductive cells
Purpose	growth and repair of body tissues	production of gametes for sexual reproduction

How does meiosis create genetic diversity?

The haploid daughter cells made by meiosis are not identical. Because the daughter cells are different, meiosis results in genetic variation.

One way that meiosis produces non-identical daughter cells occurs during prophase I. When pairs of homologous chromosomes line up at the center of the cell, they do so randomly. This means that each daughter cell gets a different, random assortment of chromosomes. The effect on genetic diversity is illustrated in the figure below.

The other way meiosis creates variation is through crossing over. Fertilization, when two haploid gametes combine, results in even more genetic variation.

Picture This

9. Identify Underline the haploid daughter cells.

Y Y S S s s y y

Diploid parent
SsYy

S s
Y y

Meiosis continues

Possibility 1
SY

Possibility 3
sY

Possibility 2
Sy

Possibility 4
sy

Potential types of haploid gametes

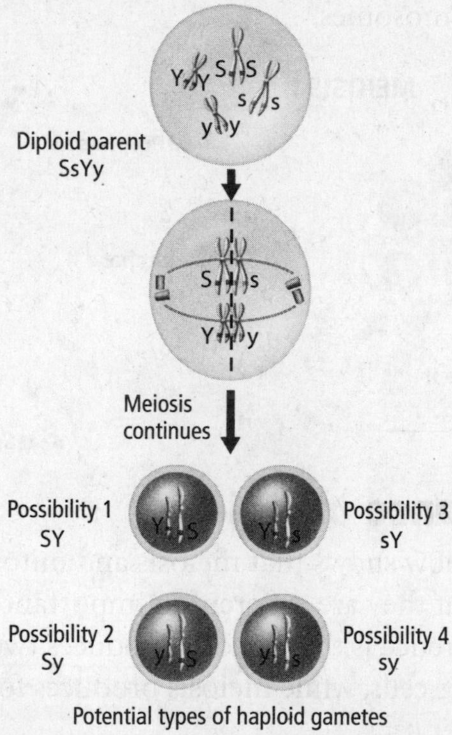

Think it Over

10. Compare How is chromosome inheritance different in sexual reproduction?

Sexual Reproduction v. Asexual Reproduction

Asexual reproduction occurs when the organism inherits all of its chromosomes from one parent. The new organism is genetically identical to its parent. Asexual reproduction does not involve meiosis.

Bacteria reproduce by asexual reproduction. Plants and some simple animals can reproduce sexually or asexually. Complex animals only reproduce sexually.

Sexual Reproduction and Genetics

section 2 Mendelian Genetics

Before You Read

Think about what you have learned about the scientific method. On the lines below, list some of the steps Mendel might have used to learn about the natural world. In this section, you will learn about Gregor Mendel's experiments.

MAIN Idea

Mendel explained how a dominant allele can mask the presence of a recessive allele.

What You'll Learn

- the law of segregation and the law of independent assortment
- how to use a Punnett square

Read to Learn

Mark the Text

Check for Understanding
As you read this section, highlight any parts you do not understand. After you have read the section, reread the parts you have highlighted.

How Genetics Began

Gregor Mendel, an Austrian Monk, lived in the 1800s. He experimented with pea plants in the monastery gardens.

Pea plants usually reproduce by self-fertilization. This means that the female gamete is fertilized by a male gamete in the same flower. Mendel discovered a way to cross-pollinate peas by hand. He removed the male gametes from a flower. He then fertilized the flower with the male gamete from a different flower.

Through these experiments, Mendel made several hypotheses about how traits are inherited. In 1866, he published his findings. That year marks the beginning of the science of **genetics**, the science of heredity. Mendel is called the father of genetics.

1. Define What is a true-breeding plant?

The Inheritance of Traits

Mendel used true-breeding pea plants—plants whose traits stayed the same from generation to generation. Mendel studied seven traits—flower color, seed color, seed pod color, seed shape, seed pod shape, stem length, and flower position. ☑

What did Mendel find when he crossed pea plants with different traits?

Mendel called the original plants the parent, or P, generation. The offspring were called the F_1 generation. The offspring of the F_1 plants were called the F_2 generation.

In one experiment, Mendel crossed yellow-seeded and green-seeded plants. All the F_1 offspring had yellow seeds. The green-seed trait seemed to disappear.

Mendel allowed the F_1 plants to self-fertilize. He planted thousands of seeds from these plants. He saw that in these offspring, the F_2 generation, three-fourths of the plants had yellow seeds and one-fourth had green seeds, a 3:1 ratio.

Mendel performed similar experiments for other traits. Each time, he observed the same 3:1 ratio.

Picture This

2. Label Fill in the boxes with the name of each generation of offspring. Draw the peas you would expect to see in the empty pods. Use shading to indicate a green pea.

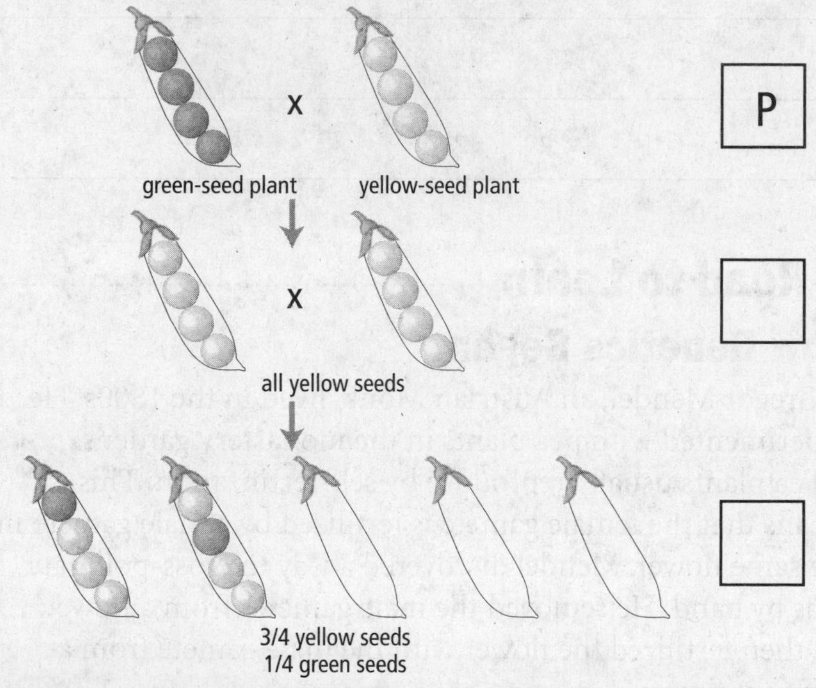

How did Mendel explain his results?

Mendel proposed that there were two forms of each trait, and each form was controlled by a factor, which is now called an allele. An **allele** (uh LEEL) is a different form of a gene passed from generation to generation. Yellow-seed plants have a different allele than green-seed plants.

Mendel proposed that each trait was controlled by two alleles. The **dominant** form is the version of the trait that appears in the F_1 generation. The **recessive** form is the version that is hidden in the F_1 generation.

Think it Over

3. Apply In Mendel's experiment with green and yellow seeds, what was the dominant trait?

____

____

How does dominance work?

When written, the dominant allele is represented by a capital letter. The recessive allele is represented by a lowercase letter.

An organism is **homozygous** (hoh muh ZI gus) if both alleles for a trait are the same. The organism is **heterozygous** (heh tuh roh ZY gus) if the alleles for a trait are different. In heterozygous organisms, only the dominant trait can be seen. Dominant alleles mask recessive alleles.

How do genotype and phenotype differ?

It is not always possible to know what alleles are present just by looking at an organism. A yellow-seed plant could be homozygous (*YY*) or heterozygous (*Yy*). An organism's allele pairs are called its **genotype** (JEE nuh tipe). The expression of an allele pair, or the way an organism looks or behaves, is called its **phenotype** (FEE nuh tipe).

What is the law of segregation?

Recall that the chromosome number is divided in half during meiosis. The gametes contain only one of the alleles. Mendel's **law of segregation** states that the two alleles for each trait separate from each other during meiosis and then unite during fertilization. When parents with different forms of a trait are crossed, the offspring are heterozygous organisms known as **hybrids** (HI brudz).

A cross which involves hybrids for a single trait is called a monohybrid cross. Mono means one. The offspring of the cross have a phenotypic ratio of 3:1.

How are two or more traits inherited?

Mendel also performed dihybrid crosses, crossing plants that expressed two different traits. Mendel crossed yellow, round-seed plants with green, wrinkle-seed plants. Round seeds are dominant to wrinkled, just as yellow color is dominant to green. He wondered whether the two traits would be inherited together or separately. Members of the F_1 generation are dihybrids because they are heterozygous for both traits.

Mendel found that the traits were inherited independently. Members of the F_2 generation had the phenotypic ratio of 9:3:3:1—9 yellow round seeds, 3 green round, 3 yellow wrinkled, and 1 green wrinkled. From experiments with dihybrid crosses, Mendel developed the **law of independent assortment,** which states that alleles distribute randomly when gametes are made.

Think it Over

4. Predict What would be the phenotype of a homozygous, recessive (*yy*) pea plant?

____

____

Think it Over

5. Apply True-breeding yellow-seeded and green-seeded plants are crossed and produce yellow-seeded offspring. Which of these plants is a hybrid?

____

____

____

Punnett Squares

In the early 1900s, Dr. Reginald Punnett developed a square to predict possible offspring of a cross between two known genotypes. Punnett squares are useful for keeping track of genotypes in a cross. ☑

Reading Check

6. Identify What is one purpose of a Punnett square?

What information does a Punnett square contain?

A Punnett square can help you predict the genotype and phenotype of the offspring. The genotype of one parent is written vertically, on the left side of the Punnett square. The genotype of the other parent is written horizontally, across the top. A Punnett square for a monohybrid cross contains four small squares. Each small square represents a possible combination of alleles in the children.

The Punnett square below shows the results of Mendel's experiment with seed color. The Punnett square shows that four different genotypes are possible—one *YY*, two *Yy*, and one *yy*. The genotypic ratio is 1:2:1.

Picture This

7. Define Circle the genotypes in the small squares that will give a yellow-seed phenotype. What will be the phenotypic ratio in the offspring?

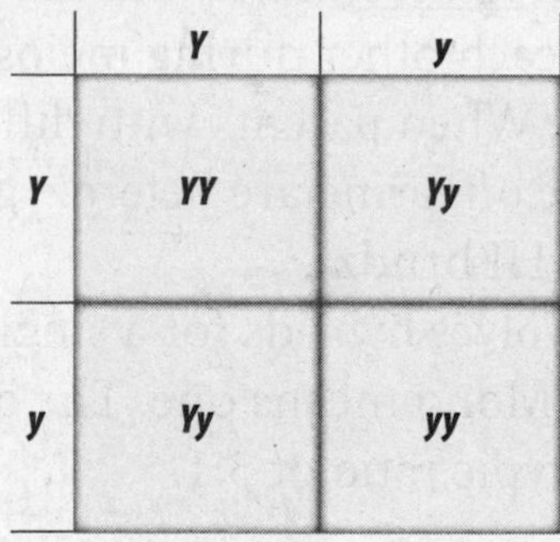

How is a Punnett square used for two traits?

Punnett squares also can be used to predict the results of a dihybrid cross. A Punnett square for a dihybrid cross is larger. It has 16 boxes to represent 16 allele combinations.

Probability

Genetics follows the rules of probability, or chance. It is like flipping a coin. The probability of flipping heads is one out of two. Because of chance, if you flip a coin 100 times, it might not land heads exactly 50 times, but it will be close.

It is the same in genetics. A cross might not give a perfect 3:1 or 9:3:3:1 ratio. The larger the number of offspring, the more closely the results will match the ratio predicted by the Punnett square.

Sexual Reproduction and Genetics

section 3 Gene Linkage and Polyploidy

Before You Read

Genetics is like a game of cards. In meiosis, chromosomes are shuffled and sorted. On the lines below, explain the chances of a player getting the same cards two games in a row. In this section, you will learn about the independent assortment of chromosomes that occurs during meiosis.

____

____

____

MAIN Idea

Crossing over of linked genes is a source of genetic variation.

What You'll Learn

- how meiosis produces genetic recombination
- how gene linkage is used to make chromosome maps
- why polyploidy is important

Read to Learn

Genetic Recombination

During meiosis, genes are combined in new ways. **Genetic recombination** occurs when crossing over and independent assortment produce new combinations of genes.

Recall that independent assortment occurs in meiosis when chromosomes separate randomly. The number of possible gene combinations due to independent assortment can be calculated using the formula 2^n, where *n* equals the number of chromosome pairs.

Pea plants have 7 pairs of chromosomes. The possible combinations of these chromosomes would be 27, or 128. Fertilization further increases the number of combinations. During fertilization, any possible male gamete can fertilize any possible female gamete. The number of combinations after fertilization would be $2^n \times 2^n$. For peas, this number is 16,384, or 128×128.

In people, the possible combinations of chromosomes are $2^{23} \times 2^{23}$—over 70 trillion. Crossing over increases genetic recombination even more.

Mark the Text

Main Ideas Highlight the main ideas under each heading. State each main point in your own words.

Applying Math

1. **Calculate** The fruit fly has four chromosome pairs. How many possible combinations of chromosomes can be produced by meiosis and fertilization?

____

____

Gene Linkage

Chromosomes contain many genes. Genes that are located close together on the same chromosome are said to be linked. This means they usually travel together during gamete formation. Linked genes do not segregate independently. They are an exception to Mendel's law of independent assortment.

Occasionally, linked genes separate due to crossing over. Crossing over occurs more frequently between genes that are far apart than between genes that are close together. ☑

Reading Check

2. Explain What event causes linked genes to separate?

What does a chromosome map show?

The relationship between crossing over and chromosome distance is very useful. The distance between two genes can be estimated by the frequency of crossing over that occurs between them. Scientists use cross-over data to create a drawing of genes along a chromosome. The drawing, called a chromosome map, shows the order of genes on a chromosome. The first chromosome maps were published in 1913 for fruit-fly crosses. One is shown in the figure below.

Picture This

3. Identify Which two genes are not likely to cross over? (Circle your answer.)

a. yellow body color and vermilion eye color

b. white eye color and vermilion eye color

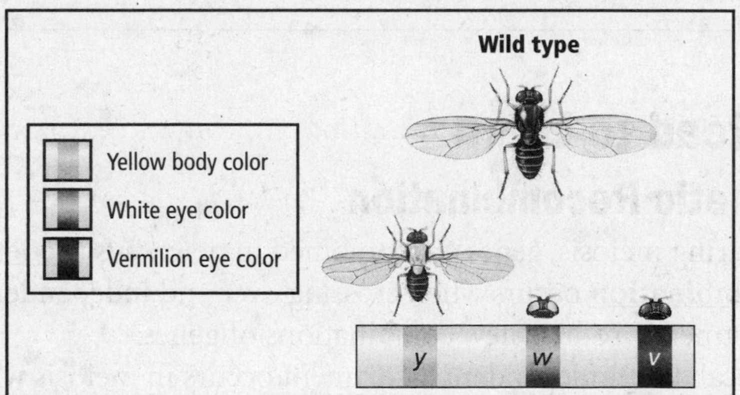

Polyploidy

Most organisms have diploid cells—cells with two chromosomes in each cell. Some species have polyploid cells. **Polyploidy** (PA lih ploy dee) means the cells have one or more extra sets of all chromosomes. For instance, a triploid organism has three complete sets of chromosomes in each cell. It is designated $3n$.

Polyploidy occurs in only a few animals, such as earthworms and goldfish. It is always lethal in humans. Polyploidy is common in flowering plants. Polyploid plants are often bigger and more vigorous. Many food plants, such as wheat ($6n$), oats ($6n$), and sugarcane ($8n$), are polyploidy. ☑

Reading Check

4. Identify Name two organisms that have polyploidy.

Complex Inheritance and Human Heredity

section 1 Basic Patterns of Human Inheritance

Before You Read

A family tree shows how people in a family are related. On the lines below, list people who might appear in a family tree. Then read the section to learn how scientists trace inheritance through several generations of a family.

____

____

MAIN Idea

The inheritance of a trait over several generations is shown in a pedigree.

What You'll Learn

- how to determine if an inherited trait is dominant or recessive
- examples of dominant and recessive disorders

Read to Learn

Study Coach

Create a Quiz After you read this section, create a quiz based on what you have learned. Then be sure to answer the quiz questions.

Recessive Genetic Disorders

During the early 1900s, Gregor Mendel's work on heredity was rediscovered. Archibald Garrod, an English doctor, was studying a disorder that results in black urine and affects bones and joints. Dr. Garrod, with the help of other scientists, discovered that the disorder was a recessive genetic disorder. This finding began the study of human genetics.

Review the table below and recall that a recessive trait is expressed when the person is homozygous recessive for that trait. A person with at least one dominant allele will not express the recessive trait. A person who is heterozygous for a recessive disorder is called a **carrier**.

Term	Description
Homozygous	An organism with two of the same alleles for a particular trait is said to be homozygous for that trait.
Heterozygous	An organism with two different alleles for a particular trait is said to be heterozygous for that trait. When alleles are present in the heterozygous state, the dominant trait will be observed.

Picture This

1. Identify Circle the term that describes the genotype of a person who expresses a recessive trait.

FOLDABLES™

Take Notes Use two sheets of paper to make a layered Foldable, as shown below. As you read, record important information you learn about human inheritance.

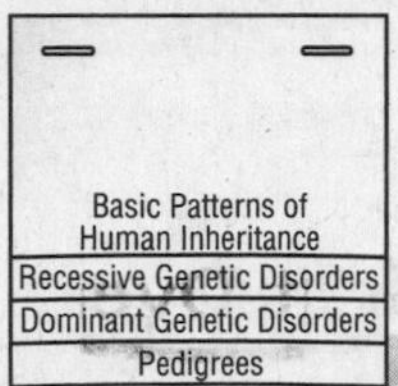

What is cystic fibrosis?

Cystic fibrosis is a recessive genetic trait. Chloride ions are not absorbed into cells but are excreted in sweat. Without the chloride ions in cells, water does not diffuse from cells. This causes the secretion of a thick mucus that affects many areas of the body. The mucus interferes with digestion, clogs ducts in the pancreas, and blocks air pathways in the lungs. Patients with cystic fibrosis often get infections because of excess mucus in their lungs.

Treatment includes physical therapy, medicine, special diets, and replacement digestive enzymes. Genetic tests are used to determine if the recessive gene is present.

What causes albinism?

Albinism is a recessive disorder found in people and animals. In humans, it is caused by the absence of the skin pigment melanin in hair and eyes. People with albinism have white hair, pale skin, and pink eyes. They need to protect their skin from the Sun's ultraviolet rays.

What is Tay-Sachs disease?

Tay-Sachs (TAY saks) disease is a recessive genetic disorder. Tay-Sachs disease (TSD) is more common among Jews whose ancestors are from eastern Europe.

People with TSD are missing an enzyme needed to break down fatty acids called gangliosides. Normally, gangliosides are made and then destroyed as the brain develops. In people with TSD, gangliosides build up in the brain, causing mental deterioration. Children born with TSD usually die by age five. Currently, there is no cure. ☑

Reading Check

2. Explain Why do gangliosides build up in the brain of people with Tay-Sachs disease?

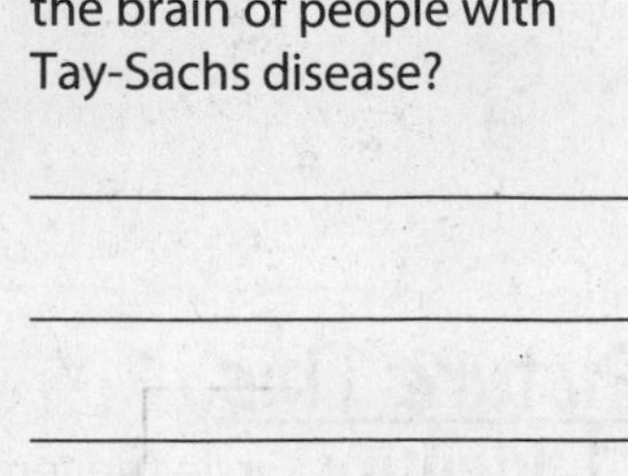

What causes galactosemia?

Galactosemia (guh lak tuh SEE mee uh) is a recessive genetic disorder. It causes intolerance of the sugar galactose. Milk contains the sugar lactose. During digestion, lactose breaks down into galactose and glucose, the sugar used by the body for energy. People with galactosemia lack the enzyme needed to break down galactose.

Dominant Genetic Disorders

Not all genetic disorders are recessive. Some are caused by dominant alleles. People who do not have the disorder are always homozygous recessive, meaning they carry two recessive genes for the trait.

What happens in Huntington's disease?

Huntington's disease is a dominant genetic disorder that affects the nervous system. It is rare. Symptoms occur when the person is between 30 and 50 years old. Symptoms are gradual loss of brain function, uncontrollable movements, and emotional disturbances. Genetic tests can tell people whether they have the gene for Huntington's disease, but there is currently no treatment or cure.

What is achondroplasia?

Achondroplasia (a kahn droh PLAY zhee uh) is a dominant genetic disorder that is also known as dwarfism. People with this disorder have a small body size and short limbs. They grow to an adult height of about 1.2 m (4 ft).

About 75 percent of people with achondroplasia have parents of average size. Because the gene is dominant, parents who are average size do not have the gene. Therefore, when average-sized parents have a child with achondroplasia, the condition occurs because of a new mutation.

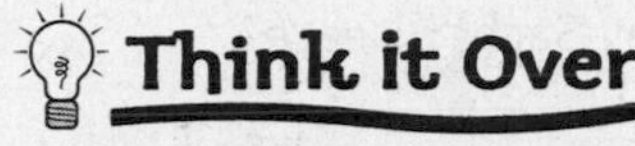

3. Explain How can scientists determine if achondroplasia developed from a new mutation?

____

____

____

____

Pedigrees

Scientists use a diagram called a **pedigree** to trace inheritance of a trait through several generations. A pedigree uses symbols to illustrate inheritance of the trait.

A sample pedigree is shown in the figure below. In the top row, the two symbols connected by a horizontal line are the parents. Their children are listed below them, oldest to youngest from left to right.

Roman numerals are used to represent generations—I for the parents, II for the children, and so on. Arabic numbers are used to represent the individuals within a generation.

Picture This

4. Evaluate Circle the carriers in the second generation.

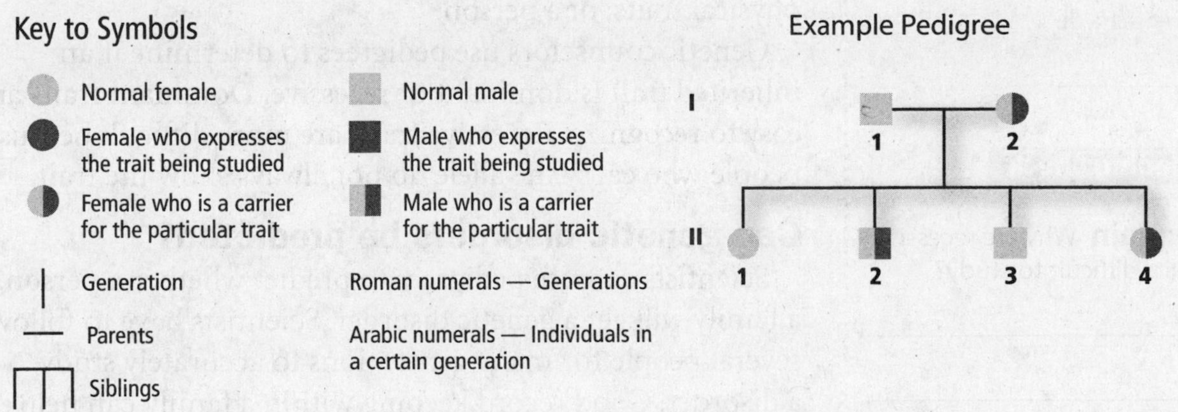

Analyzing Pedigrees

The figure below is a pedigree showing the inheritance of Tay-Sachs disease, a recessive disorder. The pedigree shows that two parents who do not have Tay-Sachs disease can have a child who has the disorder.

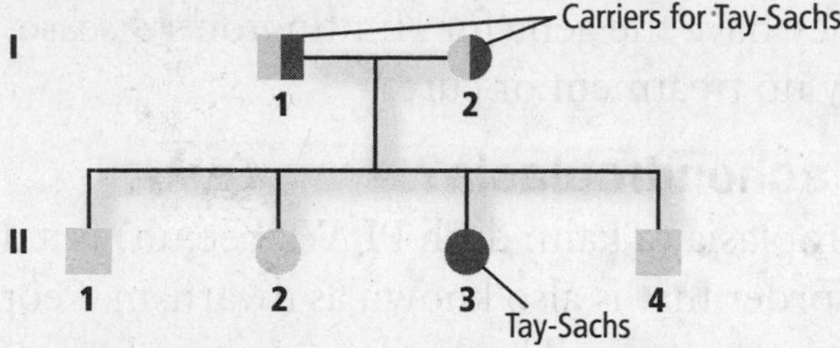

Picture This

5. Calculate What percentage of the children in this family inherited Tay-Sachs disease?

How is the inheritance of a dominant disorder shown on a pedigree?

The pedigree below shows the inheritance of the dominant disorder, polydactyly (pah lee DAK tuh lee)—extra fingers and toes. A person with dominant disorders could be homozygous or heterozygous for the trait. A person who does not have polydactyly would be homozygous recessive for the trait.

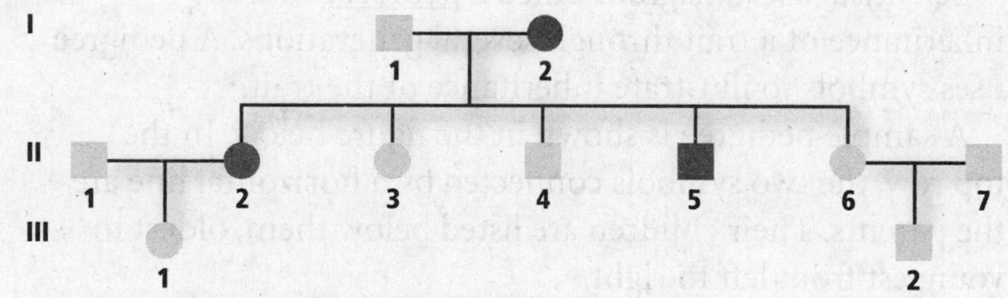

Picture This

6. Identify Do any of the grandchildren in this family have polydactyly?

How are genotypes deduced?

A pedigree can be used to learn the genotype of a person. The genotype is determined by observing the phenotypes, or physical traits, of a person.

Genetic counselors use pedigrees to determine if an inherited trait is dominant or recessive. Dominant traits are easy to recognize. Recessive traits are more difficult because people who carry the allele do not always show the trait. ☑

Reading Check

7. Explain Why are recessive traits difficult to study?

Can genetic disorders be predicted?

Scientists can use pedigrees to predict whether a person in a family will get a genetic disorder. Scientists have to follow several people for many generations to accurately study a disorder. Good record keeping within a family can help scientists predict the inheritance of a disorder.

Complex Inheritance and Human Heredity

section 2 Complex Patterns of Inheritance

Before You Read

Cats can look different from one another because of differences in their coats. On the lines below, describe differences you have seen in the coats of cats. Then read the section to learn more about complex inheritance patterns.

____

____

MAIN Idea

Complex inheritance of traits does not follow the inheritance patterns described by Mendel.

What You'll Learn

- the difference between sex-linked and sex-limited inheritance
- how environment can influence a trait

Read to Learn

Mark the Text

Highlight each question head. Then highlight the answer to the question.

Incomplete Dominance

Not all traits follow Mendel's rules. Some traits are not dominant or recessive. Sometimes, the heterozygous organism has a mixed phenotype. **Incomplete dominance** occurs when the heterozygous phenotype is an intermediate phenotype between the two homozygous phenotypes.

An example of incomplete dominance occurs in snapdragon flowers. Red-flowered snapdragons ($C^R C^R$) can be crossed with white-flowered snapdragons ($C^W C^W$) to produce offspring with pink flowers ($C^R C^W$). When heterozygous F_1 generation snapdragon plants ($C^R C^W$) self-fertilize, the offspring have a 1:2:1 ratio of red, pink, and white flowers.

Codominance

In Mendel's experiments with pea plants, heterozygous pea plants expressed only the dominant allele. **Codominance** occurs when a heterozygous organism expresses both alleles. Sickle-cell anemia is an example of codominance. People who are heterozygous for the sickle-cell trait have both normal and sickle-shaped cells. ☑

Reading Check

1. Define What is codominance?

____

____

____

What happens in sickle-cell disease?

Sickle-cell disease is common in people of African descent. Sickle-cell disease affects red blood cells and their ability to transport oxygen. Changes in the protein in red blood cells cause those red blood cells to change from a normal disc shape to a sickle or *C* shape. ☑

Reading Check

2. Describe What effect does sickle-cell disease have on red blood cells?

Sickle-cell disease is a codominant trait. People who are heterozygous for the trait make both normal and sickle-shaped cells. The normal cells compensate for the sickle-shaped cells.

How does sickle-cell disease relate to malaria?

Sickle-cell disease is found in areas of Africa where malaria occurs. Scientists have discovered that people who are heterozygous for the sickle-cell trait are resistant to malaria. Because the sickle-cell gene helps people resist malaria, they are more likely to pass the sickle-cell trait on to their offspring.

Multiple Alleles

So far you have learned about traits that result from a gene with two alleles. Some traits are controlled by a gene that has **multiple alleles**. Blood groups in humans is an example of a multiple allele trait.

How are blood types produced?

There are four blood types in people: A, AB, B, or O. The four types result from the interaction of three different alleles, as shown below. The allele I^A produces blood type A. I^B produces blood type B. The allele i is recessive and produces blood type O. Type O is the absence of AB alleles. People with one I^A and one I^B allele have blood type AB. Blood types are examples of multiple alleles and codominance.

Rh factors are also in blood. One factor is inherited from each parent. Rh factors are either positive or negative (Rh+ or Rh−); the Rh+ is dominant.

Picture This

3. Evaluate What phenotype results from a genotype of I^Bi?

Genotypes	Resulting Phenotypes
I^AI^A	Type A
I^Ai	Type A
I^BI^B	Type B
I^Bi	Type B
I^AI^B	Type AB
ii	Type O

What genes control coat color in rabbits?

The fur color of rabbits is another trait controlled by multiple alleles. In rabbits, four alleles control coat color: C, c^{ch}, c^{h}, and c. The alleles are dominant in varying degrees. The hierarchy can be written as $C > c^{ch} > c^{h} > c$.

Allele C is dominant to all other alleles and results in a dark gray coat color. Allele c^{ch} is dominant to c^{h}, and c^{h} is dominant to c. Allele c is recessive and results in an albino when the genotype is homozygous recessive.

Multiple alleles increase the possible number of genotypes and phenotypes. Two alleles have three possible genotypes and two possible phenotypes. Four alleles have ten possible genotypes and can have five or more phenotypes.

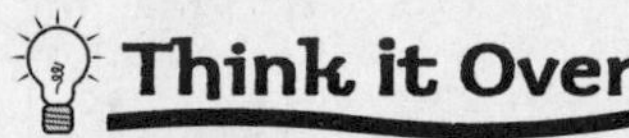

4. Evaluate What allele is dominant over c^{ch}?

a. c^{h}
b. c
c. C

Epistasis

Epistasis (ih PIHS tuh sus) occurs when one allele hides the effects of another allele. Coat color in Labrador retrievers is a trait controlled by epistasis. Labrador coats vary from yellow to black. Two different genes control coat color. The dominant allele *E* determines whether the coat will have dark pigment. A dog with genotype *ee* will not have any pigment. The dominant allele *B* determines how dark the pigment will be. If the genotype is *EEbb* or *Eebb* the coat will be chocolate. If the genotype is eebb, *eeBb,* or *eeBB* the coat will be yellow because the *e* allele hides the effects of the dominant *B* allele.

Sex Determination

Each cell in your body contains 23 pairs of chromosomes. One pair, the **sex chromosomes,** determines gender. The other 22 pairs of chromosomes are called **autosomes**.

There are two types of sex chromosomes—X and Y. A person's gender is determined by the sex chromosomes present in the egg and sperm cell. Females inherit two X chromosomes. Males inherit one X and one Y chromosome.

5. Identify A person has 22 pairs of autosomes and two X chromosomes. What is the person's gender?

____

Dosage Compensation

In humans, the X chromosome carries genes needed by males and females. The Y chromosome mainly carries genes needed to develop male characteristics. Because females have two X chromosomes and males have only one, body cells randomly turn off one of the X chromosomes. This is called dosage compensation or X-inactivation.

How is coat color determined in calico cats?

The coat color of calico cats is controlled by the random inactivation of X chromosomes. Orange patches are formed when an X chromosome carrying the allele for black coat color is turned off. Black patches are formed when an X chromosome carrying the allele for orange coat color is turned off.

What are Barr bodies?

Canadian scientist Murray Barr first observed inactivated X chromosomes, now known as Barr bodies. Barr bodies appear as dark objects in the cell nuclei of female mammals.

Think it Over

6. Draw Conclusions Why is a recessive sex-linked trait less likely to occur in females than in males?

____

____

____

Sex-Linked Traits

Traits controlled by genes on the X chromosome are called **sex-linked traits** or X-linked traits. Males who have only one X chromosome are affected more than females by recessive sex-linked traits. Females would not likely express a recessive sex-linked trait because one X chromosome will mask the effect of the recessive trait on the other X chromosome.

How is red-green color blindness inherited?

The trait for red-green color blindness is a recessive sex-linked trait. People who are color blind cannot see the colors red and green. About 8 percent of males in the United States are red-green color blind. Examine the Punnett square below to see how red-green color blindness is inherited.

Picture This

7. Predict Circle the genotype that represents a color-blind person.

X^B = Normal
X^b = Red-green color blind
Y = Y chromosome

	X^B	Y
X^B	X^BX^B	X^BY
X^b	X^BX^b	X^bY

How is hemophilia inherited?

Normally, when a person is cut, the bleeding stops quickly. Hemophilia is a recessive sex-linked disorder that slows blood clotting. Hemophilia is more common in males. Until the discovery of clotting factors in the twentieth century, most men with hemophilia died at an early age. Safe methods of treating the disorder now allow for a normal life span.

Polygenic Traits

So far you have learned about traits that are controlled by one gene with different alleles. **Polygenic traits** develop from the interaction of multiple pairs of genes. Many traits in humans are polygenic, including skin color, height, eye color, and fingerprint pattern. ☑

Reading Check

8. List an example of a polygenic trait.

Environmental Influences

The environment influences many traits. Factors such as sunlight, temperature, and water can affect an organism's phenotype. For example, the gene that codes for the production of color pigment in Siamese cats functions only under cooler conditions. Cooler parts of the cat's body, such as the ears, nose, feet, and tail, are darker. The warmer parts of the body, where pigment production is inhibited, are lighter.

Environmental factors also include an organism's actions. Heart disease can be inherited, but diet and exercise also strongly influence the disease. An organism's actions are considered part of the environment because they do not come from genes.

Twin Studies

Scientists can learn about inheritance patterns by studying twins. Twin studies often reveal how genes and the environment affect phenotype.

Identical twins have identical genes. If a trait is inherited, both identical twins will have the trait. Scientists presume that traits that are different in identical twins are strongly influenced by the environment. The percentage of identical twins who both have the same trait is called a concordance rate, as shown in the graph below. The higher the concordance rate, the stronger the genetic influence.

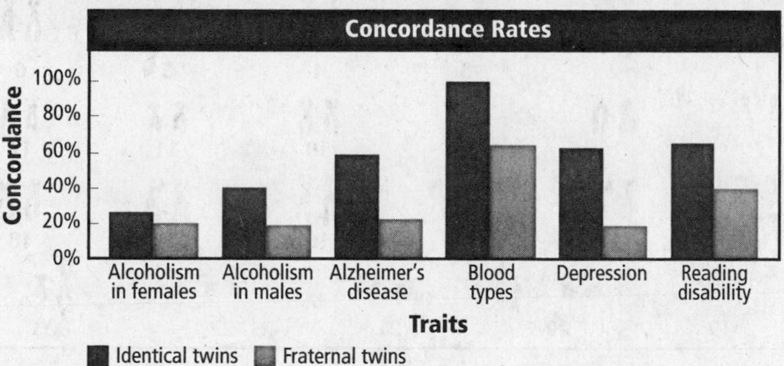

Picture This

9. Evaluate Circle the trait that shows the strongest genetic influence.

Complex Inheritance and Human Heredity

section 3 Chromosomes and Human Heredity

MAIN Idea

Chromosomes can be studied using karyotypes.

What You'll Learn

- the role of telomeres
- how nondisjunction leads to Down syndrome and other abnormalities
- the benefits and risks of fetal testing

Before You Read

Think about the traits that people in a family might share. On the lines below, list the ways that people in families resemble each other. Then read to learn more about how scientists study genetic material.

____

____

Mark the Text

Identify Main Ideas Highlight the main idea of each paragraph.

Read to Learn

Karyotype Studies

Genetics not only is the study of genes, it is also the study of chromosomes. Images of chromosomes that have been stained during metaphase are studied. The staining bands mark identical places on homologous chromosomes. The homologous chromosomes are arranged, from biggest to smallest, to produce a micrograph called a **karyotype** (KER ee uh tipe). A karyotype is shown in the figure below.

Picture This

1. Apply Examine the karyotype. Are these chromosomes from a male or female?

____

Chromosomes of a human cell

1 2 3 4 5 6
7 8 9 10 11 12
13 14 15 16 17 18
19 20 21 22 (XY)

_____ **(No. of chromosome pairs)** × 2 = _____ **(No. of chromosomes)**

Telomeres

Telomeres are protective caps at the ends of chromosomes. They are made of DNA and proteins. Scientists have discovered that telomeres might be involved in both aging and cancer.

Nondisjunction

During cell division, the chromosomes separate and move to opposite poles of the cell. This ensures that each new cell has the correct number of chromosomes.

Cell division during which sister chromatids do not separate properly is called **nondisjunction**. Nondisjunction does not often occur. ☑

Nondisjunction during meiosis results in gametes that do not have the correct number of chromosomes. When one of these gametes undergoes fertilization, the offspring will not have the correct number of chromosomes. The figure below shows nondisjunction during meiosis. Trisomy (TRI so me) means having a set of three chromosomes. Monosomy (MAH nuh so me) means having only one copy of a chromosome.

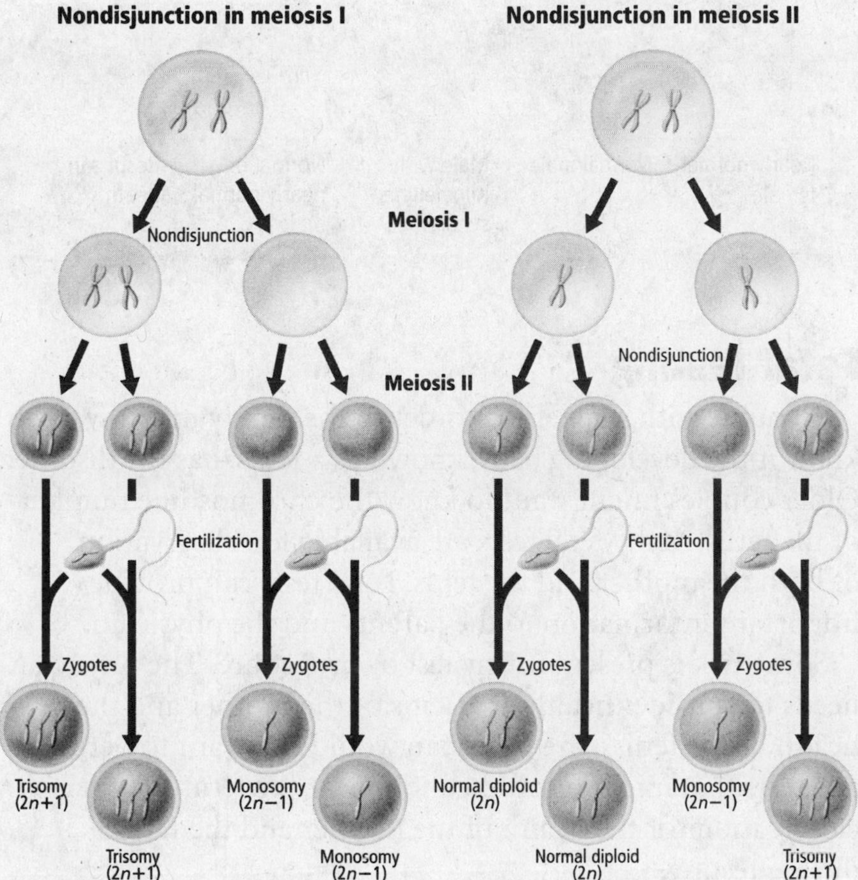

Reading Check

2. Define What happens during nondisjunction?

Picture This

3. Evaluate Does nondisjunction during meiosis produce any normal gametes? Explain.

How does nondisjunction lead to Down syndrome?

Down syndrome is usually the result of an extra copy of chromosome 21. People with Down syndrome have distinctive facial features, are short, and have heart defects and mental disability. Approximately one out of 800 children born in the United States has Down syndrome. Older women have a greater chance of having a child with Down syndrome.

Does nondisjunction occur with sex chromosomes?

People can inherit incorrect numbers of both autosomes and sex chromosomes. The results of nondisjunction in sex chromosomes are shown in the figure below. A female with Turner's syndrome has only one sex chromosome. A male with Klinefelter's syndrome has two X chromosomes and one Y chromosome.

Picture This

4. Label Circle the pictures that show a trisomy in the sex chromosomes.

Nondisjunction in Sex Chromosomes

Genotype	XX	XO	XXX	XY	XXY	XYY	OY
Example							
Phenotype	Normal female	Female with Turner's syndrome	Nearly normal female	Normal male	Male with Klinefelter's syndrome	Normal or nearly normal male	Results in death

Fetal Testing

A couple with a genetic disorder in the family might want to know if the developing baby, known as a fetus, has the disorder. Older couples might want to know the chromosome number of the fetus. Many fetal tests are available for observation of both the mother and the fetus. Fetal tests can provide important information to the parents and the physician. ☑

Some risk is present in any test or procedure. The physician needs to consider health problems of the mother and the health of the fetus. The physician would not want to perform any tests that might harm the mother or the fetus. Physicians closely monitor the health of the mother and the fetus during testing.

Reading Check

5. State the purpose of fetal testing.

____

____

____

Molecular Genetics

section 1 DNA: The Genetic Material

Before You Read

In 1953, James Watson and Francis Crick discovered the structure of DNA. But they were not the first people to ask, "How is genetic information passed from one generation to the next?" Watson and Crick's work was possible because of the work of other scientists. On the lines below, identify a task that is possible only because of the work of many people. Then read the section to learn more about how scientists discovered that DNA is the genetic code.

____

____

____

MAIN Idea

The discovery that DNA is the genetic code involved many experiments.

What You'll Learn

- the basic structure of DNA
- the structure of a eukaryotic chromosome

Read to Learn

Discovery of the Genetic Material

Mendel's laws of inheritance became well known to scientists in the early 1900s. Scientists knew that genes were carried on chromosomes. They also knew that chromosomes were made of DNA and protein. For many years, scientists tried to determine which of these two molecules—DNA or protein—carried the genetic information of a cell.

Mark the Text

Identify People As you read this section, underline the name of each scientist introduced. Highlight the sentences that explain each person's contribution to understanding DNA.

What are smooth and rough bacteria?

In 1928, Frederick Griffith conducted an experiment that led to the discovery of DNA as the genetic material. Griffith studied two strains of *Streptococcus pneumoniae.* One strain had a sugar coat and caused pneumonia. It was called smooth (S) because colonies of bacteria appear smooth. Another strain did not have a sugar coat and did not cause pneumonia. It was called rough (R) because its colonies have rough edges. Follow along with Griffith's study described on the next page. ☑

Reading Check

1. Contrast Name two ways that rough bacteria differ from smooth bacteria.

____

____

____

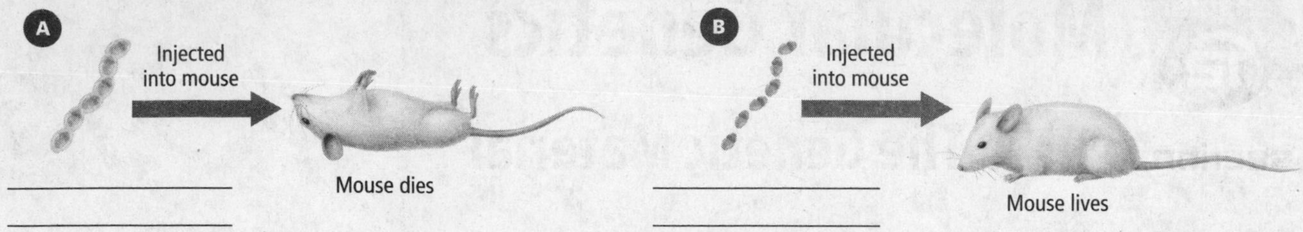

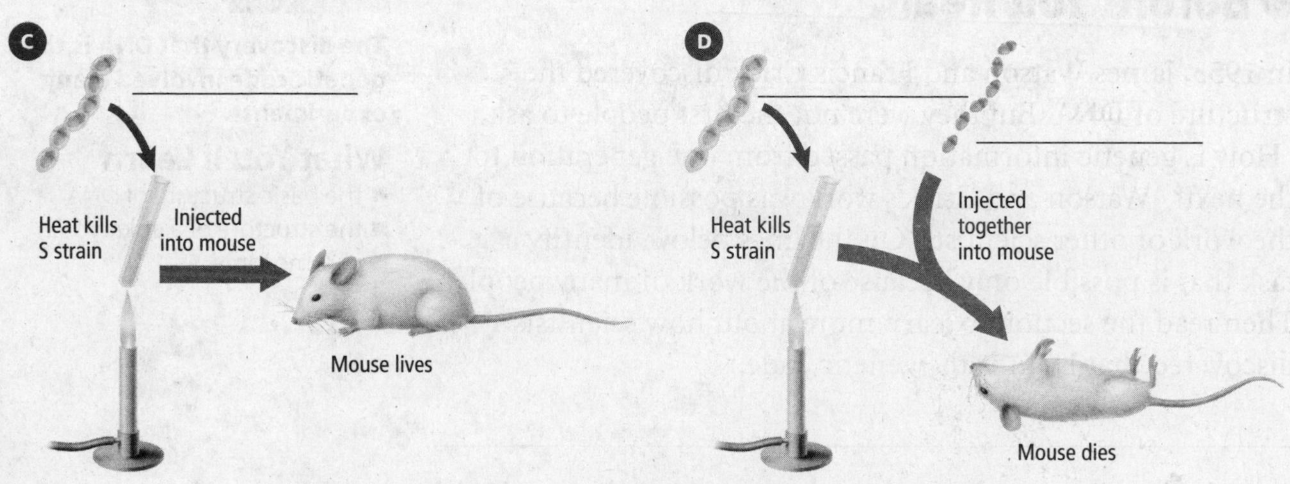

Picture This

2. Label Write the name of each strain of bacteria (*S* for smooth, *R* for rough) in each of the experiments shown.

How were bacteria transformed?

When Griffith injected live S strain into a mouse, the mouse died. When Griffith injected live R strain into a mouse, the mouse did not die. These results are shown above in Parts A and B.

Next, Griffith heated and killed the S strain. When injected, the dead S strain no longer killed the mouse. This step is shown in Part C above.

Next, Griffith mixed the heat-killed S strain with the live R strain. As shown in Part D above, when he injected the mixture into a mouse, something unexpected happened—the mouse died. Griffith studied live bacteria from the dead mouse. The smooth trait was visible. He concluded that the live R stain had changed into live S strain.

Reading Check

3. State what Avery found.

How was the transforming factor identified?

In 1931, Oswald Avery, along with other scientists, identified the molecule that transformed the R strain into S strain. Avery tested DNA, protein, and lipids from heat-killed S strain. He found that only DNA was able to change R strain into S strain. ☑

Who proved that DNA was the genetic material?

In spite of Avery's result, many scientists still questioned whether proteins or DNA were the genetic material. In 1952, Alfred Hershey and Martha Chase published results of an experiment proving that DNA was the genetic material. Hershey and Chase did an experiment with bacteriophages (bak TIHR ee uh fayjz), a type of virus that infects bacteria. The bacteriophages were made of DNA and protein. They reproduce by attaching to and injecting their genetic material into a living bacterial cell. ☑

Hershey and Chase used radioactive phosphorus (^{32}P) to label the DNA of one set of bacteriophages. They used radioactive sulfur (^{35}S) to label the protein of a second set of bacteriophages.

As shown in the figure below, Hershey and Chase mixed bacteria with viruses from the two groups. The viruses injected their genetic material into the bacteria. The viruses were separated from the bacteria.

Reading Check

4. Identify What two molecules make up a bacteriophage?

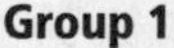

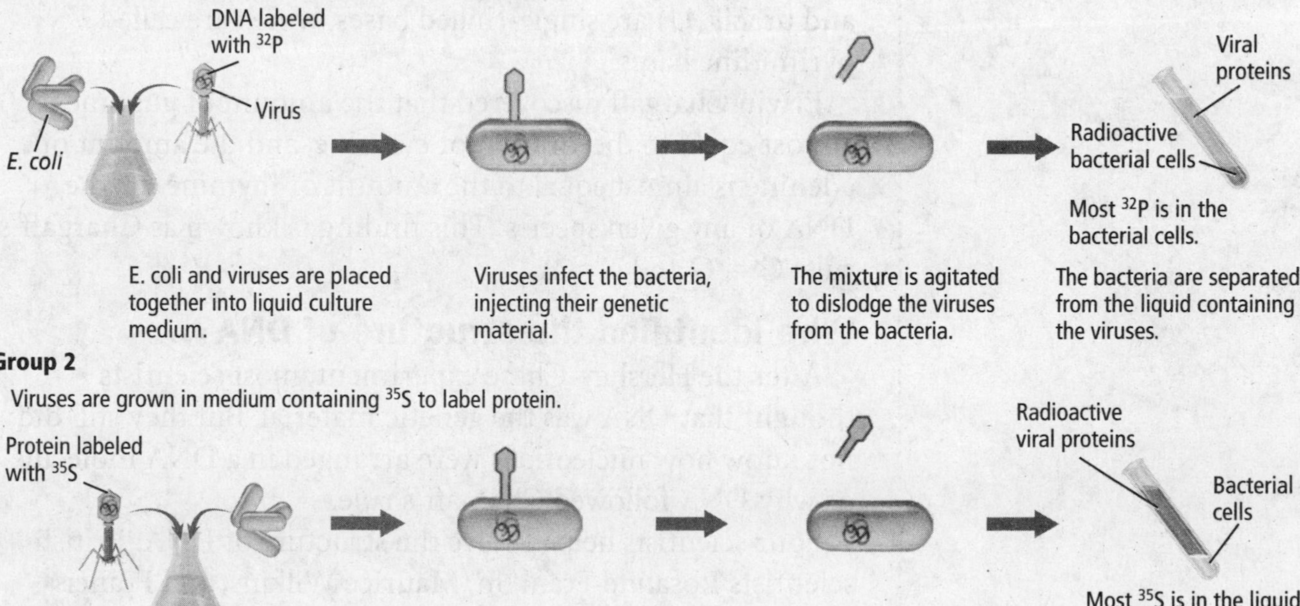

What did the viruses inject into bacteria?

Hershey and Chase found that both sets of viruses had replicated inside the bacterial cells. But only the labeled DNA had entered the bacterial cells. The labeled protein remained outside the bacterial cells. This experiment provided evidence that DNA, not protein, was the genetic material.

Picture This

5. Identify Circle the bacterial cells that have radioactive material inside of them.

DNA Structure

DNA is made of nucleotides. In the 1920s, biochemist P.A. Levene showed that each DNA nucleotide contains the sugar deoxyribose (dee ahk sih RI bos), a phosphate group, and one of four nitrogenous bases—adenine (A dun een), guanine (GWAH neen), cytosine (SI tuh seen), or thymine (THI meen).

RNA also is made of nucleotides. Each RNA nucleotide contains the sugar ribose, a phosphate, and one of four nitrogenous bases—adenine, guanine, cytosine, and uracil (YOO ruh sihl). The figure below shows the structure of a nucleotide.

Picture This

6. Identify Is the nucleotide in the picture a DNA nucleotide or an RNA nucleotide? How can you tell?

____

____

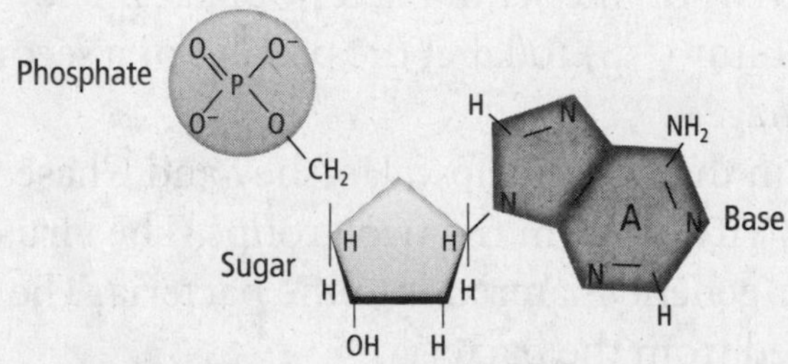

Adenine (A) and guanine (G) are double-ringed bases, which are called purine bases. Thymine (T), cytosine (C), and uracil (U) are single-ringed bases, which are called pyrimidine bases.

Erwin Chargaff discovered that the amount of guanine is almost equal to the amount of cytosine, and the amount of adenine is almost equal to the amount of thymine for the DNA of any given species. This finding is known as Chargaff's rule: C = G and T = A.

Who identified the structure of DNA?

After the Hershey-Chase experiment, most scientists thought that DNA was the genetic material. But they still did not know how nucleotides were arranged in a DNA molecule or why DNA followed Chargaff's rule.

Four scientists helped solve the structure of DNA: British scientists Rosalind Franklin, Maurice Wilkins, and Francis Crick and American scientist James Watson.

Reading Check

7. Identify What technique did Franklin use to take a picture of DNA?

____

____

What did Franklin's picture show?

Franklin worked for Wilkins at King's College in London, England. She took a picture of DNA using X-ray diffraction, a technique that involved aiming X rays at DNA. Franklin's picture, called Photo 51, showed that DNA was a **double helix**, with two strands of nucleotides twisted around each other like a twisted ladder. ☑

What did Watson and Crick propose?

Watson and Crick saw Franklin's X-ray diffraction picture. They used Franklin's picture and data as well as other mathematical data to determine the specific structure of the DNA double helix.

Watson and Crick built a model of DNA with two outside strands made of deoxyribose alternating with phosphate. The bases were on the inside of the molecule—cytosine paired with guanine, thymine paired with adenine. ☑

What is the structure of DNA?

Imagine DNA as a twisted ladder. The rails of the ladder are made of alternating deoxyribose and phosphates. The pairs of bases (cytosine—guanine or thymine—adenine) form the rungs. The rungs are always the same width because a purine base always binds to a pyrimidine base.

Now imagine two pencils lying side by side with the point of one pencil next to the eraser of the other. Like these pencils, the two strands of the sugar-phosphate chain of a DNA double helix run in opposite directions. The ends of the sugar-phosphate strands are named 5' (read "five-prime") and 3' (read "three-prime"). One strand runs 5' to 3'. The other strand runs 3' to 5'.

When was the structure of DNA announced?

In 1953, Watson and Crick published a letter in the journal *Nature* suggesting the structure for DNA and a hypothesized method of copying the molecule. In the same issue, Wilkins and Franklin published separate articles that supported Watson and Crick's structure. Scientists had solved some mysteries. However, they still did not know how DNA worked as a genetic code.

Chromosome Structure

The DNA molecule in prokaryotes is contained in the cytoplasm. The DNA forms a ring with its associated proteins.

DNA in eukaryotes is organized into chromosomes. Human chromosomes vary in length from 51 million to 245 million base pairs. The DNA fits into the nucleus of a eukaryotic cell by coiling around a group of beadlike proteins called histones. **Nucleosomes** are DNA that are tightly coiled around the histones. Nucleosomes are twisted together into chromatin fibers, which supercoil into a chromosome. ☑

Reading Check

8. Identify how the DNA bases pair.

____

____

____

Reading Check

9. Define What is a nucleosome made of?

____

____

Molecular Genetics

section 2 Replication of DNA

MAIN Idea

DNA replicates by making a complementary strand to each original strand.

What You'll Learn

- the role of enzymes in the copying of DNA
- how leading and lagging DNA strands are made

Before You Read

DNA is the instruction manual for a living thing. Each time one of your cells divides, your DNA is copied. That way, each new cell has its own copy of the instruction manual. On the lines below, list some items that come with instructions.

____

____

Mark the Text

Identify Main Ideas
As you read, underline or highlight the main ideas in each paragraph.

Read to Learn

Semiconservative Replication

Every time a cell divides, it must copy its DNA. That way, each cell has its own copy of the genetic material. When Watson and Crick presented their model of DNA, they also suggested a possible method of replication—semiconservative replication. In **semiconservative replication**, the two strands separate, serve as a pattern, and produce DNA molecules with one strand of the parental DNA and one strand of new DNA. Other scientists, armed with the knowledge of DNA's structure, began to explore ways that cells might copy DNA.

What are the steps of DNA replication?

Recall that DNA replication occurs during interphase of mitosis and meiosis. The process of semiconservative replication happens in three phases: unwinding, base pairing, and joining. ☑

Reading Check

1. **List** the three phases of semiconservative replication.

____

____

____

What happens during unwinding?

DNA replication is shown in the figure on the next page. In the first phase, an enzyme called DNA helicase unwinds the double helix. Single-stranded binding proteins hold the two strands apart. The RNA primase enzyme adds a short piece of RNA, called an RNA primer, to each strand of DNA.

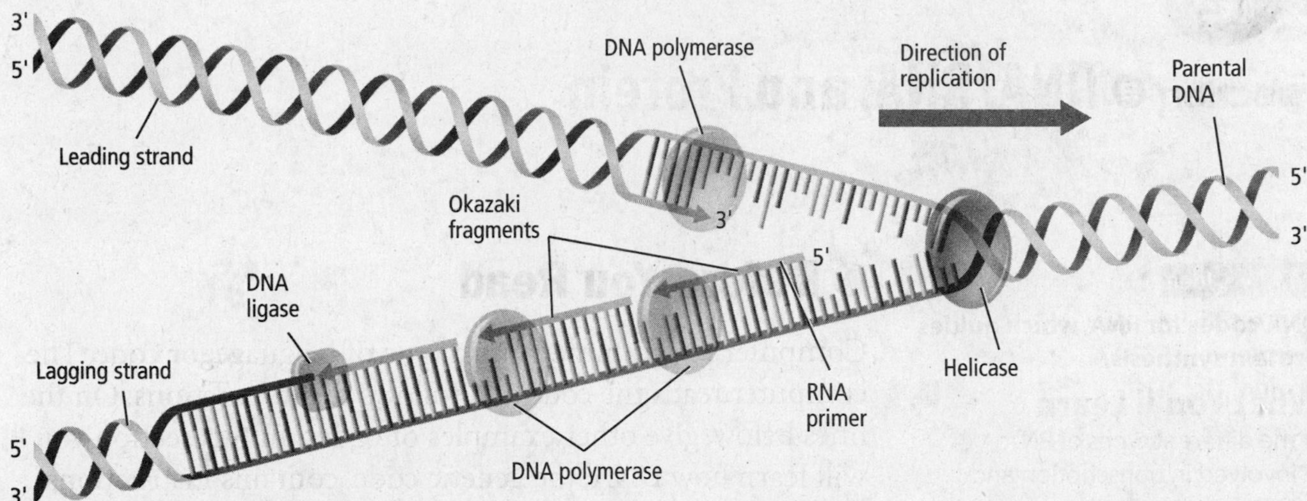

How does base pairing occur?

In the next step, the enzyme <u>**DNA polymerase**</u> helps the addition of nucleotides, bonding a new nucleotide to the parent strand and creating base pairs as it forms new strands.

Recall that one DNA strand runs 3' to 5'. The other runs 5' to 3'. The two strands are copied differently. One strand, called the leading strand, is made longer as the original DNA unwinds and nucleotides are added to its 3' end. The other strand, called the lagging strand, becomes longer as small pieces called <u>**Okazaki fragments**</u> are added in the 3' to 5' direction.

How are the fragments joined?

DNA polymerase removes RNA primers and replaces them with DNA nucleotides. When the last RNA primer has been removed and replaced with DNA nucleotides, the enzyme DNA ligase connects the DNA nucleotides.

Comparing DNA Replication in Eukaryotes and Prokaryotes

Eukaryotic DNA replication occurs at many places at the same time. The sites of DNA replication look like bubbles in the DNA strand.

Prokaryotic DNA is circular. In prokaryotes, the DNA strand is opened at one place on the circle. Replication occurs in both directions, unzipping the circle, until the whole DNA strand is copied.

Picture This

2. Identify Circle the names of the three enzymes involved in the process of semiconservative replication.

Think it Over

3. Compare Name one difference between eukaryotic DNA replication and prokaryotic DNA replication.

____

____

____

____

____

chapter 12 Molecular Genetics

section 3 DNA, RNA, and Protein

MAIN Idea

DNA codes for RNA, which guides protein synthesis.

What You'll Learn

- the different types of RNA involved in transcription and translation
- the role of RNA polymerase in the making of messenger RNA

Before You Read

Computer programmers use a type of language, or code. The computer reads the code and follows the instructions. On the lines below, give other examples of codes. In this section you will learn how DNA, the genetic code, contains instructions for making proteins.

____

____

____

Study Coach

Create a Quiz Read this section, and create a quiz with answers.

FOLDABLES™

Record Information Make a three-tab Foldable, as shown below, to organize information about the three types of RNA and how each is involved in transcription and translation.

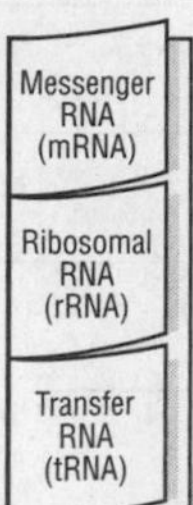

Read to Learn

Central Dogma

Proteins function as the building blocks of a cell and as enzymes. The instructions for making proteins are found in DNA. The information is read and expressed from DNA to RNA to proteins. This chain of events is called the central dogma of biology: DNA codes for RNA, which guides the making of proteins.

What are the types of RNA?

RNA, like DNA, is a nucleic acid. RNA nucleotides contain the sugar ribose instead of deoxyribose and the base uracil instead of thymine. RNA usually exists in single strands. Cells contain three main types of RNA—messenger RNA, ribosomal RNA, and transfer RNA.

Messenger RNA (mRNA) molecules are long strands of RNA nucleotides that direct ribosomes to make proteins. They travel from the nucleus to the ribosome. **Ribosomal RNA** (rRNA) molecules make up part of the ribosomes of the cell in the cytoplasm. **Transfer RNA** (tRNA) molecules transport amino acids from the cytoplasm to the ribosomes.

What happens during transcription?

During **transcription** (trans KRIHP shun), mRNA is made from DNA. As shown below, the DNA is unzipped in the nucleus. **RNA polymerase** binds where mRNA will be made. RNA polymerase makes mRNA in the 5' to 3' direction. Uracil is added to the mRNA strand instead of thymine, making a complement to the DNA strand. The mRNA strand moves out of the nucleus through nuclear pores into the cytoplasm.

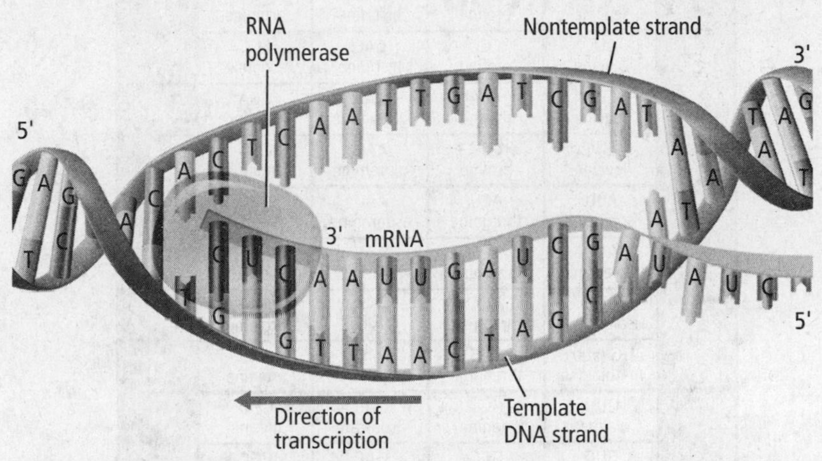

How is mRNA processed?

The mRNA code that is made during transcription has all of the DNA code. Before it leaves the nucleus, the pre-mRNA molecule undergoes processing. First, some of the sequences that interrupt the DNA code, called **introns**, are cut out. The sequences that remain in the mRNA molecule are called **exons**. Exons are the protein-coding sequences. ☑

A protective cap is added to the 5' end of the mRNA strand. A string of adenines, called a poly-A tail, is added to the 3' end.

The Code

The only way DNA varies among organisms is the sequence of bases. There are 20 amino acids that are used to make proteins, so DNA must provide at least 20 codes.

In the 1960s, scientists discovered that the DNA code is a three-base code called a **codon**. Each codon is transcribed into the mRNA code.

The genetic code for all 64 codons is shown in the table at the top of the next page. Three codons—UAA, UAG, and UGA—do not code for an amino acid. They are called stop codons because they stop the transcription process. AUG codes for methionine and is the start codon.

Picture This

1. Draw an arrow pointing to the location where the DNA is unzipped and where mRNA is made.

Reading Check

2. Define What is the name of the sequences that are removed from mRNA during processing?

____

Picture This

3. Apply Determine the amino acid chain that would result from the sequence AUG-CCC-GGA-UUA-UAG.

____

____

____

Think it Over

4. Name Where in the cell is the mRNA translated? (Circle your answer.)

a. cytoplasm
b. nucleus
c. ribosome

Dictionary of the Genetic Code

First Base	Second Base				Third Base
	U	C	A	G	
U	UUU phenylalanine	UCU serine	UAU tyrosine	UGU cysteine	U
U	UUC phenylalanine	UCC serine	UAC tyrosine	UGC cysteine	C
U	UUA leucine	UCA serine	UAA stop	UGA stop	A
U	UUG leucine	UCG serine	UAG stop	UGG tryptophan	G
C	CUU leucine	CCU proline	CAU histidine	CGU arginine	U
C	CUC leucine	CCC proline	CAC histidine	CGC arginine	C
C	CUA leucine	CCA proline	CAA glutamine	CGA arginine	A
C	CUG leucine	CCG proline	CAG glutamine	CGG arginine	G
A	AUU isoleucine	ACU threonine	AAU asparagine	AGU serine	U
A	AUC isoleucine	ACC threonine	AAC asparagine	AGC serine	C
A	AUA isoleucine	ACA threonine	AAA lysine	AGA arginine	A
A	AUG (start) methionine	ACG threonine	AAG lysine	AGG arginine	G
G	GUU valine	GCU alanine	GAU aspartate	GGU glycine	U
G	GUC valine	GCC alanine	GAC aspartate	GGC glycine	C
G	GUA valine	GCA alanine	GAA glutamate	GGA glycine	A
G	GUG valine	GCG alanine	GAG glutamate	GGG glycine	G

What happens during translation?

After the mRNA is made, it leaves the nucleus and moves to the cytoplasm. There the 5' end of the mRNA connects to a ribosome. The ribosome is where the code is read and translated to make proteins through a process called **translation**.

During translation, tRNA molecules act as interpreters of the mRNA. Each tRNA molecule is folded into a cloverleaf shape. The 3' end attaches to a specific amino acid. At the middle of the folded strand, there is a three-base coding sequence called an anticodon. Each anticodon is complementary to a codon on the mRNA. The anticodon is read 3' to 5'.

What is the role of the ribosome?

You can think of the ribosome as a factory for making proteins. Ribosomes are made of two parts—a large subunit and a small subunit. The subunits are not involved in making protein. The two parts of the ribosome come together and attach to an mRNA molecule to complete the ribosome.

How does translation work?

Transcription takes place in the nucleus. Translation occurs in the cytoplasm. Follow the process of protein production in the picture on the next page.

How is a protein made?

A tRNA molecule with the anticodon CAU carrying a methionine amino acid will move in and bind to the mRNA start codon—AUG—on the 5' end.

A tRNA, that is complementary to the mRNA codon, binds to a groove in the ribosome known as the P site. A second tRNA, that is complementary to the second mRNA codon, moves into a second groove called the A site.

The ribosome bonds the new amino acid in the A site and the amino acid in the P site. As the two amino acids join, the tRNA in the P site is released to the third site, called the E site, where it exits the ribosome. Find these sites in the figure.

The ribosome then moves along the mRNA molecule, so the tRNA in the A site shifts to the P site. A new tRNA then moves into the A site. The process continues, adding and joining amino acids in the sequence determined by the mRNA.

The ribosome continues moving along the mRNA, linking amino acids together, until it reaches a stop codon. The mRNA is released from the last tRNA by proteins called release factors, and the ribosome subunits come apart, ending protein synthesis.

5. Identify In what order does a tRNA move through the binding sites on a ribosome? (Circle your answer.)

a. E site–P site–A site
b. A site–E site–P site
c. P site–A site–E site

One Gene—One Enzyme

Scientists still needed to learn the relationship between the genes and the proteins for which they coded. In the 1940s, George Beadle and Edward Tatum conducted experiments on the mold *Neuorspora.* The mold spores they studied were mutated by exposure to X rays. Their experiments showed that a gene can code for an enzyme. Based on the results of their experiments, Beadle and Tatum proposed that each gene makes one enzyme, the "one gene—one enzyme" hypothesis.

How has the one gene—one enzyme hypothesis been revised?

Later experiments by other scientists showed that some enzymes contain more than one polypeptide chain. As a result, the hypothesis has been changed to state that each gene codes for one polypeptide.

Picture This

6. Identify Circle the tRNA in the P site.

How does translation make a protein polypeptide?

Follow along with the picture below. You will see how the ribosome translates the mRNA code into a chain of amino acids, called a polypeptide.

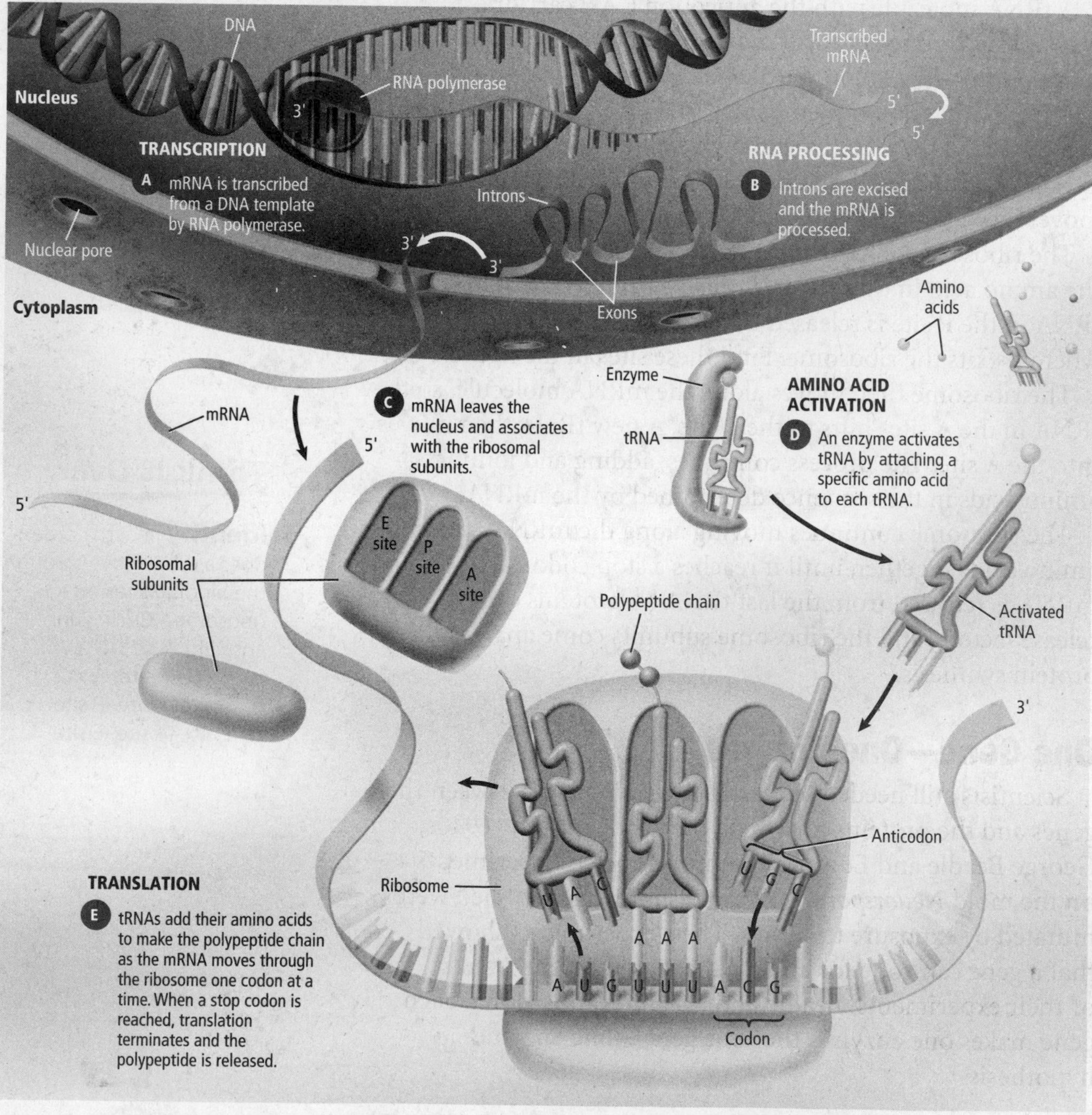

Molecular Genetics

section 4 Gene Regulation and Mutation

MAIN Idea

The cell regulates gene expression, and mutations can affect this expression.

What You'll Learn

- how bacteria can regulate their genes by operons
- how eukaryotes regulate transcription of genes

Before You Read

Like a DNA codon, the sentence, "The dog ran." contains only three-letter words. Insert one letter "Z" and the sentence changes to, "ThZ edo gra." On the lines below, write the sentence that results if you insert two letters, then three letters. Which sentence is easiest to read? In this section, you will read about how mutations affect gene expression.

____

____

____

Read to Learn

Mark the Text

Restate the Main Point Highlight the main point in each paragraph. State each main point in your own words.

Prokaryote Gene Regulation

Cells use **gene regulation** to control which genes are transcribed in response to the environment. Prokaryotes use operons to control the transcription of genes. An **operon** is a section of DNA that contains the genes for the proteins needed for a specific metabolic pathway. An operon contains an operator, a promoter, and a regulatory gene. The operator is like an on/off switch for transcription. The promoter is where RNA polymerase first binds to the DNA.

How does the *trp* operon work?

The tryptophan (*trp*) operon in the bacteria *Escherichia coli* (*E. coli*) is a repressible operon. Tryptophan synthesis occurs in five steps. Each step is triggered by a specific enzyme. The tryptophan operon contains five genes (*trpA* through *trpE*) needed to make the amino acid tryptophan. When tryptophan levels are low, RNA polymerase binds to the operator, turning on the transcription of the five enzyme genes needed for tryptophan synthesis. ☑

Reading Check

1. **Explain** What triggers each step in the synthesis of tryptophan?

____

____

How is the *trp* operon turned off?

The figure below shows what happens when tryptophan is abundant. The cell has no need to make tryptophan. Tryptophan binds to the repressor protein to activate it. The complex binds to the operator, keeping RNA polymerase from binding. The genes needed for tryptophan synthesis are not made.

Picture This

2. Identify Circle the operator sequence where the tryptophan-repressor complex is bound.

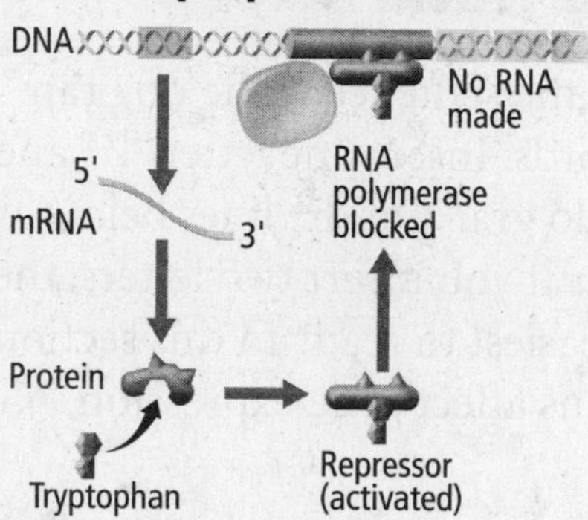

How does the *lac* operon work?

The *E. coli* lactose (*lac*) operon, an inducible operon, is shown below. The lac operon contains a promoter, an operator, a regulatory gene, and three genes that code for enzymes needed to digest the sugar lactose as food. The *lac* operon is switched on by an inducer, a molecule present in food containing lactose. The inducer binds to the *lac* repressor and inactivates it. RNA polymerase can bind to the promoter and transcription proceeds, and the lactose-digesting enzymes are made.

E. coli does not need to make lactose-digesting enzymes when lactose is not available. In that case, an inducer is not present and the regulatory gene makes a repressor protein that binds to the operator and blocks transcription.

Picture This

3. Label Circle the allolactose-repressor complex. How is the *lac* repressor different from the *trp* repressor?

____

____

____

____

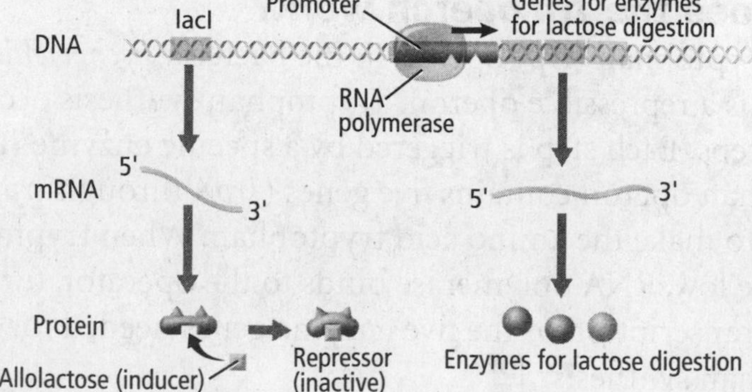

Eukaryote Gene Regulation

Eukaryotes have many more genes than prokaryotes. They also use different, more complex methods of gene regulation.

How do eukaryotes control transcription?

Proteins called transcription factors control when a gene is turned on and how much of that protein is made. Some transcription factors guide the binding of RNA polymerase to a promoter. Other transcription factors control the rate of transcription. ☑

How do Hox genes work?

Homeobox (Hox) genes code for transcription factors. Hox genes control differentiation, the process through which cells become specialized in shape and function. Hox genes are used during embryo development and are active in different zones of the embryo. They control what body part will develop in different parts of the embryo.

What is RNA interference?

Another way that eukaryotic genes are regulated is RNA interference (RNAi). Interfering RNA molecules are small segments of double-stranded RNA that bind to a protein complex that breaks down one strand of the RNA. The resulting single-stranded interfering RNA and protein complex bind to mRNA sequences and prevent mRNA from being translated.

Mutations

A permanent change in a cell's DNA is called a **mutation**. Mutations that occur in a gene sequence can change the protein that is made. Mutated proteins often do not work.

What mutations involve a single nucleotide?

Point mutations occur when a single nucleotide is changed. They can result in genetic disorders. A substitution is a kind of point mutation that occurs when one base is exchanged for another. A missense mutation is a substitution in which the DNA code is changed so that it codes for the wrong amino acid. A nonsense mutation changes the codon for an amino acid to a stop codon. Nonsense mutations often cause translation to stop early, making a protein that is too short. Muscular dystrophy is an example of a disease caused by a nonsense mutation. ☑

Reading Check

4. State two ways that transcription factors control genes.

____

____

____

Reading Check

5. Name What type of mutation occurs when a stop codon replaces an amino acid codon?

____

____

What are some other types of mutations?

Insertions and deletions occur when a nucleotide is added or lost. Insertions and deletions can cause a frameshift mutation, causing the ribosome to misread the codons. THE BIG FAT CAT ATE THE WET RAT becomes THE BIG ZFA TCA TAT ETH EWE TRA. Cystic fibrosis and Crohn's disease are both caused by frameshift mutations. Some mutations involve large pieces of DNA containing many genes. A piece of a chromosome can be deleted, moved to a different location on the same chromosome, or moved to another chromosome. Such mutations often have serious effects.

In 1991, a new type of mutation was discovered. This mutation happens when repeated sequences, called tandem repeats, increase in number. Fragile X syndrome and Huntington's disease are both caused by this type of mutation.

Think it Over

6. Evaluate Would a 3-nucleotide insertion result in a frameshift? Why or why not?

____

____

____

____

How do mutations affect protein folding?

Small mutations, like substitutions, can lead to genetic disorders. Changing one amino acid for another can change how a protein folds and, as a result, change how it functions.

What causes mutations?

Some mutations occur simply because DNA polymerase makes a mistake, adding the wrong nucleotide during DNA replication. Other mutations are caused by **mutagens** (MYEW tuh junz), which are chemicals or radiation that can damage DNA.

Some mutagens resemble nucleotides so closely, that DNA polymerase mistakes them for nucleotides and adds the mutagen into the DNA chain. Chemical mutagens are being studied for possible use in treating HIV—the virus that causes AIDS.

UV radiation from the Sun can damage DNA. It can cause thymine bases that are next to each other to bind together. This creates a kink in the DNA, and it cannot replicate. ☑

Reading Check

7. Explain How does UV light damage DNA?

____

____

____

How are mutations inherited?

Mutations in body cells, or somatic cells, are not passed on to the next generation. Sometimes these mutations do not cause problems for the cell. Other times they kill the cell. Some somatic cell mutations lead to cancer.

Mutations in sex cells are passed on to the organism's offspring. Every cell in the offspring will carry the mutation. Sometimes the mutations do not change how those cells function. Other times the mutations have serious effects.

Genetics and Biotechnology

section 1 Applied Genetics

Before You Read

Imagine that you could design the perfect dog. What color would it be? Would it be big or small? On the lines below, describe the traits your dog would have. In this section, you will learn how selective breeding produces certain traits.

____

____

____

MAIN Idea

Selective breeding is used to create animals or plants with certain traits.

What You'll Learn

- how inbreeding differs from hybridization
- how to use test crosses and a Punnett square to find the genotypes of organisms

Read to Learn

Selective Breeding

For thousands of years, people have been breeding animals and plants to have certain traits. For instance, some dogs, such as huskies, have been bred to be strong runners. Other dogs, such as Saint Bernards, have been bred to have a good sense of smell.

People have also bred plants, such as tomatoes, apples, and roses, to taste better, resist disease, or produce fragrant flowers. Selective breeding is the process used to breed animals and plants to have desired traits. As a result of selective breeding, desired traits become more common.

Study Coach

Create a Quiz After you read this section, create a five-question quiz from what you have learned. Then, exchange quizzes with another student. After taking the quizzes, review your answers together.

What is hybridization?

A hybrid is an organism whose parents each have different forms of a trait. For instance, a disease-resistant tomato plant can be crossed with a fast-growing tomato plant. The offspring of the cross would be a tomato plant that has both traits. The hybrid is disease resistant and grows quickly.

Hybridization is the process of making a hybrid organism. Hybridization is expensive and takes a long time, but it is a good way to breed animals and plants with the right combination of traits. ☑

Reading Check

1. **Name** an advantage of hybridization.

____

____

____

How is inbreeding used?

Inbreeding is another example of selective breeding. **Inbreeding** occurs when two closely related organisms that both display the desired trait are bred. Inbreeding can be used to ensure that the desired trait is passed on. Inbreeding can also eliminate traits that are not desired.

Purebred animals are created by inbreeding. Clydesdale horses are an example of a purebred animal. Clydesdale horses were first bred in Scotland hundreds of years ago. They were bred for use as farm horses that could pull heavy loads. All Clydesdales have the traits of strength, agility, and obedience.

A disadvantage of inbreeding is that harmful traits can be passed on. Harmful traits are usually carried on recessive genes. Both parents must pass on the recessive genes for the harmful traits to appear in the offspring. Inbreeding increases the chance that both parents carry the harmful traits.

Test Cross

Breeders need a way to determine the genotype of the organisms they want to cross before creating a hybrid. They use test crosses to find out the genotype of an organism. In a **test cross**, an organism whose genotype for a desired trait is unknown is crossed with an organism that has two recessive genes for the trait. ✔

✔ Reading Check

2. Explain What is the purpose of a test cross?

____

____

____

When are test crosses performed?

An orchard owner might use a test cross to find out the genotype of a white-grapefruit tree. In grapefruits, white color is a dominant trait and red color is a recessive trait. A red-grapefruit tree has two recessive genes (*ww*). A white-grapefruit tree might have two dominant genes (*WW*), or it might have one dominant gene and one recessive gene (*Ww*).

Picture This

3. Label Fill in the phenotype with the word *white* or *red* for each genotype.

Genotype	Phenotype
Homozygous dominant (*WW*)	
Homozygous recessive (*ww*)	
Heterozygous (*Ww*)	

How does a test cross reveal the genotype?

The orchard owner decides to do a test cross to find out the genotype of a white grapefruit tree. The white grapefruit tree is crossed with a red grapefruit tree. The orchard owner uses a Punnett square to understand the results of the cross.

The figure below shows a Punnett square for the test cross if the white grapefruit tree is homozygous, meaning it has two dominant genes (*WW*) for white fruit. All the offspring from the test cross will be heterozygous, meaning they will have one dominant and one recessive gene (*Ww*). All the offspring of the test cross are white grapefruit trees.

Homozygous white grapefruit (top); Homozygous red grapefruit (side)

	W	**W**
w	***Ww***	***Ww***
w	***Ww***	***Ww***

What if the test cross involved a heterozygous tree?

The figure below shows a Punnett square for the test cross if the white grapefruit tree is heterozygous (*Ww*). Half the offspring from the test cross will be white (*Ww*). Half the offspring from the test cross will be red (*ww*).

Heterozygous white grapefruit (top); Homozygous red grapefruit (side)

	W	**w**
w	***Ww***	***ww***
w	***Ww***	***ww***

Picture This

4. Evaluate If you planted 100 seeds from this test cross, about how many would be white? How many would be red?

____

____

Picture This

5. Calculate If you planted 100 seeds from this test cross, about what percentage would be white? What percentage would be red?

____

____

Genetics and Biotechnology

section 2 DNA Technology

MAIN Idea

Genetic engineering manipulates recombinant DNA.

What You'll Learn

- the difference between selective breeding and genetic engineering
- how genetic engineering can be used to improve human life

Before You Read

The tools that a chef uses to prepare food differ from the tools a mechanic uses to fix cars. On the lines below, describe a few of the tools you use at home and school. In this section, you will learn about tools scientists use to study DNA.

____

____

Mark the Text

Main Ideas As you read, underline or highlight the main ideas in each paragraph.

Read to Learn

Genetic Engineering

For many years, scientists knew the structure of DNA and knew that information flowed from DNA to RNA and from RNA to proteins. In the last few decades, scientists have learned more about how individual genes work by using genetic engineering. **Genetic engineering** is a way of manipulating the DNA of an organism by inserting extra DNA or inserting DNA from another organism.

One example of genetic engineering uses green fluorescent protein (GFP). GFP is a protein made naturally in jellyfish. GFP causes jellyfish to turn green under ultraviolet light. Scientists have inserted the DNA for making GFP into other organisms. This makes the organisms glow.

Reading Check

1. State What do scientists have to do to a gene before they can manipulate it?

____

____

____

DNA Tools

An organism's **genome** is all the DNA present in the nucleus of each cell. Genomes can contain millions of nucleotides in the gene's DNA. In order to study a specific gene, scientists isolate it from the rest of the organism's DNA. Scientists can then manipulate it. To understand how scientists do this, it is helpful to know the DNA tools scientists use. ☑

What are restriction enzymes?

Scientists have found hundreds of restriction enzymes. **Restriction enzymes** are proteins made by bacteria. Each restriction enzyme cuts, or cleaves, DNA at a specific DNA sequence.

How do restriction enzymes work?

One restriction enzyme that is often used by scientists is called *EcoRI*. *EcoRI* cuts DNA containing the sequence GAATTC. After *EcoRI* cuts DNA, it leaves single-stranded ends, called *sticky ends,* as shown in the figure below. DNA that has been cut with *EcoRI* always has the same sticky ends. DNA fragments with sticky ends can be joined with other DNA fragments with complementary sticky ends.

Not all restriction enzymes leave sticky ends. Some restriction enzymes cut straight across both DNA strands, leaving blunt ends. DNA fragments with blunt ends can be joined to other DNA fragments with blunt ends.

How is gel electrophoresis used to separate DNA fragments?

After DNA is cut with a restriction enzyme, the DNA fragments are different sizes. Scientists use **gel electrophoresis** to separate DNA fragments according to the size of the fragments.

DNA fragments are placed on the negatively charged end of a material called gel. An electric current is applied to the gel. The DNA fragments move toward the positive end of the gel. Smaller fragments move through the gel faster than larger fragments. The unique pattern made by the DNA fragment can be compared to the patterns of known DNA fragments for identification. The figure below shows a gel in which DNA has been separated by electrophoresis.

Think it Over

2. Explain Why can two different fragments of DNA cut with *EcoRI* be joined?

___

___

___

Picture This

3. Analyze Use the figure to explain to a partner how gel electrophoresis works.

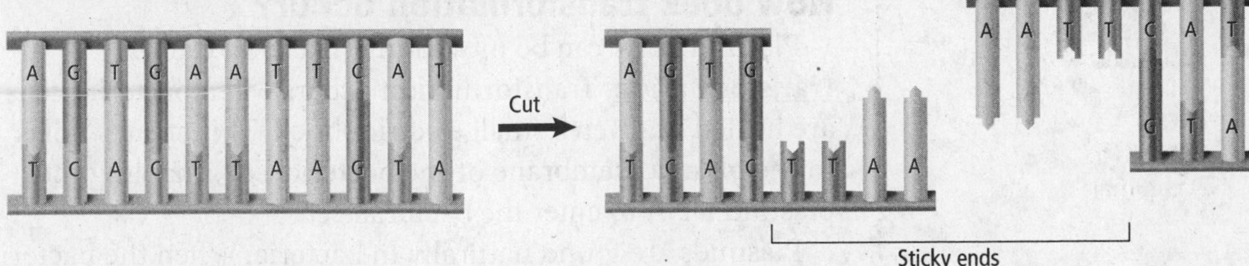

FOLDABLES™

Take Notes Make a four-tab Foldable, as shown below. As you read, take notes and organize what you learn about recombinant DNA technology.

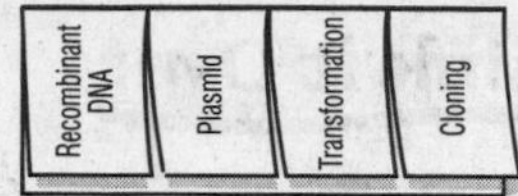

Picture This

4. Identify Circle the carrier in the figure.

Recombinant DNA Technology

Once DNA fragments have been separated using gel electrophoresis, fragments can be removed from the gel. These DNA fragments can then be combined with DNA fragments from another source, as shown in the figure below. This new DNA molecule, with DNA from different sources, is called **recombinant DNA**. Scientists use of recombinant DNA allows scientists to study individual genes.

Scientists often need to make a lot of recombinant DNA to study it. Scientists use host cells, such as bacteria, to copy the recombinant DNA. A carrier, known as a vector, is used to carry the recombinant DNA into the host cell. One commonly used vector is a small, circular, double-stranded DNA molecule called a **plasmid**. Plasmids can be cut with restriction enzymes. DNA fragments and plasmids cut with the same restriction enzyme can be combined at their sticky ends. An enzyme called **DNA ligase** is then used to join the plasmids and the DNA fragments chemically.

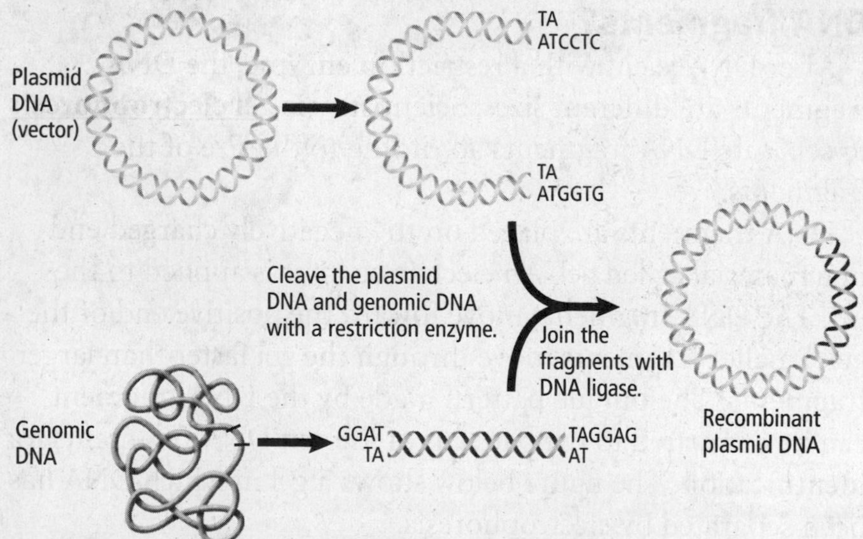

How does transformation occur?

Plasmid DNA can be moved into bacterial cells by **transformation**. Transformation occurs when bacterial cells are heated or given a small electric shock. This creates holes in the plasma membrane of the bacterial cell, enabling the plasmid DNA to enter the bacterial cell.

Plasmids are found naturally in bacteria. When the bacteria reproduce and copy their own DNA, they also copy the plasmid DNA. **Cloning** occurs when bacteria reproduce and copy recombinant DNA molecules.

What is DNA sequencing?

DNA sequencing involves finding out the exact order of the nucleotides that make up an organism's DNA. Knowing the DNA sequence of an organism gives scientists clues about how that organism's genes work. Scientists can compare genes from different organisms. Scientists can also find errors in the DNA. Long DNA molecules must be cut with restriction enzymes before they can be sequenced. ☑

How is DNA sequenced?

The figure below shows how DNA is sequenced. Scientists mix an unknown DNA fragment, DNA polymerase, and the four nucleotides—A, C, G, and T. Each of the four nucleotides are tagged with a different color of fluorescent dye.

What stops the growth of a DNA strand?

Usually, when DNA polymerase copies the DNA fragment it will put normal nucleotides on the growing strand. However, sometimes a fluorescent-tagged nucleotide will be added to the strand. Every time these tagged nucleotides are added, the new DNA strand stops growing. This produces DNA strands of different lengths. The tagged fragments are separated by gel electrophoresis. An automated DNA sequencing machine is used to detect the color of each tagged nucleotide. The sequence of the original DNA is determined from the order of the tagged fragments.

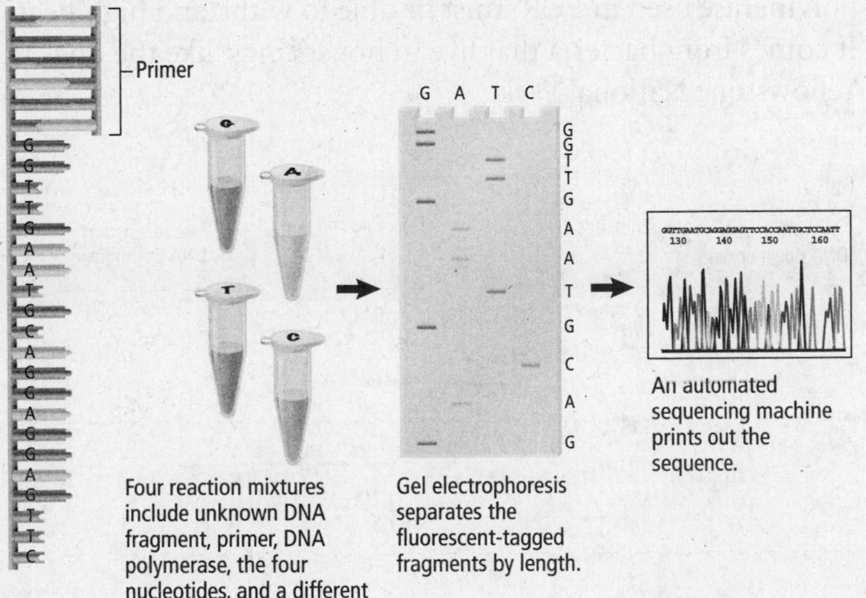

Four reaction mixtures include unknown DNA fragment, primer, DNA polymerase, the four nucleotides, and a different tagged nucleotide.

Gel electrophoresis separates the fluorescent-tagged fragments by length.

An automated sequencing machine prints out the sequence.

Reading Check

5. Determine How are restriction enzymes used?

Picture This

6. Identify Which step in the process separates DNA fragments by length?

What is polymerase chain reaction?

Polymerase chain reaction (PCR) can be used to make millions of copies of a specific region of a DNA fragment. PCR is so sensitive that it can detect a single DNA molecule in a sample. With PCR, scientists can copy a single DNA molecule many times so they can study it. ☑

PCR is a powerful tool used by scientists. Forensic scientists use PCR to identify suspects and victims of crimes. Doctors use PCR to detect diseases such as AIDS.

Reading Check

7. Explain Why is polymerase chain reaction used to make millions of copies of a DNA fragment?

____

____

____

What are the steps of PCR?

Follow the figure below as you read the steps of PCR.

Step 1 Four things are mixed in a small tube: the DNA fragment to be copied, DNA polymerase, the four DNA nucleotides—A, G, C, and T—and two short, single-stranded pieces of DNA called primers. The primers are complements to the ends of the DNA fragment to be copied. The primers are used as starting points for the DNA copies.

Step 2 The tube is placed into a thermocycler. The thermocycler heats and cools the tube over and over again. When the tube is heated, the two strands of the DNA fragment separate. When the tube is cooled, the primers bind to the ends of the separated strands of the DNA fragment.

Step 3 Each primer binds to one strand of the DNA fragment. DNA polymerase then puts the correct nucleotides between the two primers making the copies. The DNA polymerase used in PCR must be able to withstand high heat. It comes from bacteria that live in hot springs, like the ones in Yellowstone National Park.

Picture This

8. Identify Underline the two starting points for the DNA copies.

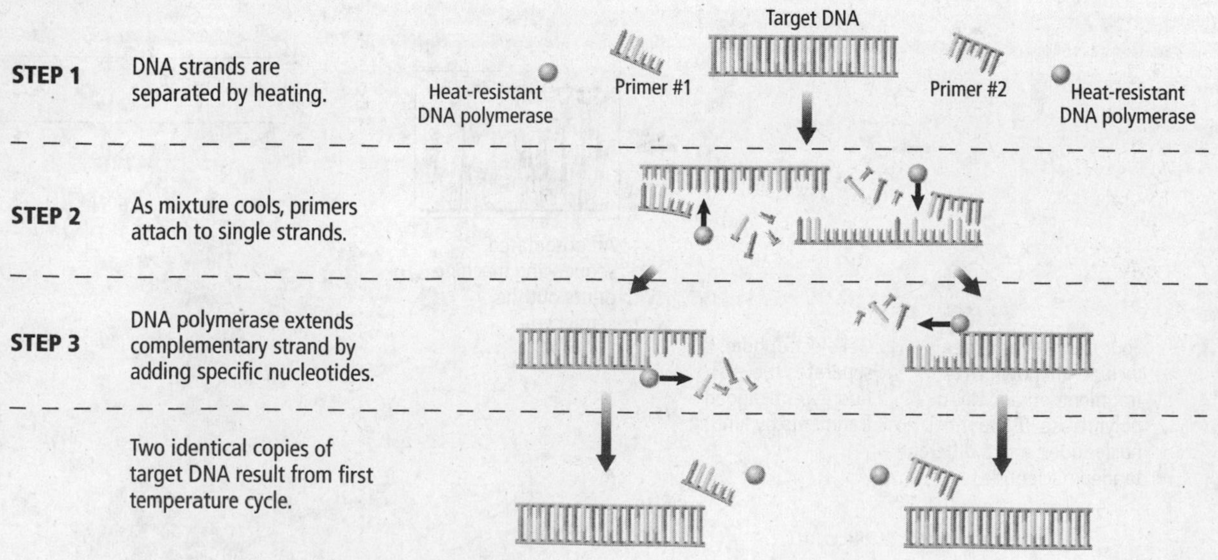

STEP 1 DNA strands are separated by heating.

Target DNA

Primer #1

Primer #2

Heat-resistant DNA polymerase

Heat-resistant DNA polymerase

STEP 2 As mixture cools, primers attach to single strands.

STEP 3 DNA polymerase extends complementary strand by adding specific nucleotides.

Two identical copies of target DNA result from first temperature cycle.

Biotechnology

Biotechnology is the application of genetic engineering to human problems. Scientists can use biotechnology to produce transgenic organisms. **Transgenic organisms** are organisms that have a gene from a different organism inserted into their DNA. Transgenic animals, plants, and bacteria are used for scientific research, in agriculture, and to treat human diseases.

How are transgenic animals used?

Most transgenic animals are made in laboratories for biological research. Some commonly studied animals are mice, fruit flies, and roundworms. Scientists use these organisms to study diseases and develop ways to treat them.

Transgenic livestock are used to improve the food supply. They also are used to improve health in people. For instance, scientists have engineered goats to make a protein that stops blood from clotting. Surgeons use this protein during operations. Several species of fish have been genetically engineered to grow faster. In the future, transgenic animals might be used as a source of organs for organ transplants in people. ☑

Reading Check

9. Identify one way scientists use transgenic animals.

______

______

______

How are transgenic plants used?

Transgenic crops are grown around the world. Farmers in at least 18 countries grow transgenic corn, soybeans, canola, and cotton on millions of acres. Farmers plant these crops because they are resistant to herbicides and insecticides. For example, scientists are now producing genetically engineered cotton. The cotton has been engineered to resist weevils, insects that harm cotton plants.

Scientists have developed other transgenic crops. They are testing these crops in fields. One of these crops is a transgenic rice that is more nutritious than normal rice. Scientists hope to use the transgenic rice to decrease malnutrition in Asian countries. Scientists are also testing crops that are designed to survive extreme weather. ☑

Reading Check

10. Explain What is one trait scientists have engineered into transgenic plants?

______

______

______

Someday, peanuts and soybeans might be developed that do not cause allergic reactions. Transgenic plants might also be used to make vaccines or biodegradable plastics.

How are transgenic bacteria used?

Scientists use transgenic bacteria to make insulin, growth hormones, and other medical substances. Transgenic bacteria have been used to protect crops from frost damage and to clean up oil spills. Garbage in some landfills is being decomposed by transgenic bacteria.

Genetics and Biotechnology

section 3 The Human Genome

MAIN Idea

Genomes contain all of the information needed for an organism to survive.

What You'll Learn

- how forensic scientists use DNA fingerprinting
- how human genome information can help diagnose diseases

Before You Read

Scientists now study genes in ways that were not invented 20 years ago. Think of the new technology in your own life. What are some new technologies you use?

____

____

Study Coach

Make Flash Cards Make a flash card for each key term in this section. Write the term on one side of the card. Write the definition on the other side. Use the flash cards to review what you have learned.

Read to Learn

The Human Genome Project

A genome is all of the genetic information in a cell. The human genome is all of the genetic information in a human cell. The Human Genome Project (HGP) was an enormous project. One goal was to learn the sequence of the billions of nucleotides that make up human DNA. Another goal was to identify all 20,000 to 25,000 human genes.

The HGP was completed in 2003. Scientists will be working for many years to understand the data.

How was the human genome sequenced?

Human DNA is organized into 46 chromosomes. To determine the human genome, each chromosome was cut. Several restriction enzymes were used to make fragments with overlapping sequences. The fragments were combined with vectors and copied. The overlapping sequences were analyzed to generate a continuous sequence.

As scientists studied the sequences in the human genome, they observed that less than 2 percent of all of the nucleotides in the human genome code for all of the proteins in the body. The rest of the DNA is made of long stretches of repeated sequences called noncoding sequences. Scientists do not yet know the function of these sequences.

Applying Math

1. Calculate What percentage of human DNA is not made of genes?

____

How is DNA fingerprinting used?

The protein-coding sections of DNA are almost identical from one person to the next. The long stretches of noncoding sections of DNA are unique to each individual. **DNA fingerprinting** uses gel electrophoresis to observe the patterns that are unique to each person.

Forensic scientists use DNA fingerprinting to identify suspects and victims in a crime. DNA fingerprinting has been used to convict criminals and free innocent people who were wrongly imprisoned. DNA fingerprinting can be used to identify soldiers killed in war and establish paternity.

When only a drop of blood or a single hair is found at a crime scene, the sample does not contain enough DNA for DNA fingerprinting. Forensic scientists use PCR to copy the DNA and make a larger sample. The DNA is then cut with restriction enzymes and separated by gel electrophoresis. The pattern of the fragments from the sample is compared with DNA samples from known sources, such as a suspect or a victim in a crime.

Identifying Genes

Once the genome has been sequenced, the next step is to identify the genes and determine their functions. Organisms, such as bacteria and yeast, do not have noncoding DNA. Scientists look for DNA sequences called open reading frames (ORFs). ORFs are made of codons—groups of three nucleotides that code for amino acids. ORFs begin with a start codon and end with a stop codon. In between the start and stop codons, ORFs contain at least 100 codons. Scientists have identified over 90 percent of genes in yeast and bacteria by looking for ORFs.

In humans and other complex organisms, the long stretches of noncoding sequence make looking for genes more difficult. Scientists use sophisticated computer programs called algorithms to identify genes.

Bioinformatics

The sequencing of DNA from humans and other organisms has created large amounts of data. It has also led to a new field of study. **Bioinformatics** is the study of how to create and use computer databases to store, organize, index, and analyze this data. Scientists are using bioinformatics to discover new ways to locate genes in DNA sequences and to study the evolution of genes. ☑

Think it Over

2. Identify What is most useful for DNA fingerprinting: protein-coding sequences or noncoding sequences? Explain.

____

____

____

____

Reading Check

3. Explain What is bioinformatics?

____

____

____

____

DNA Microarrays

In any cell at any time, some genes are expressed, meaning those genes are making proteins. The rest of the genes are silent. In a different cell or at a different time, other genes will be expressed. ☑

DNA microarrays are tiny microscope slides or silicon chips that contain tiny spots of DNA fragments. One microarray can contain thousands of genes. Scientists use DNA microarrays to study the expression of a lot of genes at once. DNA microarrays are used to study when and where genes are expressed. Microarrays can reveal how gene expression changes under different conditions. Microarrays can be used to compare cancer cells to normal cells. By finding genes that are expressed in cancer cells, scientists can learn more about cancer. They can learn better ways to treat people with cancer.

The figure below shows two DNA microarrays. Each spot represents a different gene. Spots that are white indicate the gene is being expressed. Spots that are black indicate the gene is not being expressed. The top microarray shows the genes that are expressed in a normal cell. The bottom microarray shows the genes that are expressed in a cancer cell.

Reading Check

4. Define What does it mean to say a gene is expressed?

____

____

Picture This

5. Analyze Find the genes that are expressed in the cancer cell but not in the normal cell. Circle the spots that represent those genes.

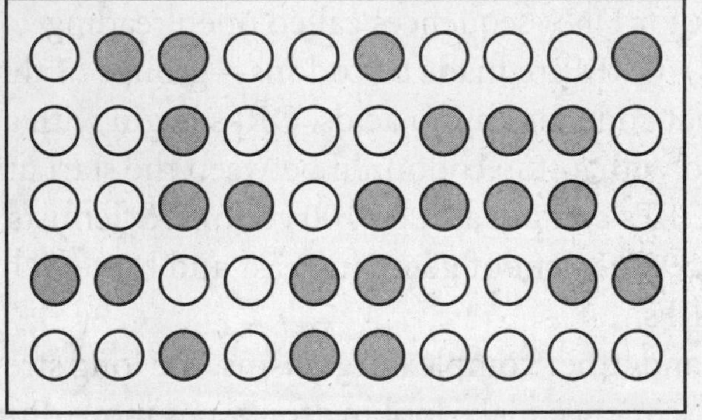
Normal Cell

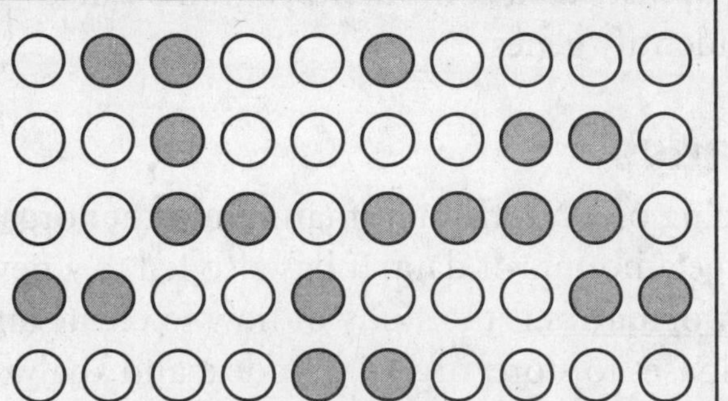
Cancer Cell

The Genome and Genetic Disorders

Over 99 percent of all nucleotide sequences are exactly the same from one person to the next. **Single nucleotide polymorphisms**, or SNPs, are variations in the DNA sequence that occur when a single nucleotide in the genome is changed. A variation is only considered an SNP if it occurs in at least 1 percent of the population.

SNPs can be useful to scientists. Many SNPs do not change how cells function, but SNPs might help scientists find other, nearby mutations that do cause genetic disease. Some SNPs occur near mutations that cause human diseases. Knowing where SNPs occur in the genome might help scientists find mutations that cause diseases.

Applying Math

6. Calculate A single nucleotide variation occurs in 7 of every 1000 people. Is this variation an SNP? Why or why not?

What is the HapMap project?

A group of international scientists is creating a list of common genetic variations in people. Genetic variations located close together on a chromosome are said to be linked. Linked variations are usually inherited together.

A **haplotype** is a section of linked variations in the human genome. The haplotype map or HapMap project is an international effort to find all the haplotypes. The project will describe what these variations are and show where they are found. The HapMap project will also describe how these variations occur among people within populations and among populations from different areas of the world.

The HapMap project will enable scientists to take advantage of how SNPs and other genetic variations are organized on chromosomes. This will help scientists find genes that cause different types of disease. The HapMap will also help scientists find mutations that affect how a person responds to medicine.

What is pharmacogenomics?

One day people might go to the doctor and have drugs specially prescribed for them based on their genes. **Pharmacogenomics** (far muh koh jeh NAW mihks) is the study of how a person's genes affect his or her response to medicine.

Researchers hope that pharmacogenomics will allow drugs to be custom made for people based on their genetic makeup. Pharmacogenomics might allow doctors to prescribe drugs that are safer, more specific, more effective, and have fewer side effects. Doctors might one day read your genetic code and prescribe drugs made especially for you. ☑

Reading Check

7. Define What is a possible benefit of pharmacogenomics?

How does gene therapy work?

Gene therapy is a way of fixing mutated genes that cause disease. Scientists insert a normal gene into a chromosome to replace the mutated gene. The normal gene can then do the work of the mutated gene.

A virus is used as a vector to transfer the normal gene to the cell. The virus releases the recombinant DNA, which contains the normal gene, into the cell. The normal gene inserts itself into the genome and begins functioning.

The Food and Drug Administration monitors new medical trials, including gene therapy. Although there have been setbacks, recent trials include work with diabetes, cancer, and retinal disease.

Picture This

8. Identify How is a normal gene inserted into a cell? (Circle your answer.)

a. by a virus releasing recombinant DNA containing the normal gene

b. by physically removing the mutated gene

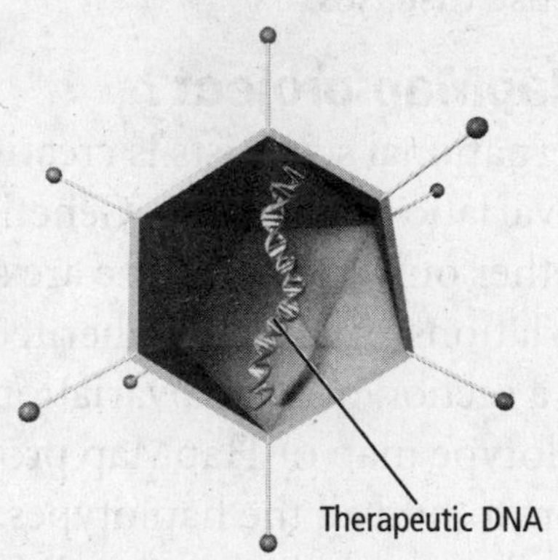

Genomics and Proteomics

Genomics is the study of an organism's genome. Following the completion of the human genome sequence in 2003, so much research has become focused on genomics that biologists call this "the genomic era."

Genomics is a powerful strategy for identifying human genes and understanding how they work. Researchers also use genomics to study plants and other organisms, such as rice, mice, fruit flies, and corn, whose genomes have been sequenced.

Genes are important because they are the way cells store information. Proteins are important because they are the machines that make cells run.

Proteomics is the large-scale study and cataloging of the structure and function of proteins in the human body. With proteomics, researchers can study hundreds or thousands of proteins at one time. ☑

Scientists use proteomics to understand human diseases. Scientists expect that proteomics will change the development of medicines to treat diseases such as diabetes, obesity, and atherosclerosis.

✔ Reading Check

9. Define What is proteomics?

____

____

____

____

The History of Life

section 1 Fossil Evidence of Change

Before You Read

Throughout Earth's history, many species have become extinct. On the lines below, name some organisms that have become extinct. Then read the section to learn more about how scientists learn about extinct species.

____

____

____

____

MAIN Idea

Fossils provide evidence of the change in organisms over time.

What You'll Learn

- how a typical fossil is made
- techniques for dating fossils
- the major events using the geologic time scale

Read to Learn

Mark the Text

Identify Definitions

As you read each section, highlight or underline the definition of each underlined term.

Earth's Early History

How did life on Earth begin? Because there were no people around to record Earth's earliest history, the answer is a mystery. Scientists who study the beginning of life on Earth must look for clues that were left behind.

Some of the clues are found in rocks. Rocks give us clues about what Earth was like in the beginning and sometimes what species lived during that time. Scientists also study other planets to uncover clues about Earth's past.

What was Earth's early land environment like?

Earth formed about 4.6 billion years ago. At first, Earth was molten—melted rock. Gravity pulled the densest elements to the center of the planet, forming Earth's core. After about 500 million years, a solid crust formed on the surface. The crust was made mostly of lighter elements.

From clues found in rocks, scientists infer that Earth's early surface was hot. Volcanoes erupted and meteorites hit the surface. It is not likely that life could have survived the heat. ☑

Reading Check

1. Explain What was Earth's early surface like?

____

____

What was Earth's early atmosphere like?

Earth's early atmosphere was probably made up of gases that were expelled by volcanoes. These gases might have been similar to those that are expelled by volcanoes today. Minerals in the oldest known rocks suggest that the early atmosphere had little or no free oxygen. Gases expelled by volcanoes do not include free oxygen.

Clues in Rocks

Earth eventually cooled, and liquid water formed on the surface, forming the first oceans. A short time later, as little as 500 million years, life appeared. Rocks provide important clues about Earth's history. The oldest clues about life on Earth date to about 3.5 billion years ago. ☑

Reading Check

2. Identify How old is the oldest evidence of life on Earth?

____

____

What is a fossil?

A **fossil** is any preserved evidence of an organism. Fossils are found in rock, ice, and amber. More than 99 percent of the species that have lived on Earth are now extinct. Only a small percentage of those species are preserved as fossils.

For an organism to be preserved as a fossil, it must be buried quickly in sediment. Organisms that live in water are more likely to form fossils than organisms that live on land because sediment in the water is constantly settling.

First the organism dies. Then sediment covers the organism. Layers of sediment build up over time. In most cases, minerals replace or fill in the pore space of the bones and hard parts of the organism. In some cases, the organism decays, leaving behind an impression of its body. The layers eventually harden into sedimentary rock, such as limestone, shale, or sandstone.

A B C D

Picture This

3. Describe the events that happen to make a fossil in sedimentary rock.

____

____

____

____

____

What can scientists learn from studying fossils?

A **paleontologist** (pay lee ahn TAH luh jist) is a scientist who studies fossils. Paleontologists use fossils as clues to learn what an organism ate and the environment in which an organism lived. Paleontologists can use fossils to put together a picture of extinct communities as if they were alive today.

How do scientists find out a rock's age?

Paleontologists use different methods to find out the age of a rock or a fossil. **Relative dating** is a method that compares rocks with rocks in other layers. Relative dating is based on the law of superposition. The **law of superposition** states that rocks form in layers, with younger layers of rock deposited on top of older layers. The oldest rocks form the bottom layer and the youngest rocks form the top layer, as shown below.

Picture This

4. Identify Label the oldest and the youngest rock layers in the picture.

What is radiometric dating?

Radiometric dating is a method used to determine the age of rocks using the decay of radioactive isotopes present in the rocks. Recall that an isotope is a form of an element that has the same atomic number but different mass number. The method requires that the **half-life** of the isotope, which is the amount of time it takes for half of the original isotope to decay, is known. The relative amounts of the isotope and its decay product must also be known. ☑

Reading Check

5. Identify What three things must be known to perform radiometric dating?

______

______

______

How does radiometric dating work?

Items such as mummies and frozen mammoths can be dated directly using carbon-14. Only materials less than 60,000 years old can be dated accurately with this isotope. Older materials do not have enough radio-isotope left.

Carbon-14 is a commonly used isotope. The decay of carbon-14 is shown in the graph below. The half-life of carbon-14 is 5730 years. After 5730 years, half of the original carbon-14 will remain. The other half will have decayed to nitrogen-14. At the half-life, there is a one-to-one ratio, or equal amounts, of carbon-14 and nitrogen-14. Scientists can determine the age of a sample by calculating the ratio of carbon-14 to nitrogen-14 in the rock.

Picture This

6. Estimate About what percent of carbon-14 is left after 11,460 y?

____

____

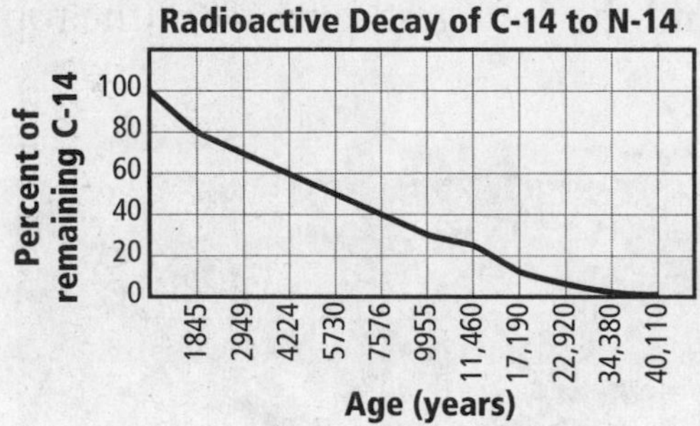

How is radiometric dating used for rocks?

Useful radioactive isotopes are found in igneous and metamorphic rocks. Sedimentary rocks are made from igneous and metamorphic sediments so they cannot be measured accurately. Scientists can determine the relative age of fossils by measuring nearby igneous and metamorphic rock in layers closely associated with them. ☑

☑ Reading Check

7. Describe How is radiometric dating used to find the age of sedimentary rock?

____

____

____

____

____

The Geologic Time Scale

The **geologic time scale** is a model that shows the major geological and biological events of Earth's history. These events include changes to Earth and to organisms.

The geologic time scale has two major divisions—Precambrian time and the Phanerozoic (fan eh roh ZOH ihk) eon. An **eon** is the longest unit of time and can include billions of years. An **era** lasts hundreds of millions of years. Eras include the Precambrian, Paleozoic, Mesozoic, and Cenozoic. Each era is further divided into one or more **periods** which last tens of millions of years. **Epochs**, which last several million years, are the smallest units of geologic time.

What occurred during the Precambrian?

The first 4 m of the geologic time ribbon makes up the Precambrian (pree KAM bree un). During the Precambrian nearly 90 percent of Earth's history occurred. It began with the formation of Earth, 4.6 billion years ago, and ended about 542 million years ago with the beginning of the Paleozoic era. ☑

Many important events occurred during the Precambrian. Earth formed and life first appeared. Autotrophic prokaryotes, such as bacteria that make organic compounds using carbon dioxide and energy from the sun or inorganic sources, enriched the atmosphere by releasing oxygen. Eukaryotic cells emerged. By the end of the Precambrian, the first animals had appeared.

During the second half of the Precambrian, glaciers might have delayed the further evolution of life. After the glaciers receded, simple organisms lived in marine ecosystems.

How did life change during the Paleozoic era?

A drastic change in the history of animal life on Earth came at the start of the Paleozoic (pay lee uh ZOH ihk) era. In just a few million years, the ancestors of most major animal groups diversified in what scientists call the **Cambrian explosion**. Fish, land plants, and insects appeared. The swampy forests were home to many types of organisms, including huge insects. The first tetrapods—animals that walk on four legs—which were the first land vertebrates, appeared. By the end of the era, reptiles appeared.

What event ended the Paleozoic era?

The Paleozoic era ended with a mass extinction. Recall that a mass extinction is an event in which many species become extinct in a short amount of time. In the mass extinction that ended the Paleozoic era, 90 percent of marine organisms disappeared. Scientists do not know why the mass extinction occurred. Most scientists agree that geological forces, including increased volcano activity, would have disrupted ecosystems or changed the climate. ☑

How did life change during the Mesozoic era?

Life continued to change during the Mesozoic (mez uh ZOH ihk) era. Mammals and dinosaurs appeared. Flowering plants evolved from nonflowering plants. Birds evolved from dinosaurs. Reptiles, including dinosaurs, were the dominate animals.

✔ Reading Check

8. Sequence What marked the beginning of the Precambrian?

____

____

✔ Reading Check

9. Identify What type of organisms became extinct during the mass extinction that ended the Paleozoic era?

____

____

What is the evidence of a meteor striking Earth?

Then, a meteorite struck Earth. The evidence for the meteorite comes from a layer of material between rocks of the Cretaceous (krih TAY shus) period and rocks of the Paleogene period. Scientists call this layer of material the **K-T boundary**. ☑

Reading Check

10. Name the rock layer found between rocks of the Cretaceous period and rocks of the Paleogene period.

In the K-T boundary, scientists have found high levels of iridium. Iridium is rare on Earth but common in meteorites. Iridium on Earth is evidence of a meteor impact.

How might a meteor strike have led to a mass extinction?

Many scientists think this meteor impact is related to the mass extinction, which eliminated all dinosaurs except birds, most marine reptiles, many marine invertebrates, and many plant species. The meteor itself did not kill these organisms, but the debris from the impact probably stayed in the atmosphere for months or years. The debris would have affected the global climate. Those species that could not adjust to the changing climate disappeared.

How did Earth change during the Mesozoic era?

Evolution in the Mesozoic era was affected by the massive geological changes of the era. As shown in the figure below, at the beginning of the Mesozoic era, approximately 225 million years ago, the continents were joined into one landmass called Pangaea.

Plate tectonics describes the surface of Earth as being broken into several large plates. Some of the plates contain continents. The plates move over a partially molten layer of rock moving the continents with them. The continents have been moving since they formed. By the end of the Mesozoic era, approximately 65 million years ago, the continents had broken apart, moved, and were in roughly the position they are now.

Picture This

11. Label the landform known as Pangaea.

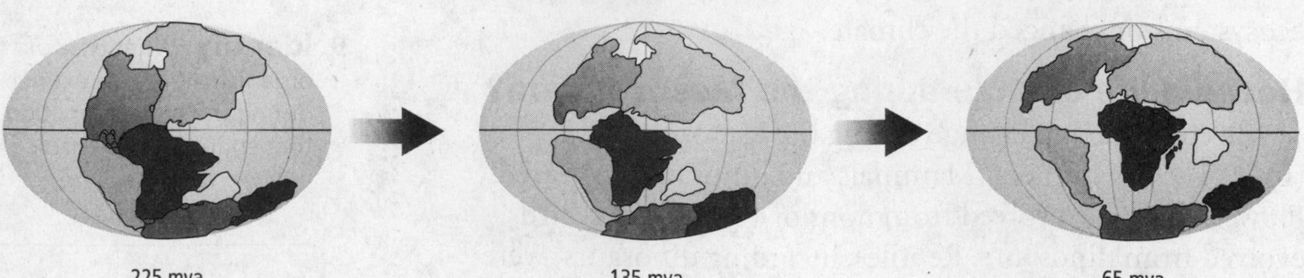

How did life change during the Cenozoic era?

The Cenozoic (sen uh ZOH ihk) era is the most recent era. Mammals became the dominant animals on land. At the beginning of the era, most mammals were small. After the mass extinction, at the end of the Mesozoic era, mammals, including primates, began to become more diverse.

When did present-day humans appear on Earth?

Present-day humans appeared near the end of the geologic time scale. Humans survived the last ice age, but many species of mammals did not. Think back to your time ribbon. The time that humans have lived on Earth takes up about two threads at the end of the ribbon.

The figure below shows the geologic time scale and gives examples of organisms that evolved during each era.

Geologic Era	Time Span	New Organisms
Precambrian	4.6 billion years ago to 542 million years ago	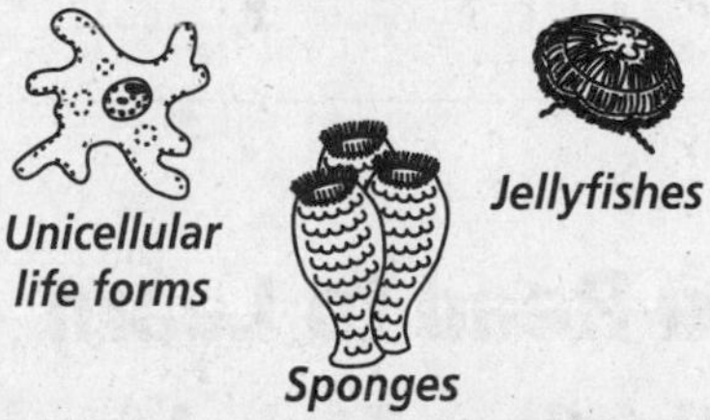 Unicellular life forms, Sponges, Jellyfishes
Paleozoic	542 million years ago to 251 million years ago	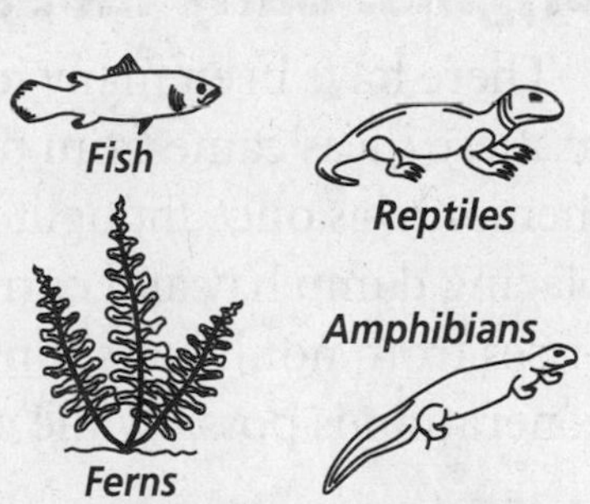 Fish, Reptiles, Ferns, Amphibians
Mesozoic	251 million years ago to 65 million years ago	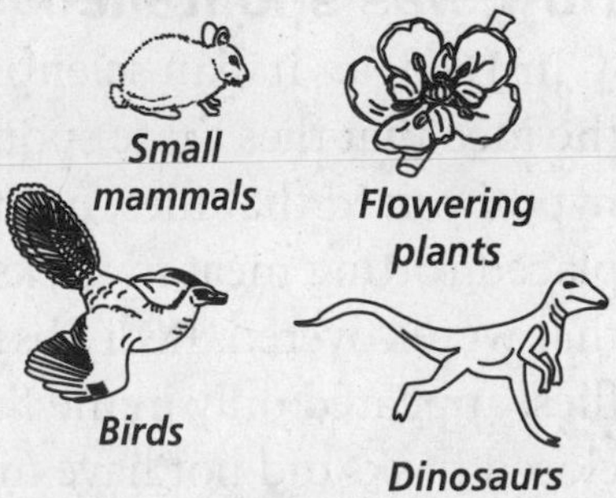 Small mammals, Flowering plants, Birds, Dinosaurs
Cenozoic	65 million years ago to present	 Large mammals, Humans

Applying Math

12. Calculate What percentage of Earth's history has included present-day humans?

____

Picture This

13. Label Circle the era when Pangaea broke apart into individual continents.

chapter 14

The History of Life

section 2 The Origin of Life

MAIN Idea

Evidence indicates that a sequence of chemical events preceded the origin of life on Earth.

What You'll Learn

- differences between spontaneous generation and biogenesis
- events that might have led to cellular life
- the endosymbiont theory

Before You Read

You want to make a sandwich but find mold growing on the bread. You don't recall seeing mold on the bread yesterday. On the lines below, explain how you think the mold got there. Then read about early ideas about the origins of life.

____

____

____

Mark the Text

Identify Scientists As you read this section, underline the name of each scientist introduced. Highlight the sentences that explain each person's contribution to understanding the history of life.

Read to Learn

Origins: Early Ideas

There have been many ideas about how life began. Many of these ideas came from people observing the world around them. It was once thought that mice could be created by placing damp hay and corn in a dark corner. This idea that life arises from nonlife is **spontaneous generation**. Spontaneous generation is possibly the oldest idea about the origin of life.

How was spontaneous generation tested?

In 1668, an Italian scientist named Francesco Redi tested the idea that flies arose spontaneously from rotting meat. He hypothesized that flies, not meat, produced other flies. Redi placed rotting meat in flasks that were opened and in flasks that were covered. Redi observed that maggots, the larvae of flies, appeared only in the flasks that were open to flies. The closed flasks did not have flies or maggots.

Despite Redi's experiment, people still believed in spontaneous generation. The microscope was beginning to be used during Redi's time. People knew that organisms too small to be seen were everywhere. Some people thought these microbes must arise spontaneously even if flies do not. ☑

☑ Reading Check

1. **Explain** What new device fueled people's belief in spontaneous generation?

____

What idea replaced spontaneous generation?

In the mid-1800s, Louis Pasteur designed an experiment, as shown in the figure below, to show that the theory of biogenesis was true even for microorganisms. The **theory of biogenesis** (bi oh JEN uh sus) is the idea that living organisms come from other living organisms. Only air was able to enter one flask containing a sterile nutrient broth. Both air and microorganisms were able to enter a second flask containing the sterile nutrient broth. Microorganisms were able to grow in the second flask but not the first flask. After Pasteur's experiment, people rejected spontaneous generation and embraced the theory of biogenesis.

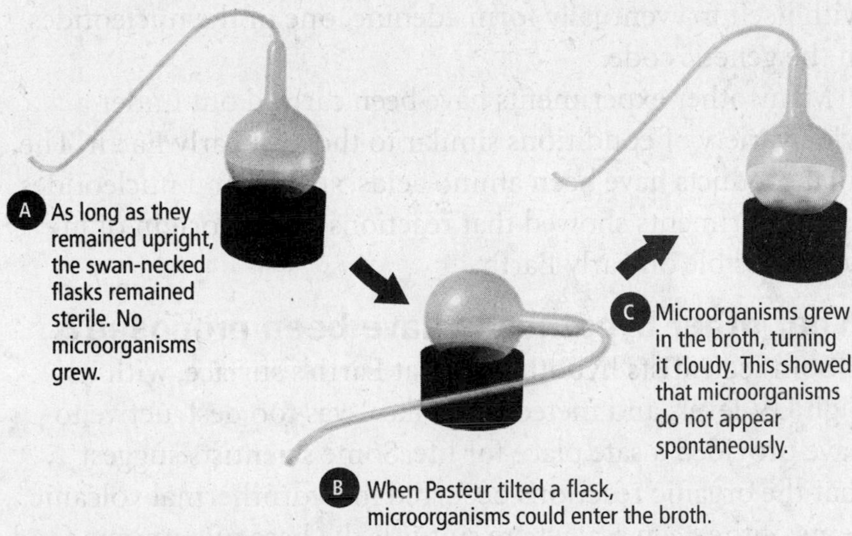

Picture This

2. Describe What did Pasteur do that allowed microorganisms to enter the flask?

____

____

Origins: Modern Ideas

Most biologists agree that life originated through a series of chemical events. During these events, complex organic molecules were made from simpler ones. Eventually, simple metabolic pathways developed. These pathways enabled molecules to be broken down. These pathways might have led to the origin of life. ☑

Reading Check

3. Determine Does biogenesis explain the origin of life?

____

How did early organic molecules form?

In the 1920s, Russian scientist Alexander Oparin suggested the primordial soup hypothesis explained the origin of life. He thought that if Earth's early atmosphere had a mix of certain gases, organic molecules could have been made from simple reactions involving those gases in the early oceans. UV light from the Sun and lightning might have provided the energy for the reactions. Oparin thought that these organic molecules would eventually lead to life.

How was Oparin's hypothesis tested?

In 1953, American scientists Stanley Miller and Harold Urey conducted an experiment showing that simple organic molecules could be made from inorganic compounds. Miller and Urey built a glass apparatus to simulate early Earth conditions. They filled the apparatus with water and gases that they thought had made up the early atmosphere. They boiled the mixture, shined UV light on it to simulate sunshine, and charged it with electricity to simulate lightning. The resulting mixture contained amino acids. Amino acids are the building blocks of proteins. ☑

Reading Check

4. Identify Why did Miller and Urey shine UV light on their apparatus?

Later, other scientists found that hydrogen cyanide could be formed from simpler molecules. Hydrogen cyanide can react with itself to eventually form adenine, one of the nucleotides in the genetic code.

Many other experiments have been carried out under a wide variety of conditions similar to those of early Earth. The final products have been amino acids, sugars, and nucleotides. The experiments showed that reactions for the origin of life were possible on early Earth.

What other hypotheses have been proposed?

Some scientists hypothesize that Earth's surface, with its high UV levels and meteorite strikes, was too destructive to have provided a safe place for life. Some scientists suggest that the organic reactions occurred in hydrothermal volcanic vents of the deep sea, where sulfur is the base of a unique food chain. Still others think meteorites might have brought the first organic molecules to Earth.

How were the first proteins made?

Proteins are chains of amino acids. The Miller-Urey experiment shows that amino acids could form on early Earth. Amino acids can bond to one another, but they can separate just as easily. Proteins might have formed when an amino acid stuck to a particle of clay. Clay would have been a common sediment in early oceans. Clay could have provided a framework for protein assembly. ☑

Reading Check

5. Explain What ocean sediment might have helped protein chains to form?

What was the first genetic code?

Another requirement for life is a genetic code—a coding system for making proteins. Many biologists think RNA was life's first coding system. RNA systems are capable of evolution by natural selection.

How do some RNA molecules behave?

Some types of RNA can behave like enzymes. These RNA molecules could have carried out some early life processes. Some scientists think clay particles could have been a template for RNA replication and that the resulting molecules developed a replication mechanism.

How did the first cells arise?

Another important step in the evolution of cells is the formation of membranes. Scientists have tested ways to enclose molecules in membranes that allow metabolic and reproductive pathways to develop. However, scientists might never know the exact steps that led to cell formation. ☑

Reading Check

6. State What functions must a membrane allow?

Cellular Evolution

Although scientists don't know what the earliest cells were like, chemicals found in rocks suggest life was present 3.8 billion years ago even though no fossils remain. Scientists recently discovered what appear to be fossilized microbes in volcanic rock that is 3.5 billion years old. This suggests that cellular activity had become established. It also suggests that early life might have been linked to volcanic environments.

Scientists think the first cells were prokaryotes, which lack a defined nucleus and most other organelles. Many scientists think prokaryotes called archaea (ar KEE uh) are the closest relatives of Earth's first cells. These microbes live in extreme environments such as hot springs and volcanic vents in the deep sea. These environments are similar to those of early Earth.

When did photosynthetic organisms appear?

Archaea are autotrophs that get their energy from inorganic compounds such as sulfur. Archaea also do not make oxygen.

Scientists think oxygen was not present in Earth's early atmosphere until about 1.8 billion years ago. Any oxygen that appeared earlier probably bonded with free iron ions. Scientists hypothesize that eventually early Earth's free iron bonded with oxygen and oxygen accumulated in the environment.

Scientists think that cyanobacteria, prokaryotes that could perform photosynthesis, evolved about 3.5 billion years ago. These organisms released oxygen into the atmosphere and eventually produced enough oxygen to support ozone layer formation. The ozone layer provided a shield from the Sun's damaging ultraviolet radiation and made conditions right for eukaryotes to develop. ☑

Reading Check

7. Sequence What came first: photosynthetic organisms or eukaryotic cells?

When did eukaryotic cells evolve?

Eukaryotic cells appeared about 1.8 billion years ago. They are larger than prokaryotes and have complex internal membranes, which enclose many organelles including the nucleus.

What is the endosymbiont theory?

American biologist Lynn Margulis proposed the **endosymbiont theory** which states that ancestors of eukaryotic cells lived together in association with prokaryotic cells. In some cases, prokaryotes might even have lived inside eukaryotes. Prokaryotes might have entered eukaryotes as undigested prey, or they might have been internal parasites. Eventually, the relationship benefitted both cells and the prokaryotes became organelles inside the eukaryotic cells, as shown in the figure below.

Evidence suggests that mitochondria and chloroplasts formed by endosymbiosis. Mitochondria and chloroplasts contain their own DNA arranged in circular patterns like the DNA of prokaryotes. Mitochondria and chloroplasts have ribosomes that are more similar to the ribosomes in prokaryotes than to those in eukaryotes. Like prokaryotic cells, mitochondria and chloroplasts reproduce by fission independent from the rest of the cell.

Scientists do not know the early steps that led to life or to its evolution. Scientists continue to test theories and evaluate new evidence as they seek answers to understand what led to life on Earth.

FOLDABLES™

Take Notes Make a four-door Foldable, as shown below. As you read, take notes and organize what you learn about the endosymbiont theory.

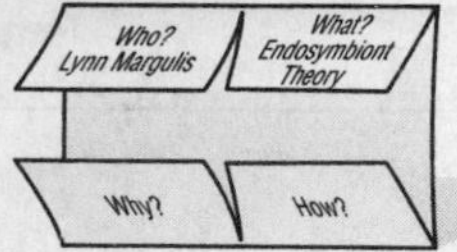

Picture This

8. Highlight the name of the structure that cyanobacteria became.

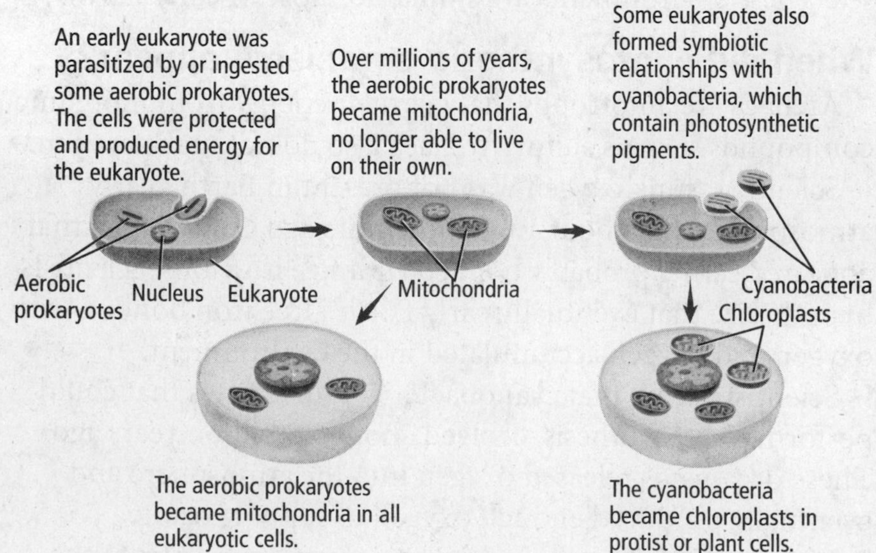

Evolution

section 1 Darwin's Theory of Natural Selection

Before You Read

In this section you will learn about Charles Darwin and his theory of natural selection. Read the first paragraph of the *Read to Learn* section. On the lines below, write the ideas about life on Earth that were common during Charles Darwin's lifetime.

____

____

____

MAIN Idea

Charles Darwin developed a theory of evolution based on natural selection.

What You'll Learn

- the evidence that led Darwin to conclude species could change over time
- the four principles of natural selection

Read to Learn

Developing the Theory of Natural Selection

When Charles Darwin boarded the HMS *Beagle*, people believed that the world was only a few thousand years old. Most people believed the plants and animals they saw had not changed. Darwin believed these things too.

Mark the Text

Identify Main Ideas Highlight every heading in the reading that asks a question. Then highlight each answer as you find it.

What did Darwin do on the HMS *Beagle*?

The mission of the *Beagle* was to survey the coast of South America. Darwin's original role on the ship was as the captain's companion. He was someone the captain could talk to during the long voyage. Darwin also served as the ship's naturalist. His job was to collect rocks, fossils, plants, and animals from the places he visited.

During the five-year voyage, Darwin read *Principles of Geology*, by Charles Lyell. Lyell's book proposed that Earth was millions of years old. The book influenced Darwin's thinking as he found fossils of marine life high in the Andes mountains. He also found fossils of giant versions of smaller living mammals. He observed how earthquakes could quickly lift rocks great distances. ☑

Reading Check

1. **Name** an observation Charles Darwin made.

____

____

____

What did Darwin find on the Galápagos Islands?

One of the places the *Beagle* sailed was the Galápagos (guh LAH puh gus) Islands off the coast of Ecuador. Darwin noticed that mockingbirds on one island were slightly different from mockingbirds on the other islands. He took careful collections of these birds. Darwin thought that the finches he saw were not related to one another and probably had representatives on mainland South America. Although he noted the differences, he did not think much about these differences at the time. Darwin also did not notice that tortoise shells were different on each island. ☑

Reading Check

2. Identify Where are the Galápagos Islands?

____

____

When Darwin returned home, he showed his specimens to naturalist John Gould. Gould told Darwin that the mockingbirds from different islands were different species. Gould also determined that the Galápagos finches did not live anywhere else in South America. They were different species too. A few of Darwin's finches are shown in the picture below. Notice the difference in beak size and shape.

The Galápagos finches most closely resembled finches from the closest mainland in South America. Darwin suspected that populations from the mainland changed after reaching the Galápagos.

Picture This

3. Explain Each of the Galápagos islands had different types of plants. How might this explain the differences in beak shape in finches?

____

____

____

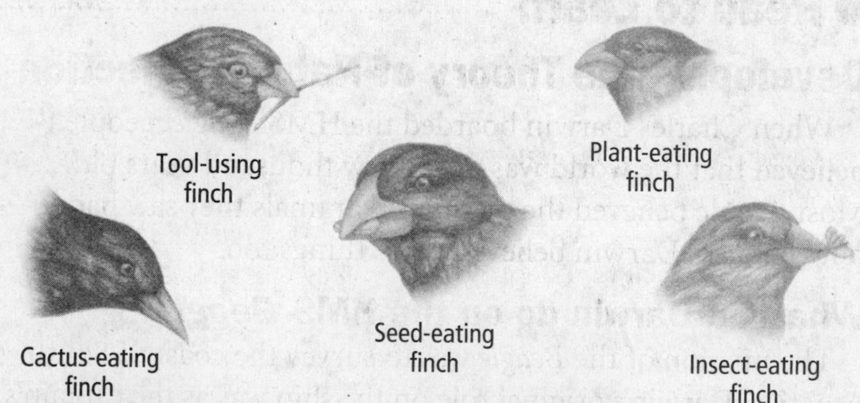

What did Darwin learn from artificial selection?

Darwin suspected that small, gradual changes might lead to new species. Darwin knew that people could create new breeds of plants and animals by breeding plants and animals that express the desired traits. Darwin called this selective breeding **artificial selection**.

Darwin thought that if people could change species by artificial selection, perhaps species could also change in the natural world. Darwin reasoned that, given enough time, gradual change could produce new species.

What principles did Darwin use?

Darwin developed a theory for how one species could change to become multiple new species. Darwin based his theory on four principles.

Principle 1 Individuals in a population show differences, even among the same species. For example, there are variations of traits among you and your classmates.

Principle 2 The variations are inherited. Traits are passed down from parents to their offspring.

Principle 3 Animals have more young than can survive on the resources in their environment. For example, the average female cardinal lays nine eggs in a summer. If each of the offspring lives one year and reproduces, in eight years the offspring would produce a million cardinals. The population of cardinals would quickly outgrow the food supply.

Principle 4 Some variations increase an organism's reproductive success, or its chance of having living offspring. Any variation that increases reproductive success will be inherited by offspring and will be more common in the next generation. For example, if pigeons with fan-shaped tails have more reproductive success than pigeons without fan-shaped tails the more pigeons in the next generation will have fan-shaped tails.

These principles formed the basis of Darwin's theory of evolution by **natural selection**, which explains how traits of a population can change over time. Darwin reasoned that given enough time, natural selection could produce a new species.

Think it Over

4. Identify Name a resource that could limit the survival of animals within a population.

The Origin of Species

Darwin began writing a book describing how natural selection could produce new species. In 1858, he learned that Alfred Russel Wallace, another English naturalist, had reached similar conclusions about natural selection. In 1858, both Wallace and Darwin presented their findings at a scientific meeting. In 1859, Darwin published his book, *On the Origin of Species by Means of Natural Selection.*

Darwin did not use the word "evolution" until the last page of his book. Today, biologists use **evolution** to describe the way a species changes over time. Darwin's theory of natural selection explains how evolution can occur.

Think it Over

5. Contrast How does natural selection differ from evolution?

chapter 15

Evolution

section 2 Evidence of Evolution

MAIN Idea

Multiple lines of evidence support evolution.

What You'll Learn

- how fossils provide evidence of evolution
- evidence of evolution from morphology
- how physiology and biochemistry provide evidence of evolution

Before You Read

To learn how different organisms might be related, scientists look for similarities and differences between the organisms. On the lines below, compare a cat and a frog. What physical features are the same? What physical features are different?

____

____

____

Mark the Text

State Main Ideas As you read, stop after every few paragraphs and put what you have just read into your own words. Then highlight the main idea in each paragraph.

Read to Learn

Support for Evolution

Science uses theories that provide explanations for how some aspects of the natural world operate. Any theory should explain available data and suggest further areas for experiments. Darwin's theory of evolution by natural selection explains the patterns scientists see in past and present organisms.

In most cases, people cannot observe evolution directly because it happens over millions of years. Fossils help us understand evolution because they are a record of species that lived long ago. The fossil record shows that some species from long ago are extinct today. Other species alive today are similar to those in fossils.

FOLDABLES™

Take Notes Make a four-door Foldable, as shown below. As you read, take notes and organize what you learn about Charles Darwin and the development of the theory of natural selection.

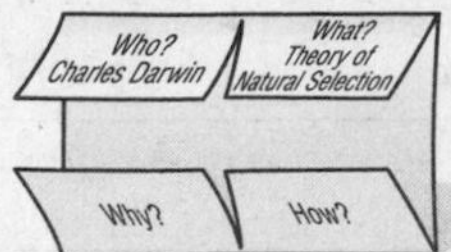

What did Darwin predict about the fossil record?

Darwin predicted that scientists would find fossils that would show organisms that were intermediate between different species. Darwin's prediction has come true. Scientists have found intermediate species for the evolution of mollusks, modern horses, whales, and humans.

What are two classes of traits?

Scientists have found fossils of *Archaeopteryx*, a dinosaur that has the teeth, claws, and a bony tail of a reptile and feathers and the ability to fly like a bird. *Archaeopteryx* is likely an intermediate organism and is evidence that birds evolved from dinosaurs.

Scientists divide traits into two classes. **Derived traits** are newly evolved traits that have not appeared in common ancestors. **Ancestral traits** are traits that are shared by species with a common ancestor. In *Archaeopteryx*, teeth are an example of an ancestral trait.

What does anatomy reveal about evolution?

The limbs of vertebrates perform different functions, but they have similar anatomy. Wings and legs have similar structures because birds and animals evolved from the same ancestor. **Homologous structures** are similar structures inherited from a common ancestor. Darwin's theory of evolution by natural selection predicts that new structures are more likely to be modifications of ancestor's structures than entirely new features. The figure below shows the homologous forelimbs of three different animals.

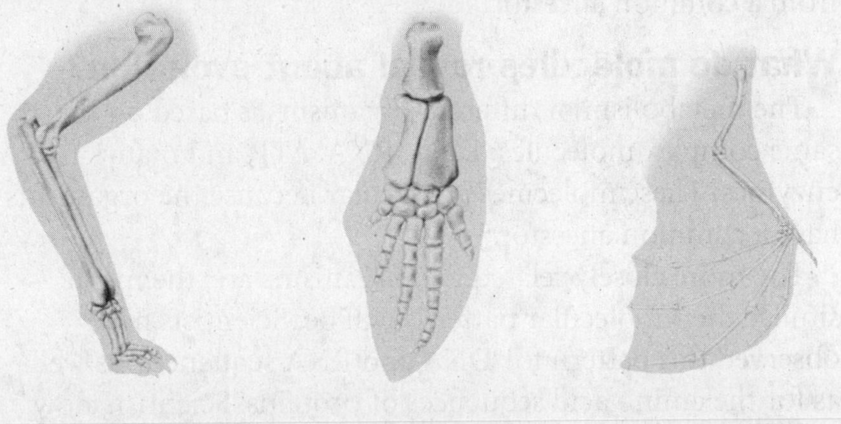

Picture This

1. Identify On the blank lines in the figure, write the function of the forelimbs in each animal.

What are vestigial structures?

In some cases, a functioning structure in one species is smaller or doesn't function in a closely related species. **Vestigial structures** are features that are reduced forms of functional structures in other organisms. Vestigial structures are reduced when structures are no longer needed. The structures become smaller over time and might eventually disappear. ☑

Reading Check

2. Define What are vestigial structures?

What are analogous structures?

Two organisms can have similar structures without being closely related. **Analogous structures** have the same function but different construction because they are not inherited from a common ancestor. Bird wings and insect wings are analogous structures. They have the same function but different anatomy.

Picture This

3. Name What analogous structure is found in both birds and insects?

____

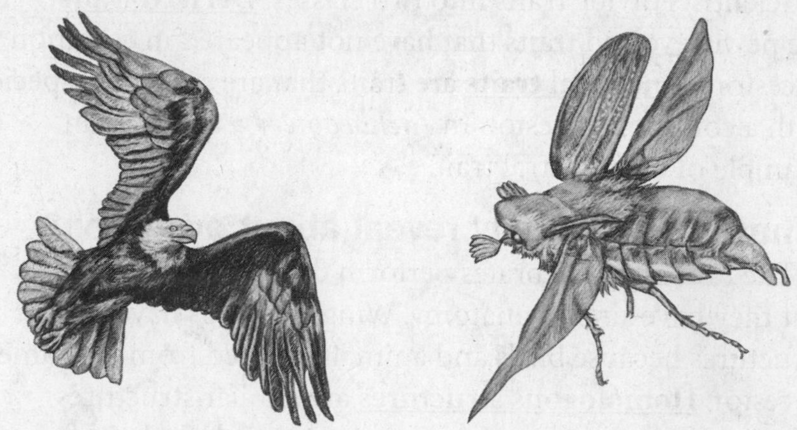

What do embryos reveal about evolution?

An **embryo** is an early stage of development in organisms. Embryos of fishes, birds, reptiles, and mammals have several homologous structures that are not present when the organisms are adults. These structures suggest that vertebrates evolved from a common ancestor.

What do molecules reveal about evolution?

The metabolism of different organisms is based on the same complex molecules: DNA, RNA, ATP, and many enzymes. These molecules are similar because the organisms have a common ancestor.

The more closely related two organisms are, the more similar their molecular patterns will be. Scientists have observed this pattern for DNA and RNA sequences, as well as for the amino acid sequences of proteins. Scientists now use similarities in DNA and RNA sequences to determine evolutionary relationships between species.

4. Define What is biogeography?

____

____

____

What does biogeography predict?

Darwin's theory of evolution by natural selection predicts that species respond to similar environments in similar ways. **Biogeography** is the study of how plants and animals are distributed on Earth. Biogeography provides evidence that similar environments can lead to the evolution of similar animals, even if the environments are far apart. ☑

Adaptation

Some traits contribute greatly to an organism's survival or reproduction. Traits that enable organisms to survive or reproduce better than organisms without those traits are called adaptations. ☑

Fitness is one way to measure the effectiveness of traits. **Fitness** is a count of offspring born to organisms with a trait compared to offspring born to organisms without that trait. Traits that enable organisms to survive or reproduce better than organisms without those traits are adaptations.

Camouflage (KA muh flahj) is an adaptation that allows an organism to blend with its surroundings. Camouflage increases fitness because it allows the organism to hide from predators.

Mimicry is an adaptation that occurs when one species looks like another species. In one form of mimicry, a harmless species evolves to look like a dangerous one. In another form of mimicry, two or more harmful species resemble one another. In both cases, predators cannot tell the species apart, so they avoid both. Mimicry increases the chance that a species will survive and reproduce.

Reading Check

5. Explain What is an adaptation?

Do all traits evolve slowly?

Bacteria that were originally killed by antibiotics such as penicillin have quickly evolved into populations of resistant bacteria. For most antibiotics, at least one species of resistant bacteria exists. Some diseases, such as tuberculosis, that doctors once believed could be controlled with antibiotics have now come back. The forms of these diseases are more harmful than the forms that were treated with antibiotics. These new forms resist treatment with today's antibiotics. ☑

Reading Check

6. Name a problem with the rapid evolution of antibiotic-resistant bacteria.

Do all traits increase fitness?

Not all features of organisms are adaptations that increase fitness. Some features arise because they are unavoidable consequences of other evolutionary changes.

For example, human babies are born helpless at an earlier stage of development than other primates. Many scientists believe that early birth is not an adaptation but is a consequence of evolution. Human babies must be small in order to squeeze through a narrow birth canal. The birth canal is narrow because human females have a narrow pelvis. The shape of the pelvis is an adaptation that enables people to walk on two legs instead of four.

Evolution

section 3 Shaping Evolutionary Theory

MAIN Idea

The theory of evolution is being refined as scientists learn new information.

What You'll Learn

- factors that influence how new species originate
- about gradualism with punctuated equilibrium

Before You Read

In this section, you will learn how our understanding of evolution has changed in the last 150 years. Other aspects of science have also changed. On the lines below, name several other scientific advances that have happened since the mid-1800s.

____

____

____

Study Coach

Make Flash Cards Write the underlined words on one side of a flash card. Write the definition on the other side of the card.

Read to Learn

Mechanisms of Evolution

Natural selection helps explain how one or two ancestors became today's diversity. Natural selection is one way that species evolve, but it is not the only way.

In the 150 years since Darwin published his findings on natural selection, scientists have learned much about evolution. They have uncovered other ways that species can change. To understand the other mechanisms for evolution, it is important first to learn about population genetics.

What is the Hardy-Weinberg principle?

In 1908, English mathematician Godfrey Hardy and German physician Wilhelm Weinberg each arrived at the same conclusion about how the laws of inheritance work in a population. The **Hardy-Weinberg principle** states that the frequency of alleles in populations does not change unless the frequencies are acted on by some factor that causes change. When the frequency of alleles remains the same, the population is in genetic equilibrium. A population in genetic equilibrium does not evolve. ☑

Reading Check

1. Explain What happens when a population is in genetic equilibrium?

____

____

How does the Hardy-Weinberg principle work?

The Hardy-Weinberg principle helps us understand when evolution can occur. Evolution occurs only when a population is not in genetic equilibrium.

Genetic equilibrium occurs when five conditions, listed in the table below, are met. When one or more of the conditions is violated, the population can change or evolve.

Populations can meet some of these requirements for long periods of time. Many populations are large enough to maintain genetic equilibrium. Other conditions do not often occur in nature. For example, one condition is that mating must be random across an entire population. But mating is rarely random. It usually occurs between closest neighbors. Because all five conditions do not usually occur in nature, most populations are able to evolve.

Conditions for Genetic Equilibrium

Condition	Violation	Consequence
The population is large.	Many populations are small.	Chance events can change population traits.
There is no immigration or emigration.	Organisms move in and out of the population.	The population can lose or gain traits with movement of organisms.
Mating is random.	Mating is not random.	New traits do not pass as quickly to the rest of the population.
Mutations do not occur.	Mutations occur.	New variations appear in the population with each new generation.
Natural selection does not occur.	Natural selection occurs.	Traits in a population change from one generation to the next.

How does genetic drift lead to evolution?

<u>Genetic drift</u> is the random change in the frequency of alleles in a population. Genetic drift usually affects small populations. Genetic drift occurs because chromosomes are sorted randomly during meiosis. The one of a parent's two alleles that passes to the offspring is determined by chance. Genetic drift is another way that a population can evolve.

Unlike natural selection, the adaptations that result from genetic drift are not always the best ones for the environment. Sometimes, important adaptations can be lost by genetic drift.

Picture This

2. Explain White rabbits blend in with the snow. Brown rabbits are more likely to be eaten by predators. Is the population of rabbits in genetic equilibrium? Explain your answer.

___

___

___

Think it Over

3. Draw Conclusions What population would be most likely to experience genetic drift? (Circle your answer.)

a. 4000 mice living in a meadow
b. 30 rabbits living on a mountaintop
c. five million people living in a city

What is the founder effect?

The **founder effect** can occur when a few individuals are separated from the rest of the population. The few individuals carry a random subset of the genes in the original population. The frequency of alleles in the subset might be different from the frequency of alleles in the original population. The founder effect is a random way that species can evolve. Unlike natural selection, the traits that result might or might not be the best available for the environment.

The founder effect often occurs on islands. New species can result from a few founders of the original population. The founder effect also occurs in people. Amish and Mennonite people live in the United States but do not usually marry outside their communities. They have many unique genes.

What happens in a genetic bottleneck?

When a large population declines in number then rebounds to a large number again, a **bottleneck** occurs. Bottlenecks reduce the total alleles in a population. The genes of the resulting population can be unusually similar. ☑

Cheetahs in Africa might have gone through a bottleneck a few thousand years ago. Cheetahs are genetically similar and appear to be inbred. Inbreeding reduces fertility and can eventually cause extinction.

What is gene flow?

A population in genetic equilibrium experiences no gene flow. No new genes enter the population and no genes leave the population. However, few populations are isolated. Gene flow occurs when individuals move among populations. This movement increases the variations in genes and reduces the differences among populations.

With whom do organisms mate?

Mating in a population is usually nonrandom. Individuals tend to mate with other individuals that live near them. This promotes inbreeding. Nonrandom mating might favor individuals that are homozygous for particular traits.

How do mutations affect genetic equilibrium?

The cumulative effect of mutations might change the allelic frequencies in a population and violate genetic equilibrium. Occasionally, a mutation provides an advantage for an organism. The mutation will become more common in future generations. Mutations are the raw material in which natural selection works.

✔ Reading Check

4. Define What happens in a bottleneck?

______

______

______

Think it Over

5. Evaluate What is a problem with inbreeding?

______

______

What are the different types of natural selection?

Recall that natural selection changes organisms to better fit their environment. There are three different ways (stabilizing, directional, and disruptive selection) that natural selection can change populations.

Stabilizing selection removes organisms with extreme forms of a trait. It is the most common form of selection. Stabilizing selection favors the average value of a trait, as shown in the figure below.

For example, human babies born with below-normal or above-normal birth weights are less likely to survive than babies born at average weights. Therefore, the average birth weight in humans remains about the same.

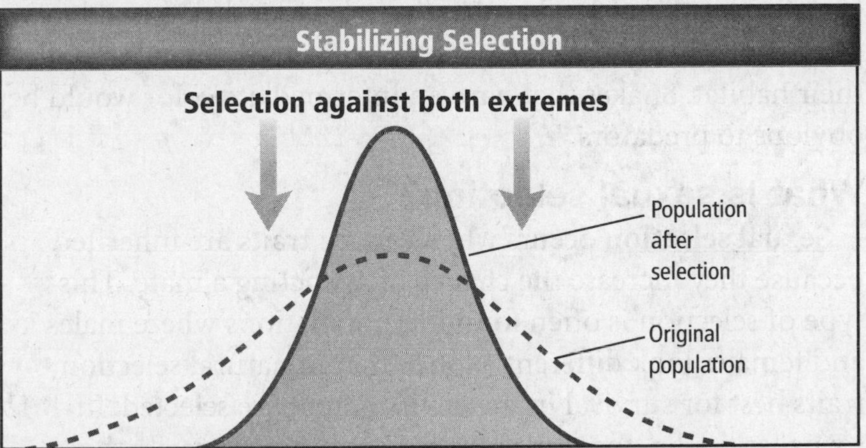

Picture This

6. Highlight the curve representing the population in which the extreme forms of a trait have been removed.

What is favored in directional selection?

Directional selection favors the extreme form of a trait. When an extreme form of a trait results in higher fitness, **directional selection** shifts the populations toward the beneficial trait. Directional selection is shown in the graph below.

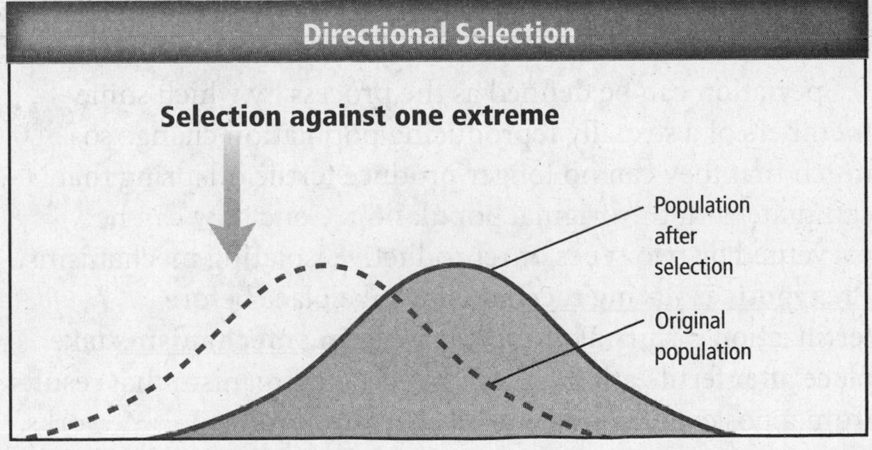

Picture This

7. Highlight the curve representing a population before it has undergone directional selection.

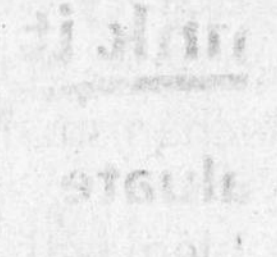

Reading Check

8. **Evaluate** In this example, what causes directional selection to occur?

____

____

What is an example of directional selection?

American biologists, Peter and Rosemary Grant have observed directional selection in Galápagos finches. In years with little water, food supplies decrease. The remaining foods are hard seeds. Birds with small beaks starve because they are unable to crack the seeds. Birds with larger beak sizes can more easily crack the seeds and survive. In years with little water, average beak size increases. When rain returns, average beak size decreases because smaller beak size is a better fit for the environment. ☑

What is disruptive selection?

Disruptive selection removes individuals with the average form of a trait. It creates two populations with extreme forms of a trait. Disruptive selection occurs in water snakes on the shores of Lake Erie. Mainland snakes live in grass habitats and have brown skin. Island snakes live on rocky shores and have gray skin. The color of both snakes helps them blend in with their habitat. Snakes that have an intermediate color would be obvious to predators.

What is sexual selection?

Sexual selection occurs when certain traits are inherited because they increase the chance of attracting a mate. This type of selection is often found in populations where males and females look different. Notice that in natural selection, traits best for survival in the environment are selected. In sexual selection, the traits selected are not necessarily those that are best for survival in the environment.

Reproductive Isolation

Genetic drift, gene flow, nonrandom mating, mutation, and natural selection are mechanisms of evolution. All these mechanisms violate the Hardy-Weinberg principle. Scientists disagree about the extent to which each of these mechanisms contributes to the evolution of new species.

Reading Check

9. **Identify** What are two types of isolating mechanisms?

____

____

____

Speciation can be defined as the process by which some members of a sexually reproducing population change so much that they can no longer produce fertile offspring that can mate with the original population. Gene flow can be prevented by two types of reproductive isolating mechanisms. **Prezygotic isolating mechanisms** take place before fertilization occurs. **Postzygotic isolating mechanisms** take place after fertilization has occurred. The organism that results from a postzygotic isolating mechanism is infertile. ☑

How do prezygotic isolating mechanisms work?

Prezygotic isolating mechanisms prevent genotypes from entering a population's gene pool. The isolation might occur geographically, ecologically, or behaviorally. The eastern meadowlark and western meadowlark exhibit a form of behavioral isolation. They have a similar appearance and live in overlapping areas. However, the two species use different mating songs and do not interbreed.

Time is another factor that can be a reproductive barrier. For example, closely related species of fireflies mate at different times of night. Different species of trout live in the same stream. Because they mate at different times of the year, they do not interbreed.

Does postzygotic isolation occur?

Postzygotic isolating mechanisms prevent offspring from surviving or reproducing. Lions and tigers are considered separate species, but they do sometimes mate. The offspring of such a mating—the liger—is sterile and cannot reproduce.

10. Apply Two closely related birds live on separate islands and do not interbreed. What type of isolation is occurring?

____

Speciation

Speciation occurs when a population reproduces in isolation. Most scientists believe that allopatric speciation is the most common form of speciation. In **allopatric speciation**, a physical barrier divides one population into two or more populations. After a long period of time, the two populations will contain organisms that can no longer successfully breed with one another. Physical barriers can include mountain ranges, wide rivers, and lava flows.

Sympatric speciation occurs when a species evolves into a new species without a physical barrier. The ancestor species and the new species live in the same habitat during the speciation process. Scientists think that sympatric speciation happens fairly often in plants. Polyploidy, a mutation that increases the number of chromosomes in an organism, might cause sympatric speciation in plants. A plant that results from polyploidy is no longer able to interbreed with the main population.

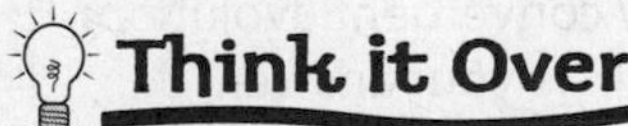

11. Explain Kaibab squirrels live on the north rim of the Grand Canyon, and Albert squirrels live on the south rim. Which form of speciation is more likely taking place? (Circle your answer.)

a. allopatric speciation

b. sympatric speciation

Patterns of Evolution

Many details of speciation are not yet known. Speciation is a long process. Observations of speciation are rare. However, evidence of speciation is visible in most patterns of evolution.

When does adaptive radiation occur?

Adaptive radiation occurs when one species evolves in a short period of time into a number of new species. Adaptive radiation can occur when a species evolves a new, useful trait or when a species arrives in a new habitat. Adaptive radiation, also called divergent evolution, can occur on a large scale. Recall that the Cretaceous period ended with a mass extinction. Soon afterward, mammals became more diverse. This example of adaptive radiation on a large scale likely produced the wide variety of mammals on Earth today. ☑

Reading Check

12. Define What happens in adaptive radiation?

____

____

____

____

How do species coevolve?

Two species can evolve together, or coevolve. Coevolution sometimes benefits both species. For example, flowers have markings that guide bees to nectar. While the bees gather nectar, they pollinate the flower. The flowers and bees have coevolved in a way that benefits both species.

What is convergent evolution?

Places far apart on Earth can have similar environments. Deserts in North America are similar to deserts in Africa. Similar environments can cause similar organisms to evolve by natural selection. In convergent evolution, unrelated organisms in different places evolve to resemble one another. Convergent evolution produces organisms with similar morphology, physiology, and behavior, even though the organisms are unrelated.

Think it Over

13. Compare Which statement describes two organisms that emerged by convergent evolution? (Circle your answer.)

a. They have similar morphology.

b. They are closely related.

How quickly do species evolve?

Early in the study of evolution, scientists thought evolution was gradual. **Gradualism** is the idea that evolution occurs in small steps over millions of years. Much evidence favors this theory.

Punctuated equilibrium is the idea that speciation occurs in sudden bursts followed by long periods of stability. Stability does not mean an organism is not changing. The organism's genes might still be changing, but the changes are not reflected in fossils of the organism.

Scientists continue to research the tempo of evolution. Some scientists think the fossil record shows that most change occurs in short bursts. Some scientists think that evolution occurs in a combination of gradual and punctuated changes. Many areas of science will contribute evidence to resolve the question of the pace of evolution.

Primate Evolution

section 1 Primates

Before You Read

Have you ever watched a monkey or an ape in a zoo or on television? On the lines below, list some humanlike behaviors you might have observed. Then read the section to learn some traits you share with your fellow primates.

____

____

MAIN Idea

Primate characteristics indicate that primates evolved from a common ancestor.

What You'll Learn

- characteristics of primates
- similarities and differences among major primate groups

Read to Learn

Characteristics of Primates

Primates are a group of mammals that includes humans, apes, monkeys, and lemurs. Primates have a high level of manual dexterity. Manual dexterity enables primates to grasp objects and move them around in their hands. Primates have well developed eyesight, long mobile arms, and large brains. Primates with the largest brains can reason.

Study Coach

Create a Quiz After you read this section, create a five-question quiz from what you have learned. Then, exchange quizzes with another student. After taking the quizzes, review your answers together.

Why is an opposable first digit important?

The hands and feet of all primates have five digits. Most have flat nails and sensitive areas on the ends of their digits. The first digit on the hands of most primates and the first digit on the feet of many primates are opposable. An **opposable first digit**, either a thumb or a toe, is set apart from the other digits. This digit can be brought across the palm or foot to touch or nearly touch the other digits. This action allows primates to grasp objects.

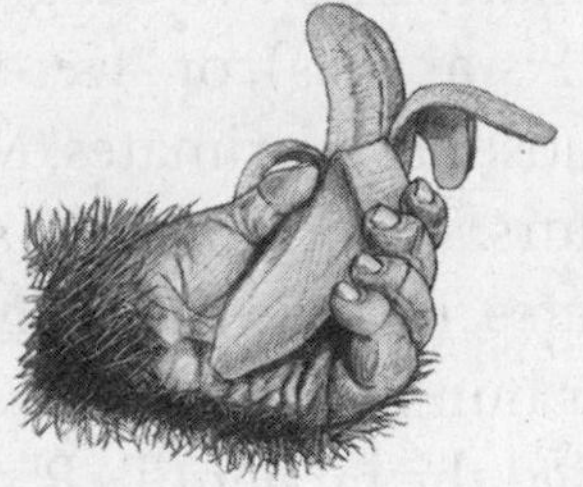

Picture This

1. Label the opposable first digit in the picture.

FOLDABLES™

Organize Information

Use two sheets of notebook paper to make a layered Foldable. As you read about primates, use your Foldable to organize the main ideas and vocabulary.

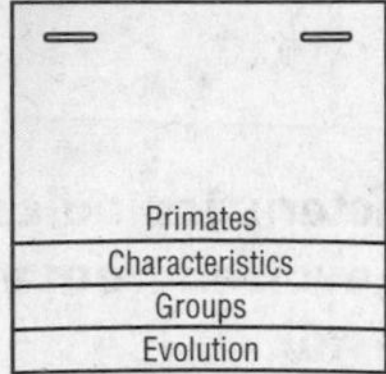

How do forward-looking eyes benefit primates?

Most primates rely more on vision than on smell. Their eyes are on the fronts of their faces. This creates overlapping fields of vision called **binocular vision**. Forward-looking eyes provide depth perception and enable primates to judge distance and movement of an object.

Most primates are **diurnal** (di YUR nul) which means they are active during the day. Many diurnal primates also have color vision. **Nocturnal** (nahk TUR nul) primates are active at night. They see only black and white colors.

With smaller snouts, primates have a reduced sense of smell. Their flattened faces aid binocular vision. Their teeth are unspecialized, enabling them to eat a variety of foods.

How do primates move?

Primates rely on hind limbs for movement. Most primates live in trees. Their flexible shoulders enable easy movement from branch to branch. On the ground, all primates except humans walk on four limbs most of the time.

What are characteristics of a primate brain?

Primates have large brains. More areas of their brains are dedicated to vision and fewer to smell. Large areas of their brains are dedicated to memory and to arm and leg movement. Many primates are able to solve problems and engage in social behaviors. Primates have complex ways of communicating with each other, including facial expressions.

What is the reproductive rate of primates?

After a long pregnancy, primates give birth usually to one infant. Newborn infants depend on their mothers for a long time. This allows infants to learn complex social interactions. The low reproductive rate, however, combined with loss of habitat and human predation, has threatened primates.

Primate Groups

Most primates are **arboreal** (ar BOHR ee uhl), or tree-dwelling, and live in tropical and subtropical forests. Scientists classify primates into two subgroups. The strepsirrhines (STREP sihr ines), or "wet-nosed" primates, are the earliest and most basic primates. Most members of this group are lemurs. Haplorhines (HAP lohr ines), or "dry-nosed" primates, include the **anthropoids** (AN thruh poydz), or humanlike primates, as well as a unique primate called the tarsier (TAR see ur). ☑

2. Identify the two subgroups of primates.

____

____

Strepsirrhines

The table below lists characteristics of some strepsirrhines. Strepsirrhines are the only primates that rely mostly on smell for hunting and social interactions. They have large eyes and ears.

Strepsirrhines live in tropical Africa and Asia. Most are found only in Madagascar and nearby islands. Scientists hypothesize that these animals evolved in isolation when Madagascar drifted away from the African mainland.

Picture This

3. Draw Conclusions Look at the picture of each animal. What feature suggests that these animals are nocturnal?

____

Strepsirrhine Group	Lemurs	Aye-Ayes	Lorises	Galagos (ga LAY gohs)
Example				
Active Period	large—diurnal small—nocturnal	nocturnal	nocturnal	mostly nocturnal
Range	Madagascar	Madagascar	Africa and Southeast Asia	Africa
Features	• vertical leapers • use long bushy tail for balance • herbivores and omnivores	• tap bark, listen, fish out grubs with long third finger	• small; slow climbers; solitary • lack tail • some have toxic secretions	• small; fast leapers • no opposable digit • long tail

Haplorhines

Tarsiers, monkeys, and apes are all members of the large group called haplorhines. Apes include gibbons, orangutans, gorillas, chimpanzees, and humans.

The tarsier lives only in Borneo and the Philippines. This small nocturnal animal has large eyes and lives in trees. It can turn its head halfway around.

Anthropoids are generally larger than strepsirrhines. They have larger brains for their body size. Most are diurnal and have color vision. Anthropoids have complex social interactions. Anthropoids are split into the New World monkeys and the Old World monkeys. New World refers to the Americas. Old World refers to Africa, Asia, and Europe.

Think it Over

4. Apply Howler monkeys live in Central and South American rain forests. Is the howler an Old World monkey or a New World monkey?

____

What traits do New World monkeys share?

New World monkeys live in the tropical forests of Mexico, Central America, and South America. New World monkeys include marmosets and tamarins. These unique primates do not have fingernails or opposable digits. Some squirrel monkeys and spider monkeys do have opposable digits.

Most of these monkeys are diurnal and live together in social bands. Their **prehensile** (pree HEN sul) tails work like a fifth limb. The tail can grasp tree limbs and support the monkey's weight.

Picture This

5. Explain how a prehensile tail might benefit a monkey.

What features distinguish Old World monkeys?

Old World monkeys live in a wide variety of habitats throughout Asia and Africa. Macaques and baboons belong to one subgroup. Colobus and proboscus monkeys belong to another subgroup.

Old World monkeys are diurnal and live in social groups, like New World monkeys. However, Old World monkeys have narrower noses and larger bodies. They spend more time on the ground. Old World monkeys do not have prehensile tails. Some do not have tails. Most have opposable digits.

How do apes differ from monkeys?

Only a few ape species exist today. They have larger brains than monkeys. Their arms are longer than their legs. Apes have barrel-shaped chests, no tails, and flexible wrists. They are highly social and make complex sounds.

Apes are classified into two subgroups: lesser apes and great apes. The lesser apes include gibbons and siamangs. The great apes include orangutans, gorillas, chimpanzees, and humans.

6. Name the ape subgroup to which you belong.

How do lesser apes travel through the trees?

Gibbons and siamangs are gymnasts of the trees. Although they can walk, they often move quickly from branch to branch using a hand-over-hand swinging motion called brachiation.

How do great apes walk?

Orangutans are the largest arboreal primates, but the large males spend more time on the ground than in trees. Orangutans are the only great ape species that lives exclusively in Asia.

Gorillas are the largest primates. Like most of the great apes, they spend most of their time on the ground. They walk on all four limbs, using their front knuckles for support. They use sticks as simple tools. In captivity, they have been taught to recognize written characters and numbers.

Chimpanzees and bonobos are also knuckle-walkers. They have well-developed communication and social systems. They are more like humans in their physical structure and behavior than any other primate. ☑

To what category do humans belong?

Humans are part of the great ape family. Humans are classified as hominins. **Hominins** are humanlike primates that appear to be more closely related to present-day humans than they are to present-day chimpanzees and bonobos. Many species of hominins have existed on Earth. However, humans are the only species of hominins that survives today. The figure below illustrates primate evolution.

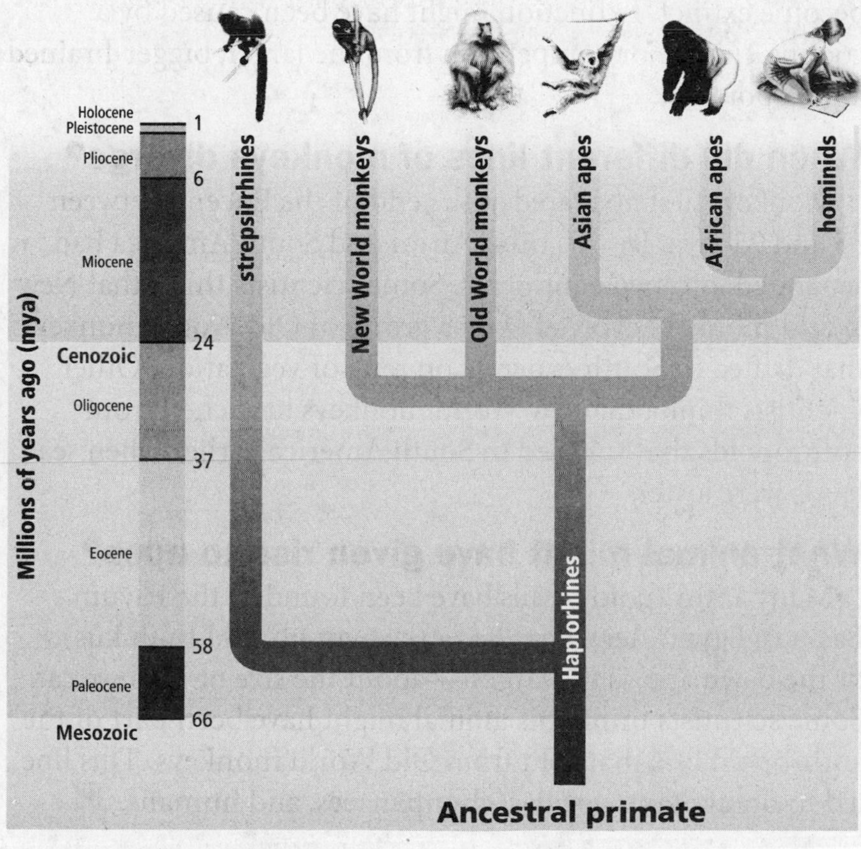

Reading Check

7. Identify the ape that is most closely related to you. (Circle your answer.)

a. gorilla
b. orangutan
c. bonobo

Picture This

8. Study the evolution of primates, and explain it to another person.

Primate Evolution

Most primates today are arboreal. Prehensile tails, long limbs, binocular vision, brachiation, and opposable toes and thumbs are all traits that help primates live in trees.

How might primates have become arboreal?

Some scientists suggest that primates evolved from ground-living animals that gathered food in top branches of shrubbery. This suggests that the flexible hand and opposable toes and thumbs might have evolved to catch insects rather than to grasp tree branches. Primates then evolved to fill other niches in trees. Other scientists suggest that the rise of flowering plants provided new niches. Arboreal adaptations then enabled primates to gather fruits and flowers of trees.

When did the first primates appear?

The first primates probably lived alongside dinosaurs about 85 mya. One of the earliest fossil primates, called *Altiatlasius* (al tee aht lah SEE us), was a small, nocturnal animal that used its hands and feet for grasping. About 50 mya, anthropoids branched off from tarsiers. By the end of the Eocene, 35 to 30 mya, anthropoids had evolved widely.

About this time, many early strepsirrhines appear to have become extinct. Extinction might have been caused by a cooling climate or competition from the larger, bigger-brained anthropoids. ☑

Reading Check

9. Name a key competitive advantage that anthropoids had over early strepsirrhines.

____

When did different lines of monkeys diverge?

Monkeys first appeared at the end of the Eocene between 35 and 25 mya. By this time, Africa and South America had separated into two continents. Some scientists think that New World monkeys evolved from a group of Old World monkeys that drifted to South America on rafts of vegetation. Other scientists think that New World monkeys branched from anthropoids that traveled to South America earlier when sea levels were lower.

What animal might have given rise to apes?

Many anthropoid fossils have been found in the Fayum Basin in Egypt. *Aegyptopithecus* (ee gypt oh PIH thuh kus), or the dawn ape, is the largest—about the size of a house cat. Some scientists think this animal might have been part of the anthropoid line that split from Old World monkeys. This line led to orangutans, gorillas, chimpanzees, and humans. ☑

Reading Check

10. Explain the importance of *Aegyptopithecus*.

____

____

____

Primate Evolution

section 2 Hominoids to Hominins

Before You Read

Humans are bipedal—they can walk upright on two legs. On the lines below, list possible advantages of walking upright, instead of walking on four limbs. Then read the section to learn why bipedalism might have evolved.

____

____

____

MAIN Idea

Hominins likely evolved in response to climate changes of the Miocene epoch.

What You'll Learn

- the evolutionary path of hominoids from *Proconsul* to *Homo*
- characteristics of various australopithecine species

Read to Learn

Mark the Text

Highlight Main Ideas As you read the section, highlight the main ideas in each paragraph.

Hominoids

Hominoids (HAH mih noydz) include all nonmonkey anthropoids—gibbons, orangutans, chimpanzees, gorillas, and humans. By comparing DNA of living hominoid species, researchers conclude that gibbons likely branched from anthropoids first. Next to the branch were orangutans, gorillas, chimpanzees and bonobos, and finally humans. Chimpanzees and bonobos are the closest living relatives to humans. The figure below shows the time line of primate evolution.

Picture This

1. Identify Circle the branch containing the human species.

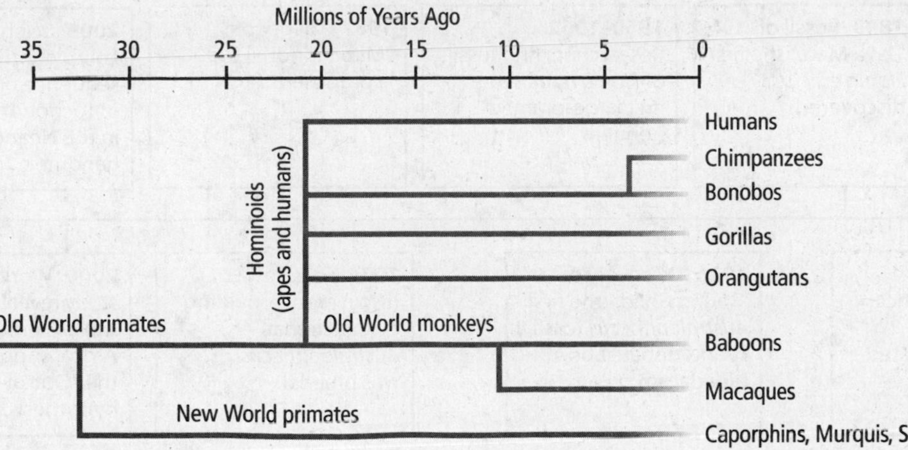

What characteristics do hominoids share?

Hominoids are the largest primates. They have the largest brains for their body sizes. Typical traits include a broad pelvis, long fingers, no tail, and flexible arm and shoulder joints. They have mostly upright postures. All hominoids, except for hominins, have longer arms than legs. Their teeth are less specialized than other primates. ☑

Reading Check

2. Identify a body characteristic of hominins that differs from other hominoids.

____

____

____

Earth's climate changed during the Miocene (24 to 5 mya). It became warmer and drier. Tropical rain forests shrank. Dry forests and savannas appeared. Many new animals evolved, including hominoids, that were able to take advantage of the new environment. Hominoids migrated from Africa to Europe and Asia.

Why is Proconsul important?

Some of the oldest hominoid fossils are from the genus *Proconsul*. *Proconsul* species had small brains. Although they lived mostly in trees, some might have been able to walk upright. Some scientists think that *Proconsul* is a human ancestor. Others suggest that humans rose from a European hominoid that returned to Africa at the end of the Miocene.

Hominins

The hominins include humans and all their extinct relatives. Hominins split off from other African apes sometime between 8 and 5 mya. The figure below highlights some important hominin discoveries.

The brains of hominins are larger than those of other hominoids and have more capacity for high-level thought. The hominin face is thinner and flatter. The teeth are smaller. Longer thumbs and more flexible wrists increase manual dexterity.

Picture This

3. Highlight each important discovery as you read about it in the section.

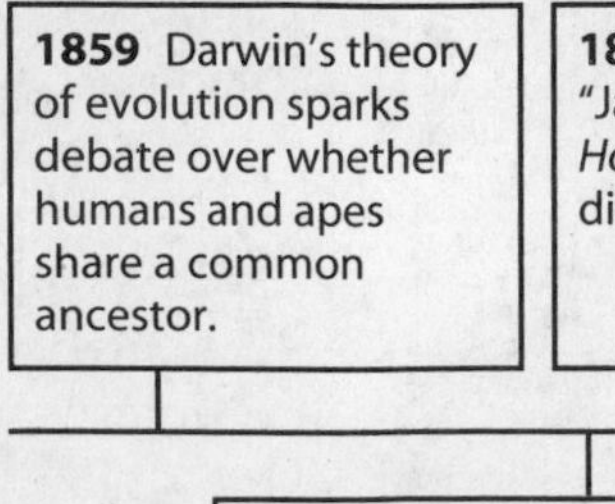

1859 Darwin's theory of evolution sparks debate over whether humans and apes share a common ancestor.

1886 Discoveries in Belgium convince scientists that Neanderthals existed.

1893 Fossil of "Java Man," the first *Homo erectus*, discovered.

1926 Discovery of "Taung Child," the first *Australopithecus* fossil, sparks debate about bipedalism.

1960–1962 Researchers find fossils of *Homo habilis*, the first large-brained hominin.

1974 Fossil of "Lucy" discovered, providing evidence that *Australopithecus* was bipedal.

1987 Theory of "Mitochondrial Eve" is proposed.

2000 Worldwide study reveals that *Homo sapiens* emerged from Africa, supporting the "Out-of-Africa" hypothesis.

2009 Scientists complete the sequence of mitochondrial DNA in the Neanderthal genome.

What structures support upright walking?

Hominins are **bipedal**—they can walk upright on two legs. The figure below shows how the structures of a biped differ from those of a quadruped, an animal that walks on all four limbs. Note how the shapes of the spine, arms, pelvis, and legs are adapted for upright walking. Also note that in quadrupeds, the spine extends from the back of the skull. In bipeds, the spine extends from the base of the skull.

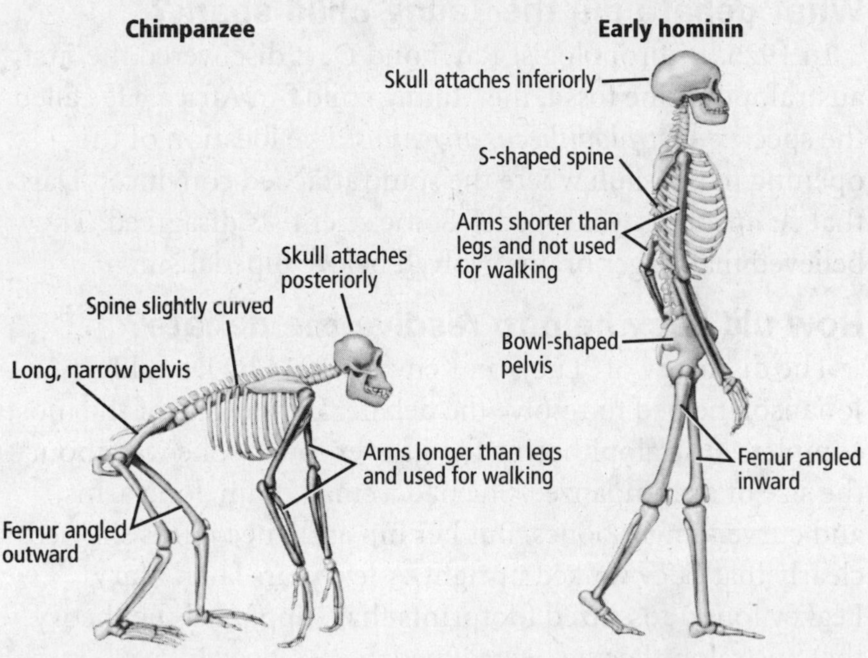

Picture This

4. Color the arms of the chimpanzee and the early hominin.

What are some disadvantages of bipedalism?

Walking upright has disadvantages. Bipedal individuals are easier for predators to see. They might not run as fast as their predators. Also, walking upright puts more strain on the hips and back. Because walking upright opposes gravity, it requires more energy than walking on all four limbs. ☑

Reading Check

5. Summarize three possible disadvantages of bipedalism.

Why did hominins become bipedal?

At the time hominins evolved, Africa was changing. Many scientists suggest that bipedalism was an adaptation to the new savanna environment. Food sources were sparse and far apart. By walking upright, individuals were able to travel longer distances and to spot food more easily.

Another theory suggests that walking on two legs freed hands for purposes such as carrying objects. Still another theory suggests that an upright posture enabled hominins to reach fruit on low tree branches.

What were the first bipedal hominins?

Bipedalism was one of the first hominin traits to evolve. Scientists look for evidence of bipedalism to identify hominin fossils. The **australopithecines** (aw stray loh PIH thuh seens) were the first truly bipedal hominins. They lived in east-central and southern Africa between 4.2 and 1 mya. They were small, only about 1.5 m tall, with apelike brains and jaws. Their teeth and limb joints were humanlike.

What debate did the Taung child spark?

In 1926, anthropologist Raymond Dart discovered the first australopithecine fossil, the "Taung child," in Africa. He called the species *Australopithecus africanus.* The location of the opening in the skull where the spine attached convinced Dart that *A. africanus* was bipedal. Some scientists disagreed. They believed that larger brains evolved before bipedalism.

Think it Over

6. Draw Conclusions Based on what you know about bipedal adaptations, where do you think the opening in the skull was in *A. africanus*?

____

____

How did Lucy help to resolve the debate?

The discovery of "Lucy" in Kenya in 1974 by Donald Johanson helped to resolve the debate. Lucy is one of the most complete australopithecine fossils ever found. She was about the size of a chimpanzee. She had a small brain, long arms, and curved finger bones. But her hip and knee joints showed clearly that Lucy walked upright. A few years later, Mary Leakey found fossilized footprints that supported the theory that australopithecines were bipedal.

How can hominin evolution be described?

Hominin fossils show a patchwork of human and apelike traits. Different body parts or behaviors evolved at different rates. This is called mosaic evolution.

Scientists have discovered many more early hominin fossils in the last 30 years. They disagree about how to classify them. For example, *A. bosei* and *A. robustus* were classified as australopithecines. Today, many scientists put these primates in a separate genus called *Paranthropus.* Paranthropoids lived alongside human ancestors but were not human ancestors themselves. ☑

Many species of hominins lived successfully for years, often overlapping with earlier species. Then they became extinct. No one knows why. By 1 mya, all australopithecines had disappeared from the fossil record. Later hominin fossils were only of humans and their close relatives.

✔ Reading Check

7. Explain what the paranthropoids were.

____

____

____

____

Primate Evolution

section 3 Human Ancestry

Before You Read

When you think of "cave men," what image comes to mind? On the lines below, describe your idea of early humans and their behaviors. Then read the section to learn about your cave-dwelling ancestors.

____

____

____

MAIN Idea

Tracing the evolution of the *Homo* genus is important for understanding humans.

What You'll Learn

- the Out-of-Africa hypothesis
- similarities and differences between Neanderthals and modern humans

Read to Learn

Mark the Text

Read for Understanding

As you read this section, highlight any sentence that you do not understand. When you finish the section, go back and reread the sentences you highlighted.

The *Homo* Genus

The genus ***Homo*** includes living and extinct humans. Members of this genus first appeared in Africa between 3 and 2.5 mya. Scientists think they evolved from an australopithecine ancestor.

Homo species had bigger brains than the australopithecines. They also had lighter skeletons, flatter faces, and smaller teeth. *Homo* species were the first to control fire and to use stone tools. As they evolved, they developed language and culture.

How did *Homo habilis* differ from its ancestors?

The first *Homo* species for which fossils exist is *Homo habilis. Homo habilis*, which means "handy man," used stone tools. The *Homo* traits of *H. habilis* included a larger brain, smaller brow and jaw, flatter face, and more humanlike teeth. Like its australopithecine ancestors, however, *H. habilis* was small, had long arms, and could climb trees. ☑

Another species, *Homo rudolfensis*, might have lived at the same time. However, few fossils of this species exist. Scientists are uncertain about how *H. rudolfensis* relates to the *Homo* line.

Reading Check

1. Describe two ways in which *H. habilis* was more like apes than like humans.

____

____

____

Why is *Homo ergaster* important?

The next species, *Homo ergaster*, appeared only briefly in the fossil record. It had a larger brain than *H. habilis.* It was also taller and lighter with longer legs and shorter arms. Scientists think that *H. ergaster* had the first humanlike nose (with nostrils facing downward).

Tools *H. ergaster* made hand axes and other tools. This species might have been a hunter or a scavenger. The tools might have been used to scrape meat off scavenged bones.

Migration *H. ergaster* appears to be the first African *Homo* species to migrate to Asia and Europe, possibly following migrating animals. Forms of *H. ergaster* in Europe and Asia are called *Homo erectus.* Scientists believe that *H. ergaster* is an ancestor to modern humans.

Think it Over

2. Explain why *H. ergaster* might have followed migrating animals.

____

____

What skills did *Homo erectus* have?

In Europe and Asia, *Homo erectus* evolved from *H. ergaster.* This species includes "Java Man," discovered in Indonesia, and "Peking Man," discovered in China. Unlike earlier species, *H. erectus* adapted to many types of environments.

H. erectus was taller than *H. habilis.* It had a bigger brain and more humanlike teeth. *H. erectus* featured a long skull, low forehead, and a thick brow ridge. This species made advanced tools, used fire, and sometimes lived in caves.

What is the significance of *Homo floresiensis*?

Most scientists believe that *H. erectus* went extinct about 400,000 years ago. Fossils discovered in 2004 on Flores island, Indonesia, suggest otherwise. This species, called *Homo floresiensis* (flor eh see EN sus), descended from *H. erectus* or another hominin. It existed until 12,000 years ago. Nicknamed "The Hobbit," *H. floresiensis* was only about 1 m tall. Basic stone tools were found with its fossils.

Reading Check

3. Describe the types of fossils that are classified as *Homo heidelbergensis.*

____

____

____

____

What traits did *Homo heidelbergensis* display?

The transition from *H. ergaster* to modern humans occurred gradually. Many fossils display a mix of traits of *H. ergaster* and modern humans. Some scientists classify these diverse fossils as *Homo heidelbergensis.* Others put them in a broader category called *Homo sapiens.* These humans had larger brains and thinner bones than *H. ergaster*, but they still had thick brow ridges and small chins. ☑

Are Neanderthals our ancestors?

The <u>**Neanderthals**</u>, or *Homo neanderthalensis*, were a species that evolved only in Europe and Asia. They likely evolved from *H. erectus* or a *Homo* species of the transition period. Neanderthals were larger than humans and had large brains. They had thick skulls and brow ridges, large noses, and heavy muscles attached to their thick bones. ☑

Neanderthals lived near the end of the Pleistocene ice age. They hunted, used fire, and made complex shelters. Evidence suggests that they cared for their sick and buried their dead.

In some areas, Neanderthals overlapped with modern humans. However, DNA tests on fossil bones show that Neanderthals were not part of the human gene pool. They were a different species. Neanderthals went extinct about 30,000 years ago.

✔ Reading Check

4. Generalize How did the Neanderthal body differ from the human body?

____

Emergence of Modern Humans

The thinner skeletons of *Homo sapiens* give them a more slender appearance than other *Homo* species. They have rounder skulls and smaller faces with an obvious chin. *H. sapiens* first appeared in what is now Ethiopia about 195,000 years ago. Early members of this species chipped stones to make hand axes and other tools. The table below compares *Homo* species.

Picture This

5. Describe the trend in brain size as evolution progressed from earlier species to more recent species.

____

____

Species	Time in Fossil Record	Characteristics
Homo habilis	2.4 to 1.4 million years ago	• average brain size: 650 cm^3 • used tools
Homo ergaster	1.8 to 1.2 million years ago	• average brain size: 1000 cm^3 • had thinner skull bones • had humanlike nose
Homo erectus	1.8 to 400,000 years ago	• average brain size: 1000 cm^3 • had thinner skull bones • used fire
Homo neanderthalensis	200,000 to 30,000 years ago	• average brain size: 1500 cm^3 • buried their dead • possibly had language
Homo sapiens	195,000 years ago to present	• average brain size: 1350 cm^3 • does not have brow ridge • has a small chin • has language and culture

What is the "Out of Africa" hypothesis?

As shown in the figure below, many hominin species overlapped until about 30,000 years ago. Then, only modern humans remained.

Some scientists believe that modern humans evolved at the same time in different areas of the world. Most scientists, however, support the "Out of Africa" hypothesis. This view suggests that humans evolved only once, in Africa, and then migrated to all parts of the world, replacing other hominins.

Picture This

6. Highlight Use a marker to highlight the direct ancestor of today's human beings.

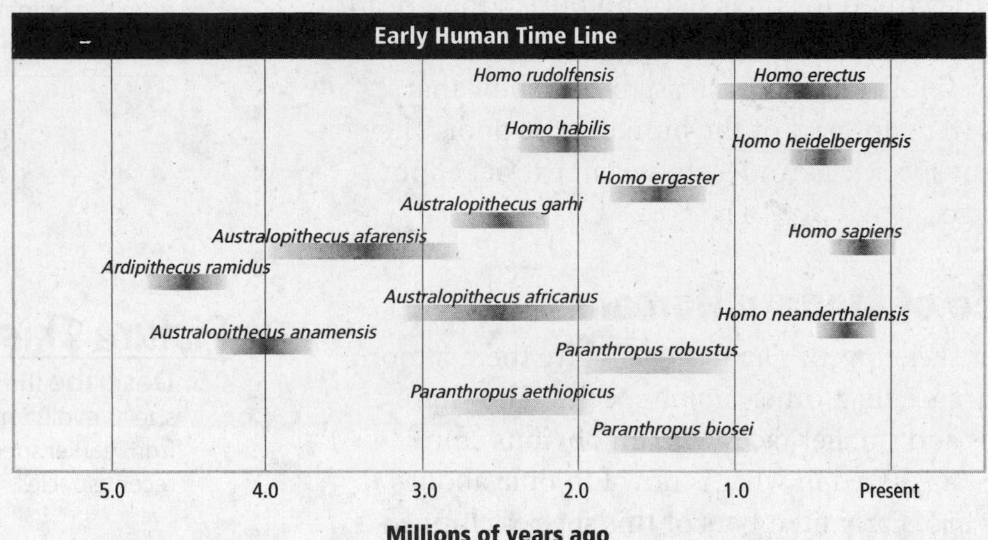

How did DNA support this hypothesis?

Mitochondrial DNA analysis of today's humans supported this hypothesis. Mitochondrial DNA changes very little over time. As a result, scientists reasoned that populations with the most variation in this DNA must have existed the longest time. They found the widest variation among Africans.

Mitochondrial DNA is inherited only from the mother. Therefore, this analysis suggested that *H. sapiens* emerged in Africa about 200,000 years ago from a hypothetical "Mitochondrial Eve."

Think it Over

7. Draw Conclusions Scientists found the widest variation in mitochondrial DNA in Africans. What conclusion could they draw from this finding?

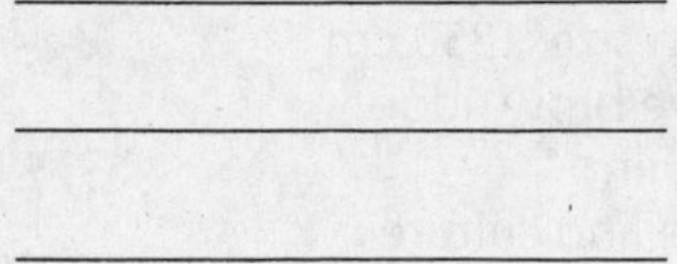

What evidence of human culture appeared?

Unlike Neanderthals, early modern humans expressed themselves using symbols and art. They drew on cave walls and decorated objects, and developed complex tools and weapons. They were the first to fish, make clothing, and raise animals. Cultural expressions such as these marked the first fully modern humans. Some people consider **Cro-Magnons** to be the first hunter-gatherers.

Organizing Life's Diversity

section 1 The History of Classification

Before You Read

On the lines below, describe how you might organize a personal collection of books or CDs. In this section, you will learn the way biologists organize living things.

MAIN Idea

Biologists use a system of classification to organize information about the diversity of living things.

What You'll Learn
- differences in methods of classifying
- how to write scientific names
- the taxa in biological classification

Read to Learn

Mark the Text

Identify Concepts Highlight each question heading in this section. Then use a different color to highlight the answers to the questions.

Early Systems of Classification

It is easier for biologists to communicate and keep information about organisms when the organisms are organized into groups. One tool biologists use to organize organisms is classification. Classification is the grouping of objects or organisms based on a set of conditions. A regular system of classification helps scientists organize and communicate information about biology.

How did Aristotle classify organisms?

More than two thousand years ago, Aristotle, a Greek philosopher, developed the first commonly accepted system of biological classification. Aristotle classified organisms as either animals or plants. Animals were classified by their habitat and their morphology. Morphology relates to the physical characteristics and structures of organisms. Animals were also classified by the presence of red blood. Aristotle's classification of "bloodless" and "red blood" animals closely matches today's classification of invertebrates and vertebrates. Plants were classified by average size and structure—as trees, shrubs, or herbs. The table on the next page shows how Aristotle might have divided some of his groups. ☑

Reading Check

1. Identify Who developed the first commonly accepted system of classification?

Picture This

2. Classify Using Aristotle's classification system, how would a daisy, a dog, and a whale be classified?

Plants		
Herbs	**Shrubs**	**Trees**
violets rosemary onions	blackberry bush honeysuckle flannelbush	apple oak maple
Animals with Red Blood		
Land	**Water**	**Air**
wolf cat bear	dolphin eel sea bass	owl bat crow

What were the limitations of Aristotle's system of classification?

Aristotle's system of classification was useful for organizing, but it had many limitations. One limitation was that Aristotle's system was based on his understanding that species are distinct, separate, and unchanging. Because of this understanding, Aristotle's classification did not account for evolutionary history or relationships. Also, many organisms have been discovered that do not fit Aristotle's classification system, such as birds that do not fly and frogs that live on land and in water. Aristotle's system was used for many centuries before it was replaced by a new system. The new system built on the knowledge humans had gained about the natural world.

How did Linnaeus classify organisms?

In the eighteenth century, Swedish botanist Carolus Linnaeus developed a branch of biology called taxonomy. **Taxonomy** (tak SAH nuh mee) is a discipline of biology concerned with identifying, naming, and classifying species based on the morphological and behavioral similarities and differences of organisms. Linnaeus's system built on the foundation of Aristotle's system of classification. Linnaeus used similarities and differences in morphology and behavior to classify birds. The morphological differences can be related to differences in where the birds lived and their behavior.

Think it Over

3. Explain How did Linnaeus's system of classification build on Aristotle's system?

What is systematics?

Taxonomy is part of a larger branch of biology called systematics. Systematics is the study of biological diversity. Scientists study diversity in the past, as well as present biological diversity.

How are scientific names written?

Linnaeus named organisms using binomial nomenclature. **Binomial nomenclature** (bi NOH mee ul • NOH mun klay chur) gives each species a scientific name that has two parts. The first part is the genus (JEE nus) name, and the second part is the specific epithet (EP uh thet), or specific name, that identifies the species. Latin is often used for binomial nomenclature because Latin is a language that is unchanging. Historically, Latin has also been the language of science.

Why do scientists use scientific names?

Biologists use scientific names because common names vary in their use. For example, the bird *Cardinalis cardinalis*, shown above, is commonly called a redbird, a cardinal, and a Northern cardinal. Binomial nomenclature is also used because common names can be misleading. A starfish is neither a star nor a fish, a great horned owl does not have horns, and a sea cucumber is not a plant.

When writing scientific names, scientists follow certain rules. The most important rules are as follows:

- The first letter of the genus name is always capitalized, but the rest of the genus name and all of the specific epithets are lowercase.
- If a scientific name is printed in a book or magazine, it is italicized.
- If a scientific name is written by hand, it is underlined.
- After the complete scientific name has been written once, the genus name will often be abbreviated to the first letter when used again. For example, the scientific name *Cardinalis cardinalis* can be written *C. cardinalis.* ☑

Picture This

4. Identify Which is the scientific name for this bird? (Circle your answer.)

a. cardinal
b. Northern cardinal
c. *Cardinalis cardinalis*

Reading Check

5. Describe How should a scientific name be written by hand?

____

____

Reading Check

6. Compare How is today's classification system different from Linnaeus's system?

How has the classification system changed?

Linnaeus's classification system made it possible to include evolutionary principles in classification in the 1800s. In the nineteenth century, important scientists, including Jean-Baptiste Lamarck, Charles Darwin, and Ernest Haeckel, introduced classification systems based on evolutionary relationships to organize biological diversity. Categories used in modern classification are based on Linnaeus's system but have been changed to show evolutionary relationships. ☑

Taxonomic Categories

Taxonomists classify organisms by dividing them into smaller groups based on more specific criteria. Taxonomic categories used by scientists are like nesting boxes—each category fits into another. The categories are arranged from broadest to most specific. A named group of organisms is called a **taxon** (plural, taxa).

What are a species and a genus?

Two of the taxa Linnaeus used were genus and species. Today, a **genus** (plural, genera) is defined as a group of species that are closely related and share a common ancestor. A species is a group of organisms that have similar characteristics such as skull shape and size. For example, the species American black bear (*Ursus americanus*) and the species Asiatic black bear (*Ursus thibetanus*), shown below, belong to genus *Ursus*. All species in the genus *Ursus* have massive skulls and similar tooth structure.

Picture This

7. Name characteristics that the American black bear and the Asiatic black bear have in common.

American blackbear

Asiatic blackbear

What is a family?

A **family** is a group of genera that have similar characteristics. All bears, both living and extinct, belong to the family Ursidae. All members of the Ursidae family have similar characteristics. For example, they walk flat-footed and have forearms that can rotate to grasp prey closely.

What are the higher taxa?

An **order** is a group of families that have similar characteristics. A **class** is a group of one or more related orders. A **phylum** (FI lum) (plural, phyla) or a **division** is a group of related classes. The term *division* is used for bacteria and plants. A **kingdom** is a group of related phyla, or divisions. The least specific of all taxa is a domain. A **domain** is a group of one or more kingdoms. The pyramid of taxa shown below will help you remember how the taxa are organized.

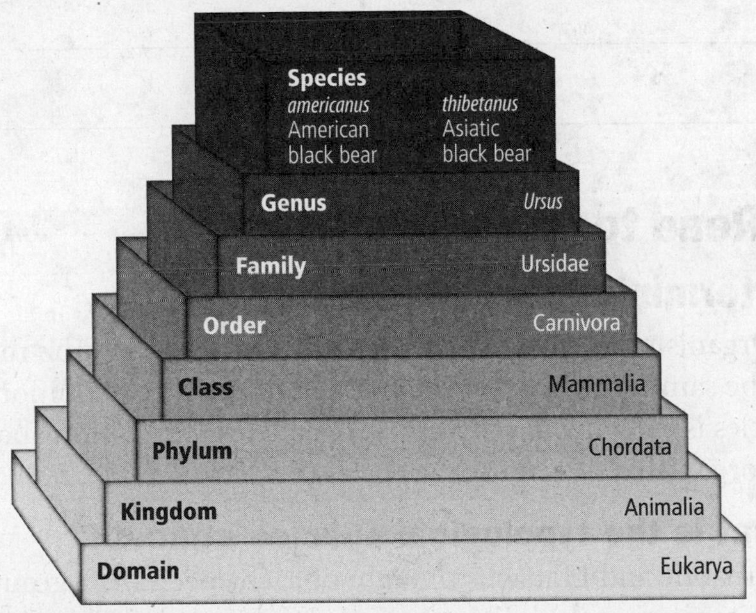

Systematics Applications

Systematicists are scientists who study classification. They provide detailed guides that enable other people to identify organisms. Many times, field guides have dichotomous (di KAWT uh mus) keys, which are keys based on a series of choices between characteristics. You can tell if a plant or animal is poisonous by using a field guide to identify it. Systematicists also work to identify new species and relationships among known species. If a known species produces a certain chemical, a close relative might produce a similar chemical.

Picture This

8. Determine At what level on the pyramid of taxa is the American black bear distinguished from the Asiatic black bear?

______

Think it Over

9. Draw Conclusions Why might it be important to know the relationships among species?

______

______

______

______

Organizing Life's Diversity

section 2 Modern Classification

MAIN Idea

Classification systems have changed over time as information has increased.

What You'll Learn

- species concepts
- methods to reveal phylogeny
- how to construct cladograms

Before You Read

On the lines below, describe your system of organizing your class notes and your method of using the information in your notes to study for tests. In this section, you will learn how scientists have used new information to make adjustments to systems and theories.

Mark the Text

Identify Main Ideas As you read, underline or highlight the main ideas in each paragraph.

Read to Learn

Determining Species

Organisms that are different species by one definition might be the same species by a different definition. The definition of species is evolving as scientists learn more information about the organisms they study.

What is the typological species concept?

Aristotle and Linnaeus thought of each species as a group of organisms with similar physical characteristics. This definition of a species is called the typological species concept. It is based on the idea that species are unchanging, distinct, natural "types." A type specimen is an individual of the species that best shows the characteristics of that species. When another specimen was found that was different from the type specimen, it was classified as a different species. ✓

Evolution causes species to change. Because there is a lot of variation among members of some species, the typological species concept has been replaced. Some of its traditions such as type specimens remain.

Reading Check

1. Define What is a type specimen?

What is the biological species concept?

In the 1930s and 1940s, the term *species* was redefined as a group of organisms that are able to interbreed and produce fertile offspring in a natural setting. This definition is known as the biological species concept. ☑

There are limitations to the biological species concept. Wolves and dogs are classified as different species, but they are known to interbreed and produce fertile offspring. Many plant species can also interbreed and produce fertile offspring. The biological species concept also does not consider extinct species or species that reproduce asexually. However, the biological species concept works for most classification, so it is often still used.

What is the phylogenetic species concept?

In the 1940s, the evolutionary species concept was proposed to go along with the biological species concept. The evolutionary species concept defines different species as two or more groups that evolve independently from an ancestral population. This concept has developed into the phylogenetic species concept.

Phylogeny (fi LAH juh nee) is the evolutionary history of a species. The phylogenetic species concept defines a species as a group of organisms that is different from other groups of organisms and that has, within the group, a pattern of ancestry and descent. When a phylogenetic species branches, it becomes two different phylogenetic species. For example, you have read that when organisms become isolated, they often develop different adaptations. Over time, these isolated organisms become different from the original group.

Reading Check

2. Describe How is species defined as the biological species concept?

______

Picture This

3. Identify Which species concept classifies according to ancestral history?

______

Species Concept	Description	Limitation	Benefit
Typological species concept	classification by the comparison of physical characteristics with a type specimen	Alleles produce a wide variety of features within a species.	Descriptions of type specimens provide detailed records of the physical characteristics of many organisms.
Biological species concept	classification by similar characteristics and the ability to interbreed and produce fertile offspring	Some organisms that are different species interbreed occasionally. It does not account for extinct species.	The working definition applies in most cases, so it is still used frequently.
Phylogenetic species concept	classification by evolutionary history	Evolutionary histories are not known for all species.	Accounts for extinct species and considers molecular data.

Has the classification of a species changed?

For more than one hundred years, Asiatic elephants have been classified as one species and African elephants have been classified as a different species. There are two populations of African elephants. One population lives in the savanna, and one population lives in the forest. Until recently, the two African populations had been classified as the same species. Scientists thought that the two African populations interbred at the borders of their habitats. Recent studies have shown that they interbreed rarely. Scientists also found large differences in the DNA and skull measurements of the two African populations. The two populations might be separate species. ☑

Reading Check

4. Summarize Why might African elephants be two species, instead of one?

______

______

______

______

______

Characters

To determine the species of an organism, scientists put together pieces of evolutionary history, also called phylogenies, using characters. **Characters** are inherited features that vary among species. Characters can be morphological or biochemical.

How are morphological characters used?

Shared morphological characters suggest that species are closely related and that they evolved from a recent common ancestor. Analogous characters do not indicate a close evolutionary relationship. Recall that analogous characters have the same function but different structure. Homologous characters might perform a different function but show similar structure that was inherited from a common ancestor.

Look at the oviraptor and the sparrow shown below. Some dinosaur fossils such as theropods show that they had feathers and large hollow spaces in their bones. Their hip, leg, wrist, and shoulder structures are similar to those of birds. These morphological characters suggest that birds are related more closely to theropod dinosaurs than to other reptiles.

Picture This

5. Compare What physical characteristics do the oviraptor and the sparrow have in common?

______

______

______

How are biochemical characters used?

Recall that chromosomes are strands of genetic material that become visible during mitosis and meiosis. The number and structure of chromosomes provide information about evolutionary relationships among species. Similarities suggest a common evolutionary history.

DNA and RNA are made up of four nucleotides. The sequence of DNA nucleotides defines the genes that give instructions to RNA for making proteins. Scientists study and understand evolutionary relationships by sequencing DNA of different organisms. They compare the sequences of a variety of organisms. Organisms that are closely related have many similar sequences of nucleotides. Therefore, they have similar proteins. ☑

Broccoli, cabbage, cauliflower, and kale look different, but they have almost the same chromosome structures, which suggest a close evolutionary relationship. Chimpanzees, gorillas, and orangutans also have similar chromosomes.

Different organisms might have many similar sequences in their DNA. However, when all of their DNA sequences are studied, major differences can be found. The more sequences they share, the more likely they are to share a common ancestor.

What are molecular clocks?

Mutations occur randomly in DNA. As time passes, mutations build up in the chromosomes. Some mutations do not affect the function of cells. The rate at which these mutations build up can be viewed as a molecular clock. A **molecular clock** is a model that uses comparisons of DNA sequences to estimate how long species have been evolving.

The rate at which mutations occur does not stay the same. The rate of a mutation is affected by many factors that include the type of mutation, where in the genome the mutation occurs, the type of protein the mutation affects, and the population in which the mutation occurs. In a single organism, genes might mutate at a different rate. This inconsistency makes molecular clocks difficult to read. Scientists are trying to find genes that mutate at a relatively consistent rate throughout a range of organisms.

Even though molecular clocks have limitations, they can be a valuable tool for helping to determine the time when a new species evolved. The molecular clock is often used along with the fossil record to identify the time of divergence.

Reading Check

6. Explain How can DNA sequences be used to determine if organisms are closely related?

______

______

______

Think it Over

7. Discuss Why does the inconsistency in the rate at which genes mutate make molecular clocks difficult to read?

______

______

______

______

______

Phylogenetic Reconstruction

Biologists often study evolutionary relationships using cladistics. **Cladistics** (kla DIHS tiks) is a way to study evolutionary relationships that rebuilds phylogenies and hypothesizes evolutionary relationships based on shared characters. The hypothesized relationships formed by cladistics suggest how different groups of organisms might have evolved. To identify possible relationships, the characters of different groups of organisms need to be known.

What are the main character types?

There are two main character types that need to be considered when using cladistics: ancestral characters and derived characters. An ancestral character is found in a variety of groups within the line of descent. A derived character is present in one group within the line of descent, but it is not found in the common ancestor. When comparing two groups of organisms, an ancestral character evolved in a common ancestor of both groups, and a derived character evolved in an ancestor of one group. ☑

For example, when comparing birds and mammals, a backbone is an ancestral character because both birds and mammals have backbones and both have ancestors with backbones. Feathers are derived characters because only birds have an ancestor with feathers. Hair is also a derived character because only mammals have an ancestor with hair.

What is a cladogram?

Scientists use ancestral characters and derived characters to make a cladogram. A **cladogram** (KLAD uh gram) is a branching diagram that shows the proposed phylogeny of a species. A cladogram is similar to a pedigree. Both have branches and show the ancestry of an individual or a group. The groups of a cladogram, called clades, have one or more related species. The branches of a cladogram show hypothesized phylogeny. The hypothesized phylogeny shown in a cladogram depends on information from DNA and RNA sequences, bioinformatics, and morphological studies. Places where branching occurs are called nodes. The common ancestor at the nodes is rarely a known organism, species, or fossil. Scientists hypothesize the ancestor's character types based on the traits of its descendants. Scientists think that the more derived characters groups share, the more recent their common ancestor. ☑

✔ Reading Check

8. Compare What is the difference between ancestral and derived characters?

____

____

____

____

____

✔ Reading Check

9. Define What is a cladogram?

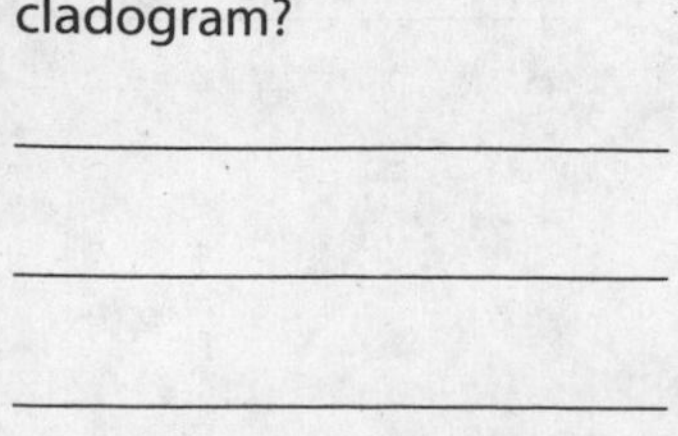

How is a cladogram made?

A cladogram of the lily, a flowering plant, is shown below. First the derived characters—vascular tissue, seeds, and flowers—were identified. Then the ancestry of a variety of species was identified based on whether the species had some or all of the derived characters. The groups that are closer to the lily in the cladogram probably share a more recent ancestor than the groups that are farther away. Flowering plants and conifers share three derived characters and are thought to have a more recent common ancestor than flowering plants and ferns have.

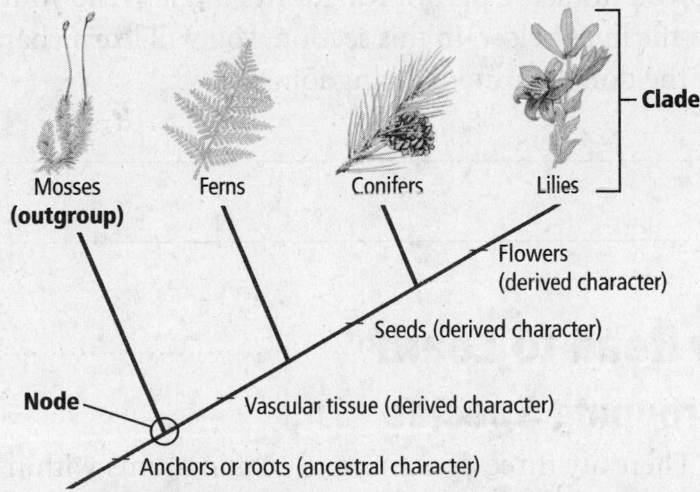

Picture This

10. Identify Which traits do flowering plants and conifers share?

____

____

____

What is a phylogenetic tree?

Phylogenetic trees are used to show the relationships among species and groups of organisms. A phylogenetic tree is a form of cladogram in which each node with descendants represents a common ancestor. The tree of life concept was introduced by scientist Ernest Haeckel. He imagined a tree with a trunk representing ancestral groups. The tree's branches showed species. Similar species were listed on nearby branches. The leaves on the branches represented individual organisms.

A tree that represented all living organisms would be gigantic. Scientists have classified about 1.75 million species. They estimate that millions more have not yet been classified. Although creating a complete tree of life is a large, difficult task, many scientists think it is important. Scientists representing many disciplines are working together to develop a comprehensive tree of life.

Organizing Life's Diversity

section 3 Domains and Kingdoms

MAIN Idea

The most widely used biological classification system has six kingdoms within three domains.

What You'll Learn

- major characteristics of the three domains
- how to classify organisms at the kingdom level

Before You Read

Kingdom Plantae includes all plants. What kinds of organisms do you think are part of Kingdom Fungi? Write your answer on the lines below. In this section, you will learn characteristics of the domains and the kingdoms.

____

____

Study Coach

Make Flash Cards Make a flash card for each kingdom in this section. Write the kingdom on one side of the card. Write the characteristics of the kingdom on the other side. Use the flash cards to review what you have learned.

Read to Learn

Grouping Species

There are three domains and six kingdoms within those domains. Organisms are classified into domains based on cell type and structure. Organisms are classified into kingdoms based on cell type, structure, and nutrition.

Recall that prokaryotes are unicellular organisms that do not have membrane-bound organelles. At one time, all prokaryotes were classified in Kingdom Monera. Today we know that there are two types of prokaryotes that are as different from each other as they are from eukaryotes. They are classified into two domains—Bacteria and Archaea.

Domain Bacteria

There is no taxonomic difference between Domain Bacteria and Kingdom Bacteria. These organisms are sometimes called eubacteria. They are prokaryotes whose cell walls contain peptidoglycan (pep tih doh GLY kan). Peptidoglycan is a polymer that contains two kinds of sugars. The amino acids on these sugars form a netlike structure that is porous and strong. ☑

FOLDABLES™

Take Notes Make a layered Foldable, as shown below. As you read, take notes and organize what you learn about the three domains and six kingdoms of living organisms.

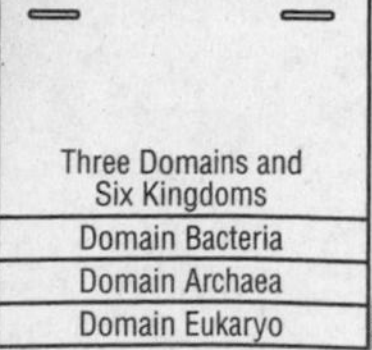

What are the characteristics of bacteria?

Bacteria are unicellular organisms that do not have a nucleus or other membrane-bound organelles. They can survive in many environments. Some bacteria are aerobic organisms, meaning that they need oxygen to live. Others are anaerobic organisms. They cannot live if atmospheric oxygen is present. Some are autotrophic organisms that make their own food. Most are heterotrophic organisms that get their nutrition from other organisms. Bacteria are found in different shapes. Bacteria are more abundant than any other organism.

Domain Archaea

There is no taxonomic difference between Domain Archaea and Kingdom Archaea. These organisms are often called archaea (ar KEE). Most scientists think that archaea are more ancient than bacteria. Archaea are prokaryotes. They are unicellular organisms that do not have a nucleus or other membrane-bound organelles. Their cell walls do not contain peptidoglycan and they have some cytoplasmic similarities to eukaryotes.

What are the characteristics of archaea?

Archaea are found in many shapes. They obtain nutrients in several ways. Some are autotrophic organisms, but most are heterotrophic organisms. Sometimes archaea are called extremophiles because they live in the most extreme environments on Earth. Extreme environments include hot springs, salty lakes, thermal vents on the ocean floor, and the mud of marshes. Little or no oxygen is found in these environments. One type of extremophile lives near thermal vents in deep ocean waters. The water temperatures can reach 98°C, almost boiling. ☑

Domain Eukarya

All eukaryotes, organisms with membrane-bound organelles, are classified in Domain Eukarya. Domain Eukarya contains Kingdom Protista, Kingdom Fungi, Kingdom Plantae, and Kingdom Animalia.

Think it Over

1. Apply What is a good definition of the term anaerobic?

____

____

____

Reading Check

2. Explain Why are archaea sometimes called extremophiles?

____

____

____

3. Compare How are plantlike protists and animal-like protists different?

____

____

____

4. Define What are hyphae?

____

____

____

____

____

What are the characteristics of Kingdom Protista?

Members of Kingdom Protista are called protists. **Protists** are eukaryotes and have membrane-bound organelles. They can be unicellular, a colony of cells, or multicellular. Protists are not similar to one another. However, they do not fit into any other kingdoms either. Protists are classified in three groups: plantlike protists, animal-like protists, and funguslike protists.

What are the characteristics of the three groups of protists?

Plantlike protists are called algae. Algae, such as kelp, are autotrophic organisms. They make their own food by performing photosynthesis.

Animal-like protists are called protozoans. Protozoans, such as amoebas, are heterotrophs. Plantlike protists and animal-like protists do not form organs like species in the plant and animal kingdoms.

Funguslike protists are slime molds and mildews. Euglenoids (yoo GLEE noyds) are protists that have both plantlike and animal-like characteristics. Euglenoids are usually grouped with plantlike protists because they perform photosynthesis.

What are the characteristics of Kingdom Fungi?

A member of Kingdom Fungi is called a fungus. A **fungus** is a eukaryote that absorbs nutrients from organic materials in its environment. Fungi are unable to move. Their cell walls contain chitin, which is a rigid polymer that gives cells structural support. Fungi also have hyphae (HI fee). Hyphae are threadlike strands that enable the fungi to grow, feed, and reproduce. More than 70,000 species of fungi have been identified. ☑

Most fungi, such as mushrooms, are multicellular. A few fungi, such as yeasts, are unicellular. Fungi are heterotrophs. Unlike other heterotrophic organisms that digest food internally, fungi secrete digestive enzymes into their food source and then absorb the nutrients directly into their cells.

Parasitic fungi include saprobes and symbionts. They grow and feed on other organisms. Saprobes get their nourishment from dead or decaying organic matter. Symbionts that live in a mutualistic relationship with algae are lichens. Lichens get their food from algae that live among the fungi's hyphae.

What are the characteristics of Kingdom Plantae?

Members of Kingdom Plantae (PLAN tuh) are called plants. There are more than 250,000 species of plants in Kingdom Plantae. Plants form the base of all land habitats.

All plants are multicellular. The cell walls of all plants contain cellulose. Most plants are autotrophs. Plants convert energy from the Sun through photosynthesis. A few plants also obtain energy from other organisms. For example, the dodder is a parasitic plant. It obtains food through suckers connected to the host plant. ☑

All plants have cells that are organized into tissues. Most vascular plants have organs such as roots, stems, and leaves. Plants cannot move. However, some plants have reproductive cells that have flagella. The flagella can move the reproductive cells through water.

✔ Reading Check

5. State How many species are in Kingdom Plantae?

What are the characteristics of Kingdom Animalia?

Members of Kingdom Animalia are called animals. All animals are heterotrophs and are multicellular. Animals are eukaryotic organisms and have membrane-bound organelles.

Animals do not have cell walls. They have cells that are organized into tissues. Most animals have tissues that are organized into organs such as skin, a stomach, and a brain. Animal organs are often organized into complex organ systems, such as digestive, circulatory, and nervous systems.

Animals range in size from a few millimeters to many meters. Animals live in water, on land, and in the air. Most animals are able to move. A few animals such as coral cannot move in their adult form.

Is there an exception to the classification system?

If you have ever had a cold or the flu, you have had a virus. A virus is a nucleic acid that is surrounded by a protein coat. Viruses do not have cells, and they are not cells. Viruses are not considered to be living. Because they are not living, they are not usually placed in the biological classification system. ☑

Virologists, scientists who study viruses, have created a special classification system to group viruses. Viral classification is based on a variety of factors.

✔ Reading Check

6. Explain Why are viruses an exception to the classification system?

Picture This

7. Name What is an example of an animal that cannot move as an adult?

____

What characteristics define differences in the six kingdoms?

The characteristics of living things are summarized in the table below. The table shows the similarities and differences in cell type and structure, nutrition, habitat, and mobility. As you review the table, think of organisms that fit into each kingdom.

Kingdom Characteristics

Kingdom	Cell Type and Structure	Nutrition	Habitat	Mobility
Bacteria	prokaryotes with cell walls made of peptidoglycan	most are heterotrophic; some are autotrophic	live in many environments	can move
Archaea	prokaryotes with cell walls that are not made of peptidoglycan	most are heterotrophic; some are autotrophic	live in extreme environments	can move
Protista	unicellular and multicellular eukaryotes	autotrophic and heterotrophic	live in moist environments	can move
Fungi	unicellular and multicellular eukaryotes with cell walls made of chitin	heterotrophic	live in many environments	cannot move
Plantae	multicellular eukaryotes with cell walls made of cellulose	most are autotrophic and perform photosynthesis; some are heterotrophic	live in water and on land	cannot move
Animalia	multicellular eukaryotes without cell walls	heterotrophic	live in water, on land, and in air	most can move; some cannot move, such as adult coral

Bacteria and Viruses

section 1 Bacteria

Before You Read

When you hear the word *bacteria*, what comes to mind? On the lines below, describe places you think bacteria might live. Then read the section to learn about some surprising places where bacteria thrive.

MAIN Idea

Bacteria are prokaryotic cells.

What You'll Learn

- how archaea differ from bacteria
- how prokaryotes can survive environmental challenges
- ways that bacteria benefit humans

Read to Learn

Study Coach

Make an Outline Make an outline of the information you learn in this section. Start with the headings. Include the boldface terms.

Diversity of Prokaryotes

Scientists think the first organisms on Earth were small, unicellular organisms called prokaryotes (proh KE ree ohts). Today, prokaryotes are the most numerous organisms on Earth. Prokaryotic cells do not have organelles or a nucleus. Instead, their DNA is found in a region of the cell. Prokaryotes are grouped into two domains—Bacteria and Archaea.

Where are bacteria found?

Bacteria live almost everywhere except in the most extreme environments. They have strong cell walls which contain peptidoglycan.

Where do archaea live?

Archaea live in extreme environments. One type lives in hot, acidic environments such as sulfur hot springs, thermal vents on the ocean floor, and around volcanoes. A second type lives in salty environments such as the Dead Sea. This type photosynthesizes using a substance other than chlorophyll. A third type cannot live in an environment that has oxygen. They use carbon dioxide during respiration and give off methane gas as waste. They live in swamps and in human intestines. They make the gases that are released from the lower digestive tract.

1. Apply Prokaryotes that live in the Great Salt Lake belong to which domain? (Circle your answer.)

a. Bacteria
b. Archaea

How do bacteria and archaea differ?

The cell walls of bacteria contain peptidoglycan. The cell walls of archaea do not contain peptidoglycan. Also, the two groups of organisms have different lipids, ribosomal proteins, and RNA. They are as different from each other as they are from eukaryotes.

Prokaryote Structure

A prokaryotic cell shares some characteristics with all cells such as DNA and ribosomes. Prokaryotic cells do not have membrane-bound organelles such as mitochondria and chloroplasts. The figure below shows the structure of a prokaryotic cell.

Picture This

2. Highlight each structure in the figure as you read about it.

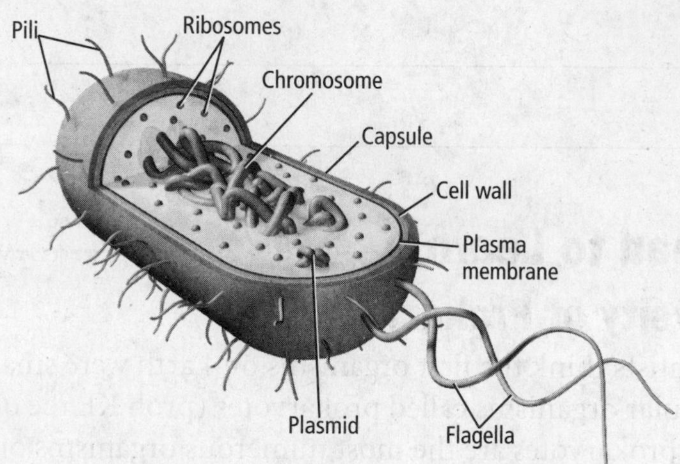

How are the chromosomes arranged?

The genes of a prokaryotic cell are on a circular chromosome in an area of the cell called the **nucleoid**. Many prokaryotes also have at least one smaller, circular piece of DNA. It is called a plasmid.

What are the functions of the capsule?

Some prokaryotic cells form a **capsule** by secreting a layer of polysaccharides around the cell wall. The capsule keeps the cell from drying out and helps it attach to surfaces. It also protects the cell from white blood cells and antibiotics.

3. Explain how pili are helpful during reproduction.

How do pili help a prokaryotic cell?

Some prokaryotes have pili on their outer surface. **Pili** (singular, pilus) are hairlike structures that are made of protein. Pili help a prokaryotic cell attach to a surface. Pili can also serve as a bridge between prokaryotes. Copies of plasmids can cross the bridge, providing new genetic characteristics. Resistance to antibiotics can be transferred this way.

How do prokaryotes benefit from their size?

Prokaryotes are small, even when viewed with a microscope. Small cells have a larger surface-area-to-volume ratio than larger cells. As a result, nutrients and other important substances can diffuse to all parts of the cell easily.

Identifying Prokaryotes

Scientists can identify prokaryotes by their shape, cell walls, and movement.

What shapes do prokaryotes display?

There are three main shapes of prokaryotes. Those shaped like spheres are called cocci (KAHK ki) (singular, coccus). Bacilli (buh SIH li) (singular, bacillus) are rod shaped. Spiral-shaped spirilli (spi RIH li) (singular, spirillium) are called spirochetes (SPI ruh keets). ☑

✔ Reading Check

4. Describe the three shapes of prokaryotes.

____

____

Why is the Gram stain test important?

All bacteria have peptidoglycan in their cell walls. Biologists add dyes to the bacteria cells to identify the two major types of bacteria—those with and those without an outer layer of lipid. The dye technique is called a Gram stain. Bacteria without a lipid layer have a lot of peptidoglycan and appear dark purple. They are called gram positive. Bacteria with a lipid layer have less peptidoglycan and appear light pink. They are called gram negative. Some antibiotics attack the cell walls of bacteria. The Gram stain identifies the type of cell wall so doctors can prescribe the right antibiotic. ☑

✔ Reading Check

5. Apply Why do doctors use Gram stains?

____

____

____

How do prokaryotes move?

Some prokaryotes do not move. Others use a flagellum (plural, flagella) to move toward light, oxygen, or sources of nutrients. Others glide over a layer of secreted slime.

Reproduction of Prokaryotes

Prokaryotes reproduce either by binary fission or by conjugation. The figure on the next page shows both.

What is binary fission?

Binary fission is the division of a cell into two cells with identical genes. In binary fission, the prokaryote's chromosome replicates. The cell gets longer as the chromosome copies separate. A new plasma membrane and cell wall form, separating the cell into two identical cells.

Picture This

6. Label Write the labels *genetic information exchanged* and *identical cells formed* next to the appropriate reproduction diagram in the figure.

How do prokaryotes reproduce by conjugation?

In **conjugation**, two prokaryotes attach to each other and exchange genetic material. As shown in the figure below, the two cells attach using their pili. The transfer of genetic material from one cell to the other creates new gene combinations. This increases the diversity of prokaryotes.

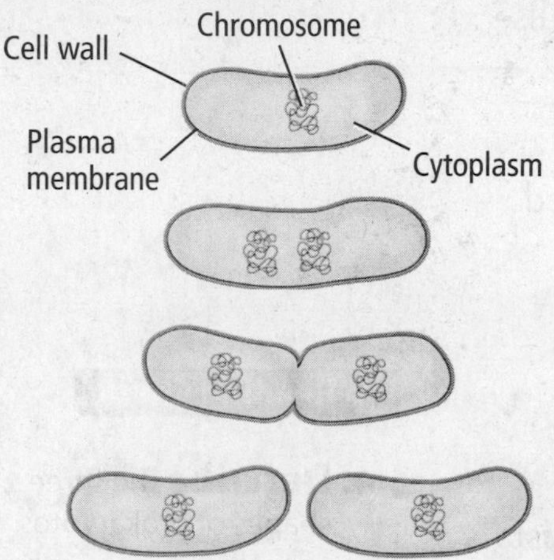

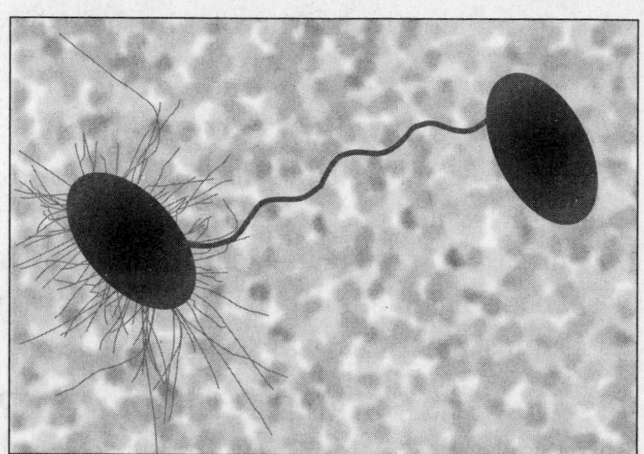

Metabolism of Prokaryotes

Obligate aerobes are bacteria that need oxygen to live. Obligate anaerobes cannot live in the presence of oxygen. They obtain energy by fermentation. Facultative anaerobes can live with or without oxygen. Besides how they use oxygen, prokaryotes are classified by how they obtain energy.

How do heterotrophs obtain energy?

Heterotrophs cannot make their own food. They need to take in nutrients. As shown in the figure on the next page, many heterotrophic bacteria are saprotrophs. They obtain nutrients by decomposing organic materials associated with dead organisms or organic waste.

Reading Check

7. Compare What is the main difference between heterotrophs and autotrophs?

____

____

____

In what ways are photoautotrophs like plants?

Autotrophs (AW tuh trohfs) can make their own food. Photoautotrophs, or cyanobacteria, carry out photosynthesis. Like plants, these bacteria live in areas where there is light, such as shallow ponds and streams, in order to make organic molecules to use as food. Also like plants, they are at the base of some food chains and they release oxygen into the environment ☑

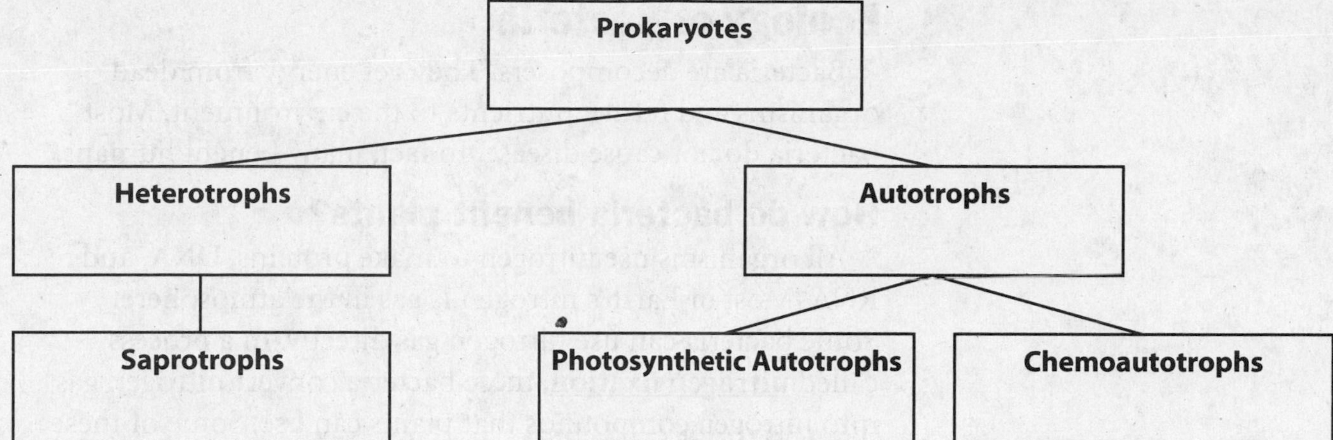

Scientists once thought that these organisms were eukaryotes and called them blue-green algae. Later discoveries showed that they were prokaryotes. Scientists call these organisms cyanobacteria. Cyanobacteria might have been the first organisms to release oxygen into Earth's early atmosphere.

What do chemoautotrophs use to make food?

Chemoautotrophs do not need light. They use the process of chemosynthesis to break down and release inorganic materials that contain nitrogen or sulfur. They help cycle nitrogen and other inorganic materials through ecosystems.

Survival of Bacteria

Bacteria have several ways that they can survive if their environment becomes unfavorable.

When environmental conditions are harsh, some types of bacteria produce a structure called an **endospore**. A spore coat surrounds a copy of the bacterial cell's chromosome and a small part of the cytoplasm. The bacterial cell dies, but the endospore can survive for long periods. An endospore might be able to survive conditions that would kill a bacterium such as extreme heat, cold, or dehydration. When conditions improve, the endospore grows into a new bacterial cell.

How do mutations benefit bacteria?

Mutations are changes or random errors in a DNA sequence. They lead to new genes, new gene combinations, new characteristics, and genetic diversity. Because bacteria reproduce quickly, gene mutations occur quickly. If the environment changes, some bacteria might have the right combination of genes that will enable them to survive and repopulate.

Picture This

8. Label each type of heterotroph and autotroph in the figure with the way it obtains nutrients.

Think it Over

9. Draw Conclusions What might happen if bacteria did not mutate and the environment changed?

Ecology of Bacteria

Bacteria are decomposers. They get energy from dead organisms and return nutrients to the environment. Most bacteria do not cause disease. In fact, many benefit humans.

How do bacteria benefit plants?

All organisms use nitrogen to make proteins, DNA, and RNA. Most of Earth's nitrogen is gas in the atmosphere. Some bacteria can use nitrogen gas directly. In a process called **nitrogen fixation**, these bacteria convert nitrogen gas into nitrogen compounds that plants can use. Some of these bacteria live in soil. Others live in root nodules of plants. Nitrogen is passed on to organisms that eat the plants. ☑

✔ Reading Check

10. Describe how nitrogen fixation benefits plants.

____

____

____

____

Why are bacteria important to humans?

Your body is covered with harmless bacteria called normal flora. Normal flora help prevent harmful bacteria from infecting your body and causing disease.

Some *Escherichia coli* (*E. coli*) bacteria can cause food poisoning. Other *E. coli* bacteria live symbiotically in the digestive tracts of humans and other mammals. These *E. coli* make vitamin K, which humans use for blood clotting. In exchange, the *E. coli* get a warm place with food to live.

Bacteria are used to make many foods such as cheese, yogurt, and pickles. Bacteria break down the covering of cocoa beans during the production of chocolate. Some vitamin pills are made with the help of bacteria. Several common antibiotics were originally made by bacteria.

Picture This

11. Explain Suppose a friend showed you this table and said, "See, bacteria are bad." How would you respond?

____

____

____

How do bacteria cause disease?

The small percentage of bacteria that cause disease do so in two ways. Some bacteria multiply at an infection site and can spread to other parts of the body. Other bacteria secrete a toxin or other substances such as the acid that causes tooth decay. The table below lists some diseases caused by bacteria.

Category	Human Diseases
Sexually transmitted diseases	syphilis, gonorrhea, chlamydia
Respiratory diseases	strep throat, pneumonia, whooping cough, anthrax
Skin diseases	acne, boils, infections of wounds or burns
Digestive tract diseases	gastroenteritis, many types of food poisoning, cholera
Nervous system diseases	botulism, tetanus, bacterial meningitis

Bacteria and Viruses

section 2 Viruses and Prions

Before You Read

Have you ever heard of mad cow disease? On the lines below, write what you know about mad cow disease. In this section, you will read about what causes mad cow disease.

____

____

MAIN Idea

Viruses and prions invade cells and can alter cellular functions.

What You'll Learn

- the general structure of viruses
- how viruses and retroviruses replicate
- how prions cause disease

Read to Learn

Viruses

A **virus** is a non-living strand of genetic material within a protein coat. Some are harmless, while others cause disease in living things. The table below lists some diseases in humans caused by viruses. Viral diseases such as HIV and genital herpes transmitted through sexual contact have no cure or vaccine.

The origin of viruses is not known. One theory, however, is that viruses came from parts of cells. Scientists found that viruses are similar to genes in cells. These genes somehow became able to exist outside the cell.

Mark the Text

Restate the Main Point As you read the section, highlight the main point in each paragraph. Then restate each main point in your own words.

Human Diseases Caused by Viruses

Category	Disease
Sexually transmitted diseases	AIDS (HIV), genital herpes
Childhood diseases	measles, mumps, chicken pox
Respiratory diseases	common cold, influenza
Skin diseases	warts, shingles
Digestive tract diseases	gastroenteritis
Nervous system diseases	polio, viral meningitis, rabies
Other diseases	smallpox, hepatitis

Picture This

1. Contrast How is a cold different from strep throat?

____

____

____

____

How are viruses structured?

The figure below shows the structure of three viruses. Adenovirus causes a cold. Influenza virus causes the flu. A bacteriophage (bak TIHR ee uh fayj) is a virus that infects bacteria. Notice that all three viruses have the same basic structure. Viruses have an outer layer called a **capsid** that is made of protein. Inside the capsid is genetic material, which could be DNA or RNA but not both. Viruses are classified as either DNA or RNA based on the type of genetic material they contain.

Smallpox, caused by a DNA virus, infected humans for thousands of years. A successful worldwide vaccination program eliminated the disease.

FOLDABLES™

Take Notes Make a vocabulary Foldable to take notes and organize information about viruses and prions.

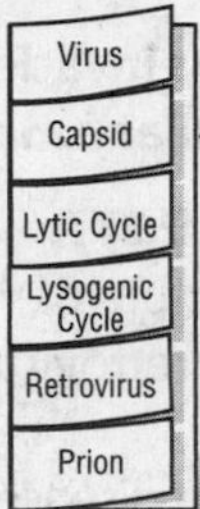

Viral Infection

A virus attaches to the host cell using receptors on the plasma membrane of the host. Different species have receptors for different types of viruses. As a result, many viruses cannot pass from one species to another.

Once attached, the genetic material enters the cytoplasm of the host cell. The virus then replicates by either the lytic cycle or the lysogenic cycle.

How do viruses replicate in the lytic cycle?

In the **lytic cycle**, the host cell makes many copies of the viral RNA or DNA. The viral genes instruct the host cell to make a protein coat around the copies of genetic material, forming capsids. These new viruses leave the cell by exocytosis or by cell lysis—bursting of the cells. The new viruses are then free to infect other cells.

Viruses that replicate by the lytic cycle cause active infections. This means that symptoms start in one to four days. Colds and flu are examples of active infections.

Picture This

2. Highlight the names of the basic structures that all three viruses in the figure have in common.

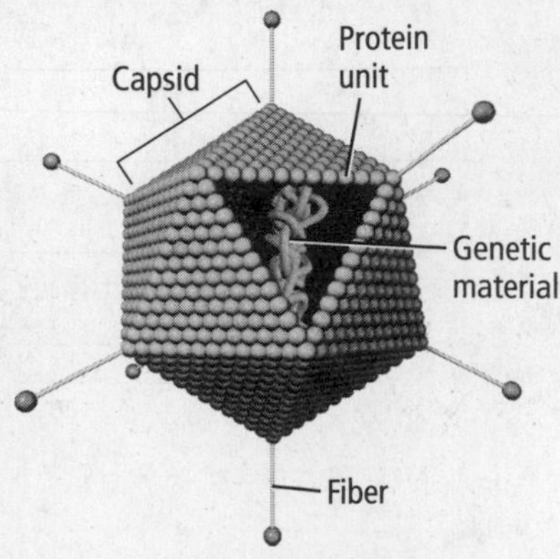

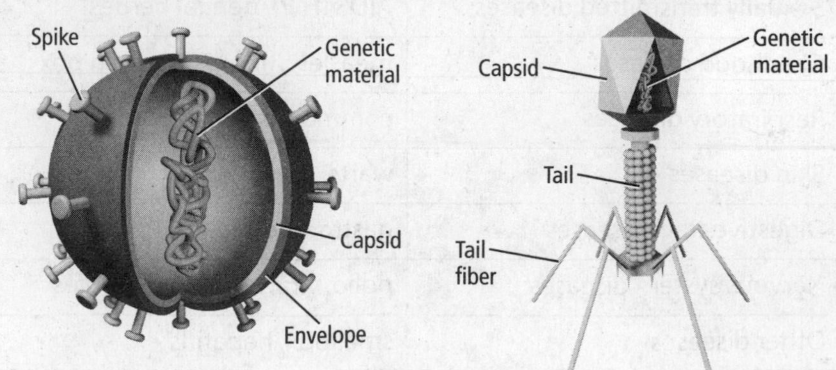

How does the lysogenic cycle differ?

In the **lysogenic cycle**, shown in the figure below, the viral DNA enters the nucleus of the host cell. The viral genes become a permanent part of the host chromosome. The genes might stay inactive for months or years until something activates them. When activated, the viral genes replicate by the lytic cycle and cause active infection.

Picture This

3. Label the active period and the dormant period in the virus's cycle.

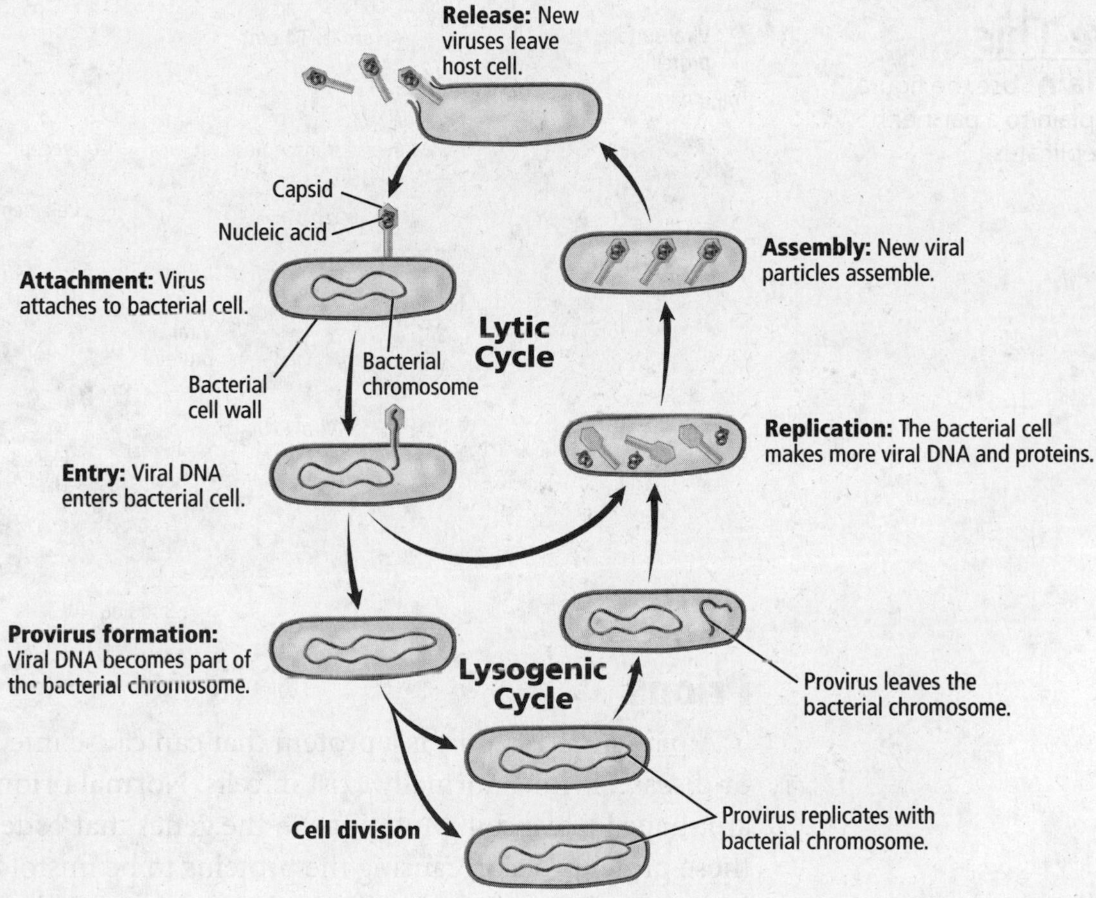

Retroviruses

Some viruses are made of RNA instead of DNA. A **retrovirus** is a type of RNA virus with a replication cycle that has many parts. Human immunodeficiency virus (HIV) is a retrovirus.

What is the structure of HIV?

The structure of HIV is similar to other viruses. It has a protein capsid. Around the capsid is a lipid envelope that it got from the plasma membrane of the host cell. In the core of a retrovirus is RNA and an enzyme called reverse transcriptase. This enzyme makes DNA from the viral RNA. ☑

Reading Check

4. Identify What does reverse transcriptase make?

____

____

How does HIV replicate?

Study the figure below to learn about HIV replication. HIV attaches to a host cell and releases its RNA. Reverse transcriptase makes DNA from the virus's RNA. Then the DNA moves into the nucleus of the host cell and becomes part of a chromosome. The viral DNA might stay inactive for years. Once it is activated, the virus makes RNA from the viral DNA. The host cell then makes new HIV particles.

Picture This

5. Explain Use the figure to explain to a partner how HIV replicates.

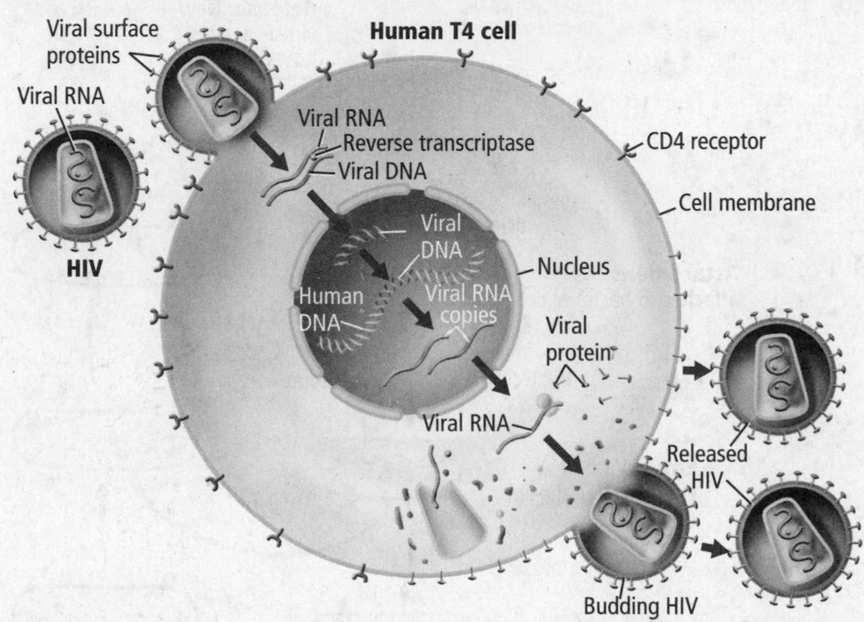

Prions

A **prion** (PREE ahn) is a protein that can cause infection or disease. Prions normally exist in cells. Normal prions are shaped like a coil. Mutations in the genes that code for these proteins occur, causing the proteins to be misfolded. Mutated prions are shaped like a piece of paper folded many times. Mutated prions can cause normal proteins to mutate. Abnormal prions infect and burst nerve cells in the brain, leaving spaces that make the brain look like a sponge. ☑

Reading Check

6. Explain how normal prions become harmful.

____

What is a disease caused by prions?

Prions cause diseases such as "mad cow." Abnormal prions are found in the brains and spinal cords of cattle. Scientists think that spinal cords might be cut during butchering, infecting the meat, and then infecting humans that eat the meat. Other examples caused by prions include Creutzfeldt-Jakob disease (CJD) in humans, scrapie (SKRAY pee) in sheep, and chronic wasting disease in deer and elk.

Protists

section 1 Introduction to Protists

Before You Read

Suppose you want to determine if an organism is more like a plant or more like an animal. On the lines below, write two questions you would ask about the organism to help you decide. Read the section to learn about the animal-like, plantlike, and funguslike characteristics of protists.

____

____

____

MAIN Idea

Protists are subdivided based on their method of obtaining nutrition.

What You'll Learn

- how protists might have evolved into organisms with organelles
- why the organization of Kingdom Protista might change

Read to Learn

Protists

Protists are more easily classified by what they are not than by what they are. Protists are not animals, plants, or fungi. Members of the Kingdom Protista are a diverse group of more than 200,000 members. All protists share one important trait—all are eukaryotes. This means that their cells contain membrane-bound organelles.

Mark the Text

Identify Class Characteristics Highlight each protist class as you read about it. In another color, highlight the characteristics of that class.

How are protists classified?

Some scientists classify protists by the way they obtain nutrition. Using this method, protists are divided into three groups: animal-like protists, plantlike protists, and funguslike protists. The table on the next page summarizes these groups.

What are animal-like protists?

A **protozoan** (proh tuh ZOH un) (plural, protozoa or protozoans) is a one-celled, animal-like protist. It is classified as animal-like because it is a heterotroph. This means that it must eat other organisms. Protozoans, such as the amoeba, usually eat bacteria, algae, or other protozoans to obtain energy.

Think it Over

1. **Explain** why a protist with chloroplasts would not be classified as an animal-like protist.

____

____

____

Why are algae considered plantlike?

Plantlike protists are known as algae (AL jee) (singular, alga). They are plantlike because they make their own food through photosynthesis. Algae can have one or many cells.

How do funguslike protists differ from fungi?

Both funguslike protists and fungi absorb nutrients from other organisms. These protists are not true fungi because they have centrioles, and fungi do not. Centrioles are small, cylindrical organelles used in mitosis. Also, the cell walls of fungi and funguslike protists contain different materials.

Picture This

2. Synthesize You can determine the group to which a protist belongs by its method of

a. moving.
b. obtaining nutrients.
c. reproducing.

Protist Group	Examples	Characteristics
Animal-like protists (protozoans)	amoebas, ciliates	animal-like because they eat other organisms; some are parasites
Plantlike protists (algae)	giant kelp, algae	plantlike because they make their own food through photosynthesis
Funguslike protists	slime mold, water mold	funguslike because they feed on decaying organic matter and absorb nutrients through cell walls

Where do protists live?

Protists usually live in damp or water environments, such as ponds, streams, oceans, decaying leaves, and damp soil. Some live in symbiotic or parasitic relationships with other organisms. **Microsporidia** (MI kroh spo rih dee uh) are microscopic protozoans. They cause disease in insects, and some can be used as insecticides.

Origin of Protists

The theory of endosymbiosis suggests that eukaryotes, including protists, formed when a large prokaryote engulfed a smaller one. The two lived symbiotically. Eventually, the two organisms evolved into a single, more highly developed organism. Some scientists suggest that the mitochondria and chloroplasts in some eukaryotes were once individual organisms. Protists might have been the first eukaryotes billions of years ago. ☑

As scientists learn more about protists' evolutionary history, the organization of the Kingdom Protista will likely change. Scientists think that protists evolved from a common ancestor. Mitochondria became part of protist cells early in evolution. Chloroplasts entered later. Algae are the only protists with chloroplasts. Thus, algae are the only protists that perform photosynthesis.

Reading Check

3. Identify two structures within protists that might have been individual organisms at one time.

____

____

____

Protists

section 2 Protozoans—Animal-like Protists

Before You Read

Have you ever tried to grab an object that was sinking to the bottom of a swimming pool? On the lines below, describe how you captured the object. Read the section to learn how tiny amoebas surround food particles in their watery home.

____

____

MAIN Idea

Protozoans are animal-like, heterotrophic protists.

What You'll Learn

- the structures and organelles of protozoans
- the life cycles of protozoans

Read to Learn

Ciliophora

Protozoans are grouped into many phyla. One way that scientists group protozoans is by the way they move. Members of the phylum Ciliophora (sih lee AH fuh ruh) are known as ciliates (SIH lee ayts). They are animal-like protists with short, hairlike projections called cilia (singular, cilium). Cilia can cover their whole bodies, or only part of the membrane. Ciliates use their cilia to propel themselves through water. Some also use cilia to pull in food. Ciliates are common in oceans, lakes, and rivers. Many are also found in mud.

What are paramecia?

Members of the genus *Paramecium* (per uh MEE see um) (plural, paramecia) are one-celled protozoans. As you read about the structures of a paramecium, identify them in the figure on the next page.

A membrane called a **pellicle** encloses a paramecium. Beneath the pellicle is a layer of cytoplasm called ectoplasm. Within the ectoplasm are the **trichocysts** (TRIH kuh sihsts). These long, cylinder-shaped bodies discharge spinelike structures. Scientists are not sure of the purpose of trichocysts. A paramecium might use them for defense, as a reaction to injury, to anchor itself, or to capture prey. ☑

Study Coach

Create a Quiz As you read this section, write quiz questions based on what you have learned. After you write the questions, answer them.

FOLDABLES™

Take Notes Make a four-door Foldable, as shown below. As you read, take notes and organize what you learn about the classification of protists, based upon the way they move.

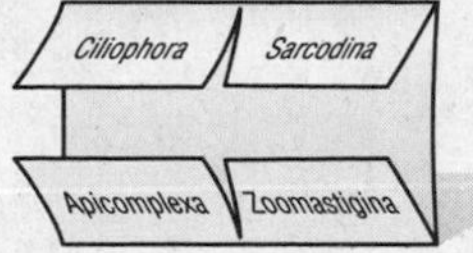

How does a paramecium feed and digest food?

Cilia completely cover the paramecium. A paramecium uses its cilia to feed and move. Cilia along the oral groove guide food into the gullet. At the end of the gullet, food is enclosed in a food vacuole. Enzymes within the food vacuole break down food into nutrients. The nutrients then diffuse into the cytoplasm. Wastes exit through the anal pore.

What is the function of contractile vacuoles?

Water constantly enters the paramecium by osmosis. The **contractile vacuoles** (kun TRAK tul • VAK yuh wohlz) collect the extra water and expel it, along with wastes, from the cell. This helps maintain homeostasis in the cell.

Picture This

1. Highlight the structures of the paramecium that are involved in feeding and digestion.

Structure of a Paramecium

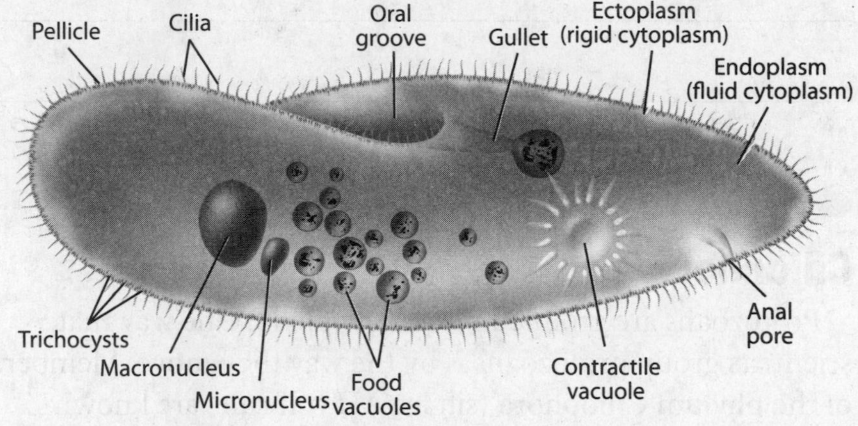

Conjugation

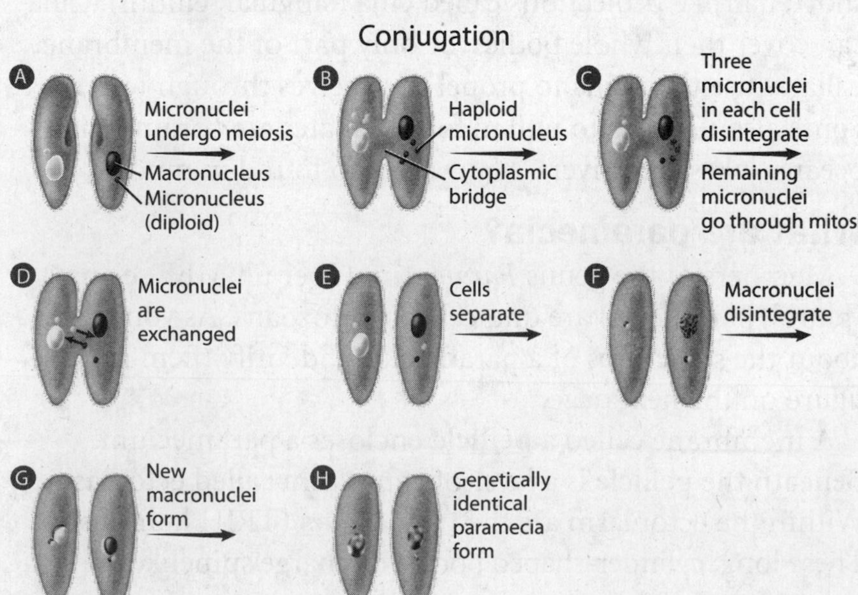

How do ciliates reproduce?

All known ciliates have two types of nuclei—the macronucleus and the micronucleus. A cell can have more than one of each type. Both nuclei contain the genetic information for the cell. The macronuclei control the everyday life functions such as feeding and maintaining water balance. The micronuclei are used for reproduction.

Ciliates reproduce asexually by binary fission. They do not divide by mitosis. Instead, the macronucleus grows longer and splits. Most ciliates exchange genetic information in a sexual process called conjugation. Conjugation is not sexual reproduction because new individuals do not result from it.

During conjugation, two paramecia form a bridge of cytoplasm. Their diploid micronuclei undergo meiosis. After three of the new micronuclei dissolve, the remaining micronucleus undergoes mitosis. One micronucleus from each connected cell is exchanged. The two paramecia then separate. The macronucleus dissolves in each paramecium. The micronuclei then combine to form a new diploid macronucleus with a new combination of genetic information.

Think it Over

2. Compare How is conjugation different from binary fission?

______

______

______

______

Sarcodina

Sarcodines (SAR kuh dinez) are members of the phylum Sarcodina (sar kuh DI nuh). They are animal-like protists that use pseudopods for feeding and movement. A **pseudopod** (SEW duh POD) is a temporary extension of cytoplasm. The figure below shows how an amoeba forms pseudopodia to take in food. Most amoebas live in salt water, but some live in streams, the muddy bottoms of ponds, and in damp moss and leaves. Other amoebas are parasites that live inside a host animal.

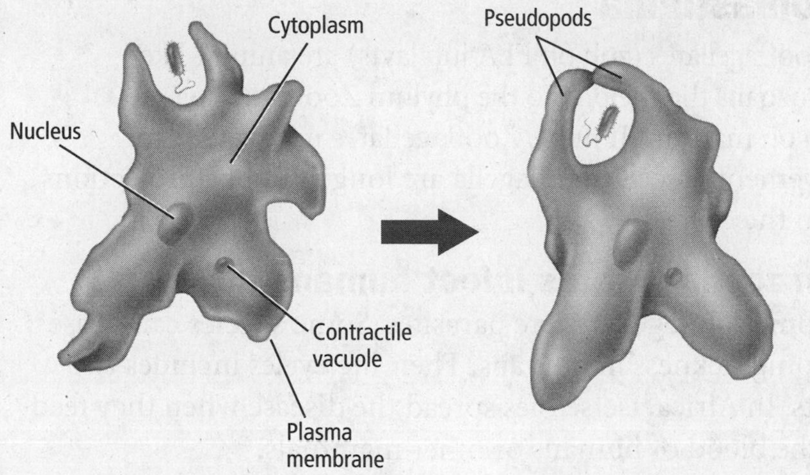

Picture This

3. Explain What happens to the pseudopods from the left figure to the right figure?

______

______

What are the structures of an amoeba?

Amoebas have an outer plasma membrane and an inner thickened membrane called ectoplasm. Inside the outer membrane is cytoplasm. The cytoplasm contains a nucleus, food vacuoles, and sometimes contractile vacuoles. Amoebas do not have an anal pore like paramecia. Amoebas excrete wastes and take in oxygen through their outer membranes by diffusion. Some species have a **test**—a hard, porous covering similar to a shell that surrounds the plasma membrane. ☑

Reading Check

4. Explain how an amoeba excretes wastes without an anal pore.

____

____

____

How does an amoeba reproduce?

An amoeba reproduces asexually. It divides into two identical offspring. In harsh living conditions, some amoebas form cysts that can survive until conditions improve.

Apicomplexa

Members of phylum Apicomplexa (ay puh KOM pleks uh) are called sporozoans (spo ruh ZOH unz) because they can reproduce through spores. Spores are reproductive cells that form without fertilization and produce a new organism. Sporozoans lack contractile vacuoles and structures for movement. Like in amoebas, respiration and excretion occur by diffusion.

All sporozoans are parasites. Organelles at one end can penetrate the host tissues. The sporozoan then obtains its nutrition from the host.

A sporozoan has both sexual and asexual stages in its life cycle. It often needs two or more hosts to complete its life cycle. Members of the genus *Plasmodium* can cause malaria in humans. Malaria is a serious disease passed to humans by mosquitoes.

Zoomastigina

Zooflagellates (zoh oh FLA juh layts) are animal-like protozoans that belong to the phylum Zoomastigina (zoh oh mast tuh JI nuh). Zooflagellates use flagella for movement. Recall that flagella are long whiplike projections from the cell.

Think it Over

5. Draw Conclusions How does a tsetse fly become infected with zooflagellates?

____

____

____

Can zooflagellates infect humans?

Some zooflagellates are parasites. Some species can cause sleeping sickness in humans. Their life cycles includes two hosts. In Africa, tsetse flies spread the disease when they feed on the blood of humans or other mammals.

Protists

section 3 Algae—Plantlike Protists

Before You Read

You probably come into contact with algae every day. Algae are used in foods, household cleaners, and other products. Some algae are used to make liquid foods thicker. On the lines below, list some foods that might contain algae. Read the section to learn about algae and their many uses.

____

____

MAIN Idea

Algae are plantlike, autotrophic protists.

What You'll Learn

- the characteristics of several phyla of algae
- the secondary photosynthetic pigments of some algae
- the differences between diatoms and most other algae

Read to Learn

Study Coach

Restate the Main Point As you read this section, stop after each paragraph, and put the main ideas into your own words.

Characteristics of Algae

Algae (singular, alga) are plantlike protists. Like plants, they have photosynthetic pigments. These pigments enable the algae to make their own food using energy from the Sun. This process is called photosynthesis. Algae are not plants because they lack roots, leaves, and other plant structures.

The pigments are found in the algae's chloroplasts. In many algae, the primary pigment is chlorophyll. This is the same pigment that gives plants their green color. Many algae also have secondary pigments. These pigments allow algae to absorb light energy in deeper water. Because of these secondary pigments, algae occur in a variety of colors.

Diversity of Algae

Scientists use three characteristics to classify algae. They group algae by pigment types, method of food storage, and composition of the cell wall. Algae can have one cell or many cells. Some one-celled algae are called phytoplanktons—meaning "plant planktons." Phytoplanktons provide the base of the food web in their aquatic ecosystem. They also produce much of the oxygen in Earth's atmosphere. ☑

Reading Check

1. **List** two reasons why phytoplankton are important.

____

____

____

Reading Check

2. Describe How do the halves of a diatom fit together?

What are diatoms?

Diatoms are one-celled algae of the phylum Bacillariophyta (BAH sih LAYR ee oh FI tuh). Diatoms have shells with two halves. One half fits inside the other, forming a box. ☑

Diatoms produce food using chlorophyll and secondary pigments called carotenoids. Carotenoids make diatoms appear golden yellow. Diatoms store their food as oil. The oil makes them a nutritious food source for other marine life. The oil also helps diatoms float close to the surface, where they can absorb energy from the Sun.

How do diatoms reproduce?

The figure below shows how diatoms reproduce both sexually and asexually. In asexual reproduction, the two shell halves separate. They create a new half that can fit inside the old one. This process produces smaller diatoms. When a diatom is about one-quarter of the original size, sexual reproduction begins. The diatom produces gametes that fuse to form a zygote. The zygote develops into a full-sized diatom, and the reproduction cycle repeats.

How do humans use the remains of diatoms?

When diatoms die their silicon shells fall to the ocean floor. The sediment is known as diatomaceous earth. This sediment can be collected and used as an abrasive and a filtering agent. It gives toothpaste its gritty texture.

Picture This

3. Identify Diatom asexual reproduction begins with

a. meiosis.

b. mitosis.

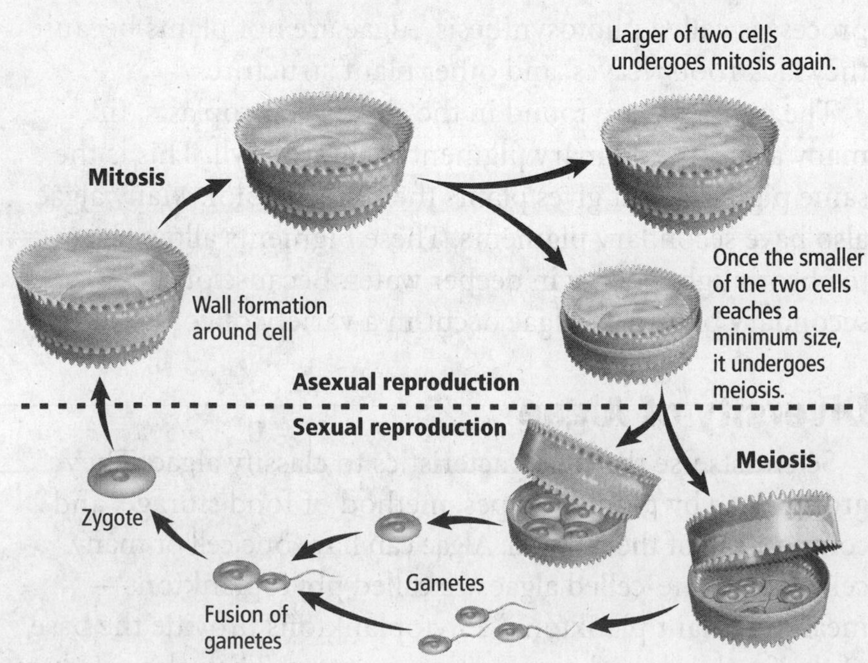

How do dinoflagellates move?

Dinoflagellates (DI nuh fla juh layts) are plantlike protists of the phylum Pyrrophyta (puh RAH fuh tuh). Most are one cell and have two flagella at right angles to one another. The beating flagella cause dinoflagellates to spin as they move through the water.

Bioluminescent (BI oh lew muh NE sunt) dinoflagellates emit light. Most dinoflagellates live in salt water as part of the phytoplankton. Some dinoflagellates are photosynthetic autotrophs. Others are heterotrophs. They can be carnivorous (meat eating), parasitic, or symbiotic.

Algal blooms When dinoflagellates have plenty of food and favorable conditions, they reproduce in great numbers. These population explosions are called algal blooms. Algal blooms deplete the nutrients in the water. When food supplies decrease, dinoflagellates die. As they decompose, they deplete oxygen in the water, killing other sea life. ☑

Red tides Some dinoflagellates have red pigments. Their blooms are called red tides because they make the ocean look red. Some species produce a poisonous substance. People can die from eating shellfish that have eaten these species.

How do euglenoids differ from plants?

Euglenoids (yoo GLEE noydz) are members of the phylum Euglenophyta (yoo gluh NAH fuh tuh). They are plantlike protists with only one cell. Most live in shallow freshwater. Like plants, most euglenoids carry out photosynthesis. Like animals, some euglenoids consume other organisms.

Unlike plants, euglenoids do not have a cell wall. Instead, they are covered by a flexible, tough outer pellicle, similar to a paramecium. They have an eyespot to detect light and flagella to move toward food or light. The contractile vacuole expels water to maintain homeostasis.

What pigments color the chrysophytes?

Members of the phylum Chrysophyta (KRIS oh fuh tuh) are called chrysophytes (KRIS oh fytz). They get their yellow or golden brown color from carotenoids. Most chrysophytes are one-celled, but some form a group of cells that join together in close association called a **colony**. Chrysophytes have two flagella attached at one end. All perform photosynthesis. They usually reproduce asexually, and they live in both freshwater and salt water.

Reading Check

4. Explain what triggers dinoflagellates to bloom.

______

______

______

Think it Over

5. Draw Conclusions A euglenoid has an eyespot that detects light. Why is the ability to detect light important to the euglenoid?

______

______

______

What phylum includes brown algae?

Brown algae belong to the phylum Phaeophyta (FAY oh FI tuh). Brown algae have many cells and a pigment called fucoxanthin (fyew ko ZAN thun) that gives them their brown color. A common example of this phylum is kelp. Most species of brown algae live along rocky coasts in cool areas. The kelp has a bulblike structure called a bladder. The bladder is filled with air, helping the kelp float near the surface where it can use sunlight for photosynthesis.

How are green algae like plants?

Green algae, of phylum Chlorophyta (kloh RAH fuh tuh), share some traits with plants. Both contain chlorophyll, have cell walls, and store their food as carbohydrates. ☑

Most green algae live in freshwater. Some live on damp ground, in snow, and even in the fur of some animals.

Green algae can have one cell or many cells. The one-celled species Volvox forms colonies that look like a hollow ball. Gelatinlike strands of cytoplasm hold the colony together. The flagella of all cells beat at the same time to move the colony. Smaller daughter colonies form balls inside the larger colony. When the daughter cells have matured, they digest the parental cell and become free-swimming.

Reading Check

6. Summarize What are three characteristics of both green algae and plants?

______

______

______

What pigment gives red algae their color?

Red algae, of phylum Rhodophyta (roh DAH fuh tuh), get their red color from the pigment phycobilins. This pigment enables red algae to absorb the light that can penetrate deep water. As a result, red algae can live in deeper water than other algae. Some red algae help form coral reefs.

Picture This

7. Draw Conclusions How do you think sea lettuce got its nickname?

______

______

______

Uses for Algae

The high protein content of algae makes them a nutritious food for animals and people. As shown in the table below, algae are used in many foods and in other products.

Type of Algae	Uses
Red algae	soups, sauces, sushi, and pie fillings; scientific gels; preservatives in canned meat and fish; thickening agent for puddings and shampoos
Brown algae	stabilizers for syrups, ice cream, and paints; eaten with meat or fish and in soups
Green algae	sea lettuce eaten in salads, soups, relishes, and with meat or fish
Diatoms	used for filtering beverages, oils, water supplies; used as abrasives

Life Cycle of Algae

Algae can alternate between forms that produce spores and forms that produce gametes. They can reproduce sexually or asexually. Green algae can also reproduce through fragmentation. In fragmentation, the green alga breaks into pieces, and each piece grows into a new individual.

What is the life-cycle pattern of algae?

The life cycles of many algae show a pattern called alternation of generations. The figure below illustrates this pattern for sea lettuce. **Alternation of generations** is a life cycle of algae that takes two generations to complete. One generation reproduces sexually. The other generation reproduces asexually. Algae alternate between a diploid (2*n*) form and a haploid (*n*) form. Each form is a generation. ☑

Reading Check

8. Compare How do the two generations differ in the way they reproduce?

___

___

How does alternation of generations work?

Follow the arrows in the figure below to review how the two generations alternate. The haploid form is called the gametophyte generation because it produces gametes. Gametes from two individuals join to form a zygote with two complete sets of chromosomes. The zygote develops into the sporophyte (2*n*). Some cells in the sporophyte divide by meiosis and become haploid spores (*n*). Spores develop into gametophytes that continue the cycle.

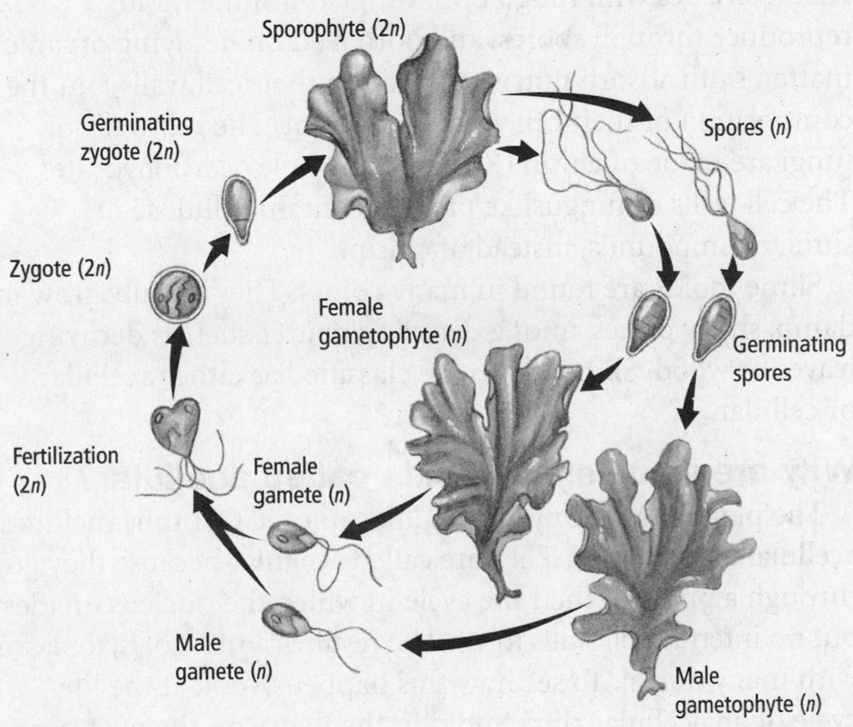

Picture This

9. Label both the haploid and diploid forms of the sea lettuce.

Protists

section 4 Funguslike Protists

MAIN Idea

Funguslike protists obtain their nutrition by absorbing nutrients from dead or decaying organisms.

What You'll Learn

- the characteristics of cellular and acellular slime molds
- the life cycle of cellular and acellular slime molds

Before You Read

Have you seen commercials about household cleaners that get rid of molds and mildews? From what you have observed, where are molds and mildews most often found? On the lines below, list the conditions that might make a good environment for molds and mildews. Read the section to learn how these funguslike protists thrive and reproduce.

____

____

Study Coach

Make an Outline Make an outline of the information you learn in this section. Start with the headings. Include the boldface terms.

Read to Learn

Slime Molds

Slime molds are funguslike protists. They share some characteristics with fungi. Both fungi and slime molds reproduce through spores, and both feed on decaying organic matter. Both absorb nutrients through their cell walls, but the composition of their cell walls is different. The cell walls of fungi are made of chitin (KI tun), a complex carbohydrate. The cell walls of funguslike protists contain cellulose or similar compounds, instead of chitin.

Slime molds are found in many colors. They usually grow in damp, shady places among decaying matter such as decaying leaves or wood. Slime molds are classified as either acellular or cellular.

Think it Over

1. Draw Conclusions From what you know about acellular slime molds, what do you think the prefix "a" means? (Circle your answer.)

a. with
b. without

Why are some slime molds called acellular?

The phylum Myxomycota (mihk soh mi COH tuh) includes acellular slime molds. They are called acellular because they go through a phase in their life cycle in which the nucleus divides, but no internal cell walls form. The result is a mass of cytoplasm with many nuclei. To see how this happens, look at the life cycle of an acellular slime mold in the figure on the next page.

Life Cycle of Acellular Slime Molds

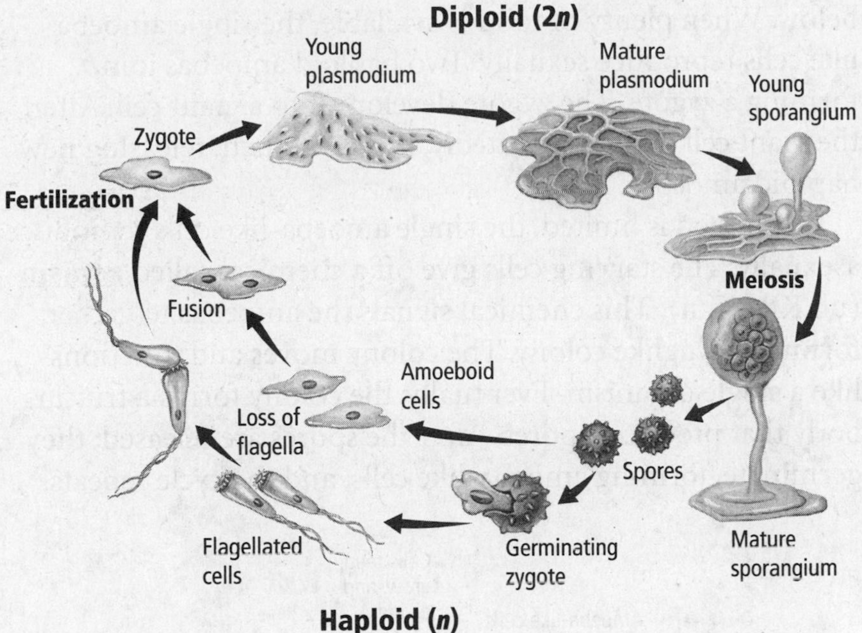

How do acellular slime molds reproduce?

As you can see in the figure above, acellular slime molds begin as spores, usually in harsh conditions. Spores form without fertilization and can produce a new organism. When water is present, the spore produces cytoplasm and flagella. The flagella propel the cell until it contacts a surface, then draw into the cell. The cell forms pseudopods that allow it to move like an amoeba. Both the flagellated cell and the amoeba-like cell are haploid (n) gametes. ☑

When two gametes join, the next phase begins. The nuclei of the fertilized cells divide repeatedly, forming a plasmodium. A **plasmodium** (plaz MOH dee um) is a mass of cytoplasm with many diploid nuclei but no separate cells. This is the feeding stage. The organism creeps through decaying leaves or wood to feed. When food or moisture becomes limited, the slime mold produces spores that are spread by the wind. When the spores are in water, the cycle repeats.

How do cellular slime molds differ?

The phylum Acrasiomycota (uh kray see oh my COH tuh) contains cellular slime molds. These funguslike protists spread over moist soil, feeding on bacteria. Unlike acellular slime molds, they spend most of their life cycle as single amoeba-like cells. They also have no flagella.

Picture This

2. Label Place a *1* beside the structure that forms without fertilization. Place a *2* beside the structure that forms when the fertilized cell divides repeatedly.

Reading Check

3. Explain how spores produce a new organism.

What is the life cycle of cellular slime molds?

The life cycle of cellular slime molds is shown in the figure below. When plenty of food is available, the single amoeba-like cells reproduce sexually. Two haploid amoebas join, forming a zygote. The zygote develops into a giant cell. After the giant cell divides repeatedly, it breaks open, releasing new haploid amoebas.

When food is limited, the single amoeba-like cells reproduce asexually. The starving cells give off a chemical called **acrasin** (uh KRA sun). This chemical signals the amoebas to gather, forming a sluglike colony. The colony moves and functions like a single organism. Eventually, the colony forms a fruiting body that produces spores. After the spores are released, they germinate, forming amoeba-like cells, and the cycle repeats.

Picture This

4. Explain Use the figure to explain the life cycle of cellular slime molds to a partner.

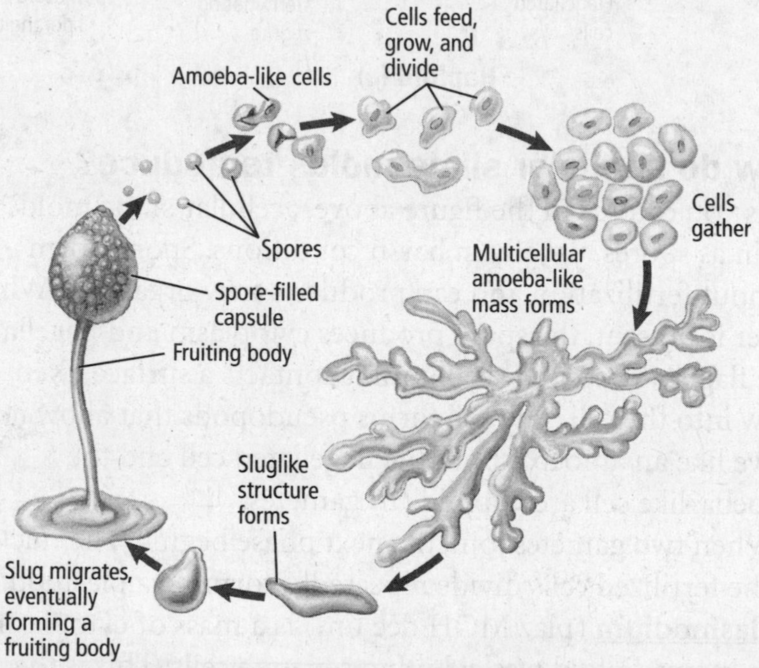

Water Molds and Downy Mildew

Water molds and downy mildews are members of phylum Oomycota (oo oh my COH tuh). Most live in water or damp places. Some absorb nutrients from the water or soil around them. Others obtain nutrients from other organisms.

Like fungi, water molds enclose their food with a mass of threads and absorb nutrients through their cell walls. Water molds differ from fungi in two main ways. Their cell walls are made of different materials, and the reproductive cells of water molds have flagella. ☑

Reading Check

5. State how water molds obtain nutrients.

Fungi

section 1 Characteristics of Fungi

Before You Read

On the lines below, describe places where you have seen mushrooms growing. In this section you will learn how mushrooms and other fungi live.

____

____

MAIN Idea

Fungi have unique characteristics.

What You'll Learn

- how decomposers and other fungi obtain nutrients
- three types of asexual reproduction in fungi

Read to Learn

Characteristics of Fungi

Members of the kingdom Fungi (FUN ji) (singular, fungus) are some of the oldest organisms on Earth. More than 10,000 species of fungi have been identified. Mushrooms on your pizza and mold growing on an old loaf of bread belong to this kingdom. All fungi are eukaryotic—their cells have a true nucleus and they have membrane-bound organelles. Also, all fungi are heterotrophs, which means that they cannot make their own food. Instead, they get nutrients from organic matter.

Mark the Text

Read for Understanding As you read this section, highlight any sentence that you do not understand. Reread the highlighted sentences to make certain that you understand their content. Ask your teacher to help you with anything that you still do not understand.

Are multicellular fungi really plants?

Most fungi, such as mushrooms, have many cells. Multicellular fungi can look like plants. Scientists have determined, however, that fungi are different enough to belong in their own kingdom.

What are unicellular fungi?

A fungus composed of one cell is called a yeast. Yeasts grow in soils, on plant surfaces, in sugary substances, and in the human body. Yeasts are used to make bread and beer.

Major Features in Fungi

Scientists have compared the features of plants and fungi. The features that distinguish fungi from plants include their cell walls, their body filaments, and their cross-walls.

FOLDABLES™

Take Notes Make a layered Foldable, as shown below. As you read, take notes and organize what you learn about the characteristics of fungi.

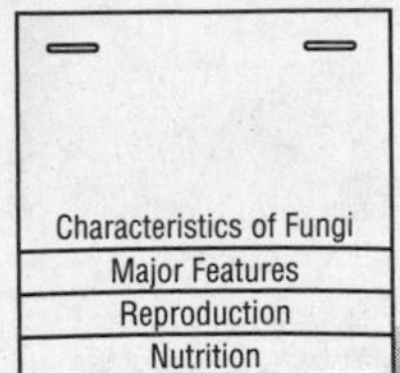

How do cell walls differ in plants and fungi?

The cell walls of plants are made up of cellulose. The cell walls of fungi are made up of **chitin** (KI tun)—a polysaccharide. Recall that polysaccharides are complex carbohydrates composed of many simple sugars.

Think it Over

1. Apply When you see a mushroom growing in the woods, what part of the fungus are you seeing? (Circle your answer.)

a. mycelium
b. fruiting body

Of what are the bodies of fungi made?

Fungi are made up of **hyphae** (HI fee) (singular, hypha), which are tubelike filaments. Hyphae grow from their tips and branches to form a netlike mass called a **mycelium** (mi SEE lee um) (plural, mycelia). The mycelium grows underground. It provides the fungus with nutrients. The part of the fungus above ground is the **fruiting body**. The fruiting body's function is reproduction. It is also composed of hyphae.

How do materials move through the septa?

In many fungi, cross-walls called **septa** (singular, septum) divide hyphae into cells. Pores in the septa allow materials, such as nutrients and cytoplasm, to flow between cells. Some fungi are aseptate—they have no septa. In these fungi, materials flow freely through the hyphae.

Nutrition in Fungi

Unlike humans, fungi digest their food before they consume it. As shown below, the digestive enzymes of fungi break down food into smaller molecules outside their bodies. The fungi then absorb the nutrients through their cell walls. All fungi feed on organic matter, but they obtain nutrients in different ways.

Which fungi are decomposers?

Saprophytic fungi are decomposers. They recycle nutrients from dead organisms or organic wastes back into food webs.

Picture This

2. Label two septa in the diagram.

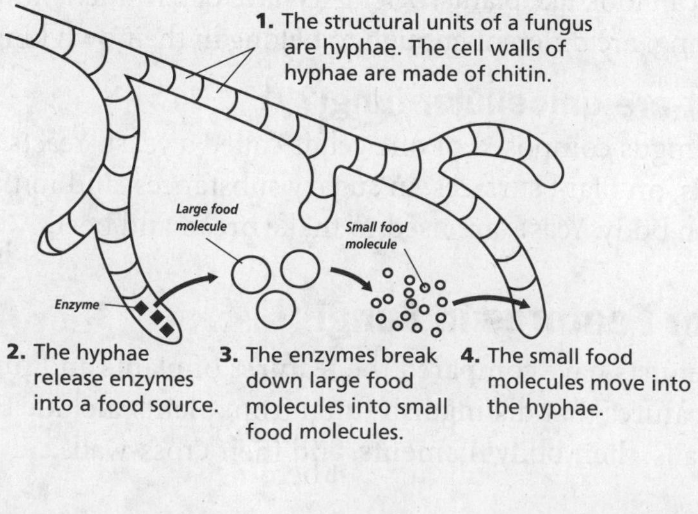

How do parasitic fungi get nutrients from hosts?

A parasitic fungus absorbs nutrients from a living host. Special hyphae called **haustoria** (haws toh REE ah) grow into the host's tissues and absorb nutrients.

What fungi live in cooperative relationships?

Fungi that live in cooperative relationships with other organisms are mutualists. For example, a fungus that covers the root of a plant takes sugar from the plant while helping the plant absorb minerals and water.

Reproduction in Fungi

Fungi can reproduce sexually or asexually. Budding and fragmentation are forms of asexual reproduction.

How do yeasts reproduce?

Yeasts reproduce asexually by budding. New cells develop while attached to the parent cells. The plasma membrane of a new cell pinches off to partially separate the bud from the parent cell. At maturity, the bud separates from the parent cell.

How does fragmentation occur?

Reproduction by fragmentation occurs when a piece of the mycelium of a fungus is broken off. If the piece lands in good growing conditions, the hyphae will grow into new mycelia.

How do fungi reproduce by spores?

Spores are part of both asexual and sexual reproduction. A **spore** is a reproductive cell with a hard outer coat. Spores develop into new organisms without the fusion of eggs and sperm. In sexual reproduction, fungi produce spores by meiosis.

How are spores adapted for survival?

Fungi produce trillions of spores, increasing the chance that some will find favorable growing conditions. Spores weigh so little that they can be spread by the wind and small animals. Their hard outer coats protect them from weather extremes. ☑

What is one way that fungi are classified?

Fungi are classified by the type of fruiting body, or sporophore, they produce. For example, some fungi produce spores in a sac or case called a **sporangium** (spuh RAN jee uhm) (plural, sporangia). The sporangium protects the spores and keeps them from drying out before they are released.

Think it Over

3. Contrast How are mutualists different from other parasitic fungi?

____

____

____

____

☑ Reading Check

4. List three features of spores that help them survive.

____

____

____

Fungi

section 2 Diversity of Fungi

MAIN Idea

Fungi are grouped into four major phyla.

What You'll Learn

- the traits that make each fungi phylum different from the others
- how members of each fungi phylum reproduce

Before You Read

Do you group similar clothes together in drawers? On the lines below, explain what you do with a unique piece of clothing, such as a bathing suit, that doesn't fit your pattern of organization. In this section, you will learn what scientists do with fungi that don't seem to fit into a major phylum.

____

____

Mark the Text

Identify Phyla Circle each phylum name as you come to it. Then highlight the characteristics of the phylum as you read about each one.

Read to Learn

Classification of Fungi

Biologists group fungi into phyla based on their structures and methods of reproduction. The four major phyla are Chytridiomycota, Zygomycota, Ascomycota, and Basidiomycota.

Fungi probably appeared on land the same time as plants. However, scientists found molecular evidence that fungi are more closely related to animals than to plants. Fungi and animals might share a common ancestor.

Chytrids

The phylum Chytridiomycota (ki TRIHD ee oh mi koh tuh) are often called chytrids (KI trihdz). Some chytrids are saprophytes, or decomposers. Others are parasites. Most are aquatic chytrids and have a unique feature—their spores have flagella. For this reason, scientists first thought that chytrids were a type of protist. ☑

Reading Check

1. Contrast How are chytrids different from all other fungi?

____

____

____

Studies of chytrid molecules, however, suggest a closer relationship to fungi than to protists. Also, like other fungi, chytrids have chitin in their cell walls. Scientists think that chytrids might have been the first fungi. Chytrids could be the evolutionary link between protists and fungi.

Common Molds

Molds found on bread and other foods belong to the phylum Zygomycota (zi goh mi KOH tuh). Most molds live on land. Many are mutualists—they live in cooperative relationships with plants.

How do molds obtain nutrients?

Molds form special types of hyphae called stolons and rhizoids. **Stolons** (STOH lunz) are hyphae that spread across the surface of food. **Rhizoids** (RIH zoydz) drill into food and absorb the nutrients. Rhizoids also anchor the mycelium and produce digestive enzymes. Zygomycetes can also be found on decaying plants and animal remains. ☑

How do molds reproduce asexually?

Zygomycetes reproduce both asexually and sexually, as shown in the figure below. Asexual reproduction begins when upright hyphae, called sporangiophores, form sporangia at their tips. Each sporangium releases thousands of spores. As the wind spreads the spores, some fall into favorable growing conditions and produce new hyphae.

What causes molds to reproduce sexually?

Sometimes growing conditions in the environment become unfavorable. For example, food can become scarce. Unfavorable growing conditions can cause molds to reproduce sexually.

Reading Check

2. Describe three main functions of rhizoids.

____

____

Picture This

3. Identify the structure that produces spores in both sexual and asexual reproduction. (Circle your answer.)

a. zygospore
b. sporangium
c. gametangium

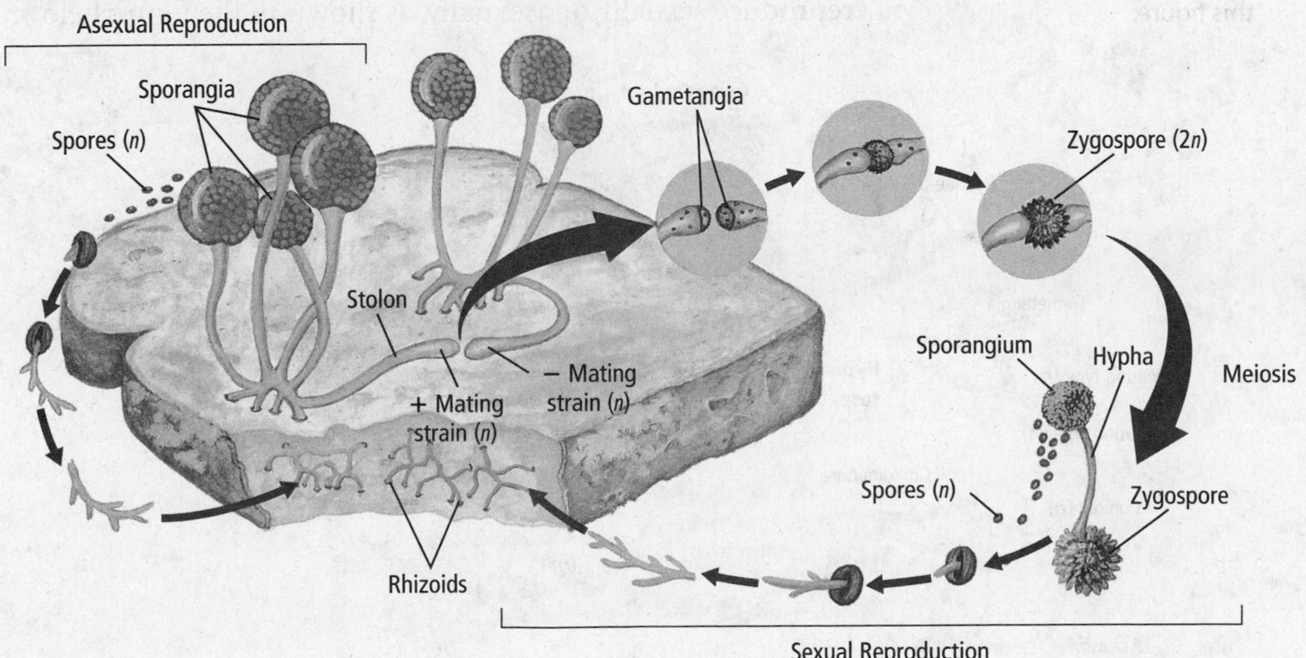

How does sexual reproduction take place?

Fungi are not divided into males and females. Instead, they have plus (+) and minus (–) mating strains. The hyphae of a plus mating strain and a minus mating strain grow together. Each hypha produces a reproductive structure called a gametangium (ga muh TAN jee um) (plural, gametangia). The haploid nuclei of the gametangia fuse to form a diploid zygote. The zygote develops a tough outer coat and becomes a dormant zygospore.

Think it Over

4. Infer What adaptive advantages are evident in zygospores?

____

____

____

How long can a zygospore remain dormant?

A zygospore can remain dormant for months. When conditions in the environment improve, the zygospore germinates. It undergoes meiosis to produce hyphae with sporangia. Each haploid spore in the sporangia can grow into a new mycelium.

Can molds cause disease?

Molds can cause infection called zygomycosis in humans and animals. The molds can infect respiratory passages, lungs, intestines, and skin. Cases of zygomycosis are rare.

Sac Fungi

The phylum Ascomycota (AS koh mi koh tuh) is the largest phylum of fungi containing over 60,000 species. Its members are often called sac fungi or ascomycetes. Unicellular yeasts belong to this phylum, as do many multicellular species. Sac fungi can reproduce sexually or asexually, as shown in the figure below.

Picture This

5. Identify Add an appropriate title to this figure.

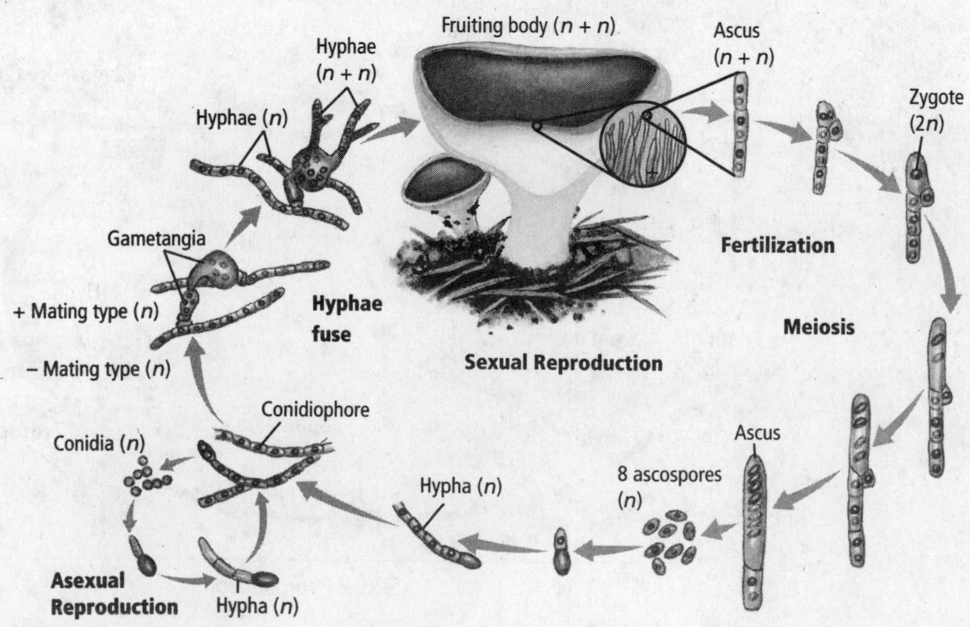

How do sac fungi reproduce asexually?

In asexual reproduction, they form special hyphae called conidiophores. **Conidiophores** (koh NIH dee uh forz) produce chains or clusters of spores called conidia at their tips. Wind, water, and animals spread the spores.

How do sac fungi reproduce sexually?

In sexual reproduction, the hyphae from plus and minus mating strains fuse. One nucleus from each strain pairs off in a separate cell. The cell contains two haploid nuclei.

As the hyphae continue to grow, they develop reproductive structures called **ascocarps**. Within an ascocarp, the haploid nuclei fuse to form a zygote. The zygote divides by meiosis and then by mitosis. The result is eight haploid nuclei. These nuclei develop into spores in a saclike **ascus**. The spores produced in the ascus are called **ascospores**. When growing conditions are favorable, the ascospores develop a mycelium.

Think it Over

6. Draw Conclusions Why do you think ascomycetes are called sac fungi?

____

____

____

Club Fungi

The members of the phylum Basidiomycota are mushrooms, or club fungi. They are called basidiomycetes (buh SIH dee oh mi see teez). Basidiomycetes can be saprophytic, parasitic, or mutualistic. Saprophytic species are major decomposers of wood.

How do club fungi reproduce?

Club fungi rarely reproduce asexually. They reproduce sexually by forming **basidiocarps** (buh SIH dee oh karpz), or fruiting bodies. The mushrooms you see in a grocery store or growing in the woods are the basidiocarps.

Basidiocarps grow quickly because their cells enlarge rather than divide. The underside of a mushroom's cap is made up of club-shaped hyphae called **basidia** (buh SIH dee uh). Basidia produce haploid spores called **basidiospores** by meiosis. Wind, water, and animals spread the basidiospores, Mushrooms can produce up to a billion basidiospores. ☑

Reading Check

7. Explain why mushrooms tend to grow quickly.

____

____

____

Other Fungi

Members of the phylum Deuteromycota are a diverse group. They share only one unique trait—they appear to lack a sexual stage in their life cycle. As a result, they are referred to as imperfect fungi. Recent studies, however, continue to lead scientists to reclassify these fungi into other phyla.

Fungi

section 3 Ecology of Fungi

MAIN Idea

Lichens have distinct characteristics.

What You'll Learn

- the features of mycorrhizal relationships
- the positive and negative effects of fungi on humans

Before You Read

Think about a recent or historical natural disaster. On the lines below, describe how you think the event affected the environment. In this section, you will learn how fungi can help the environment recover from a natural disaster.

____

____

____

Study Coach

Make Flash Cards Make a flash card for each question heading in this section. On the back of the flash card, write the answer to the question. Use the flash cards to review what you have learned.

Read to Learn

Fungi and Photosynthesizers

Lichens and mycorrhizae are mutualists. They are a combination of fungi and other organisms. Both organisms benefit from the symbiotic relationship.

What is a lichen?

A **lichen** (LI ken) is a symbiotic relationship between a fungus and an organism capable of photosynthesis. The fungus is usually an ascomycete. The photosynthetic organism is either a green alga or a cyanobacterium. The figure on the next page summarizes the benefits of this relationship to both organisms and to the ecosystem.

The photosynthetic organism provides food for both organisms. The fungus provides a dense web of hyphae in which the algae or cyanobacterium can grow.

How do lichens get what they need to live?

Lichens need only light, air, and minerals to grow. They can be found in the harshest environments. The fungus absorbs moisture and minerals from the air and from rainwater. Some fungi produce chemicals that keep organisms from eating the lichen.

Think it Over

1. **Recall** What is photosynthesis?

____

____

____

Where do lichens live?

Most lichens live in temperate or arctic areas. On the tundra, grazing animals feed on lichens that cover the ground.

Lichens can survive droughts. When there is little water, lichens dry out and stop photosynthesizing. Dry pieces can break off, blow away, and form new colonies where they land. When the rain returns, lichens absorb the moisture and start photosynthesizing again.

Recall that pioneer species can grow on rocks and in thin soil. Lichens are often a pioneer species after lava flows or other natural disasters clear the land. Lichens help plants return to the cleared area. The fungi produce acids that break down rocks into soil. The lichens trap soil and nitrogen that plants need.

How do lichens serve as bioindicators?

A **bioindicator** is a living organism that is one of the first organisms to respond to changes in environmental conditions. Lichens absorb water and minerals directly from the air and rain. As a result, they are very sensitive to air pollution. When lichens begin to die, it is a sign that pollution is rising in the area.

2. Explain how lichens can reproduce in drought conditions.

Picture This

3. Explain why it is important for fungi to partner with an organism that can photosynthesize.

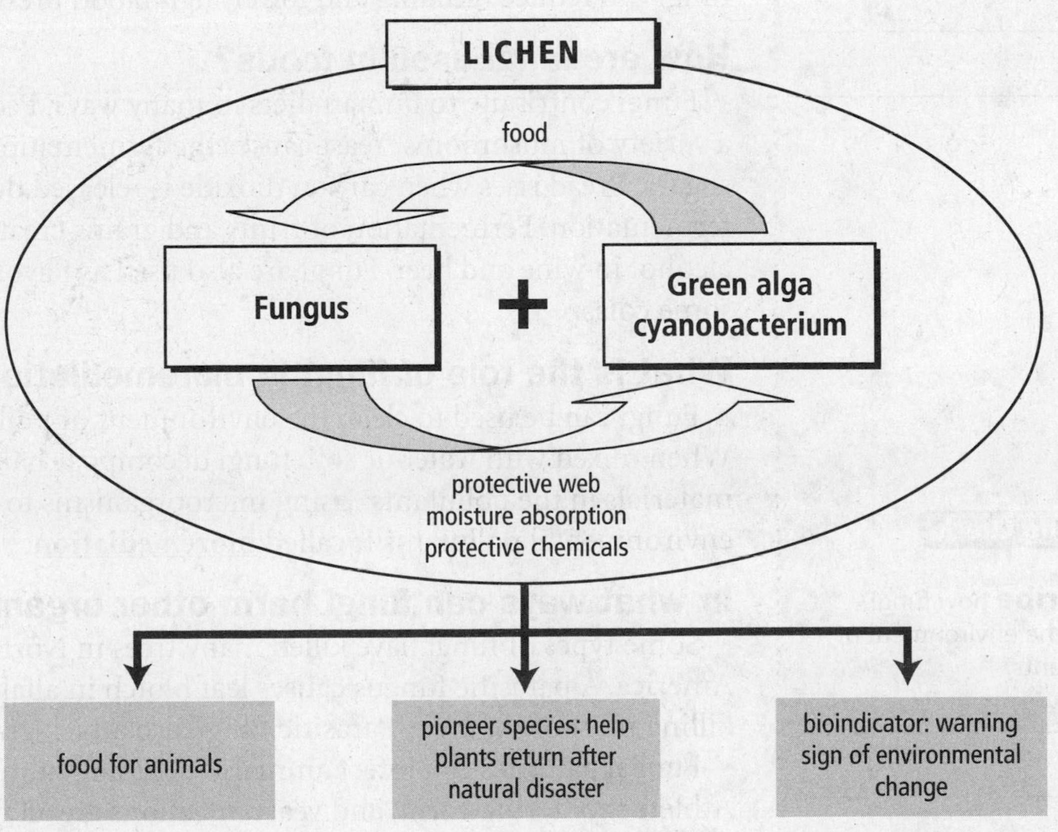

What are mycorrhizae?

A **mycorrhiza** (my kuh RHY zuh) (plural, mycorrhizae) is a symbiotic relationship between a fungus and the root of a plant. The fungus absorbs minerals for the plant roots. The fungus also creates a larger surface area for the plant roots, which helps the plant absorb more water and minerals. In return, the plant provides food for the fungus.

Why are mycorrhizae important?

Between 80 and 90 percent of plants have mycorrhizae. Many crops, including corn and potatoes, depend on mycorrhizae to help them stay healthy. Some plants, like the orchid, cannot survive without mycorrhizae.

Fungi and Humans

Fungi are mostly beneficial to humans. Their role as decomposers is especially important. Fungi recycle nutrients from dead organisms back into food webs.

Think it Over

4. Predict how Earth's surface could be different without decomposers.

____

____

____

How are fungi used in medicine?

Fungi have many medical uses. A type of fungi is the source of penicillin, a life-saving antibiotic. Chemicals found in some fungi can reduce bleeding and lower high blood pressure.

How are fungi used in foods?

Fungi contribute to human diets in many ways. People eat a variety of mushrooms. Yeast causes the fermentation of sugars. Bread rises when carbon dioxide is released during fermentation. Fermentation of fruits and grains creates the alcohol in wine and beer. Fungi are also used as flavoring in some colas.

What is the role of fungi in bioremediation?

Fungi can be used to clean the environment of pollutants. When mixed with water or soil, fungi decompose harmful materials in the pollutants. Using microorganisms to remove environmental pollutants is called **bioremediation**. ✔

✔ Reading Check

5. Describe how fungi clean the environment of pollutants.

____

____

____

In what ways can fungi harm other organisms?

Some types of fungi have killed many trees in North America. A parasitic fungus causes leaf blotch in alfalfa plants, killing most of the crop. Parasitic fungi also attack grapes.

Fungal parasites can infect animals, including humans. Athlete's foot, ringworm, and yeast infections are all fungal infections in humans.

Introduction to Plants

section 1 Plant Evolution and Adaptations

MAIN Idea

Adaptations allow plants to live on land.

What You'll Learn

- how plants and green algae are alike
- the importance of vascular tissue to plant life on land
- the alternation of generations in plants
- the divisions of the plant kingdom

Before You Read

In your mind, picture different plants that you have seen. Scientists classify living things by their characteristics. On the lines below, write at least four characteristics of plants.

____

____

____

Read to Learn

Plant Evolution

Plants are necessary for human survival. Much of the oxygen in the atmosphere comes from plants. Humans use plants for food. Many of the things that make our lives comfortable, such as clothing and furniture, come from plants. When you think of a plant, you might picture a tree, a shrub, or a houseplant.

Biologists describe plants as multicellular eukaryotes with tissues and organs. The tissues and organs have specialized structures that perform various functions. For example, most plants have tissues where photosynthesis occurs. Organs such as roots that anchor plants to the soil or to another object also have specific functions.

Primitive land plants first appeared about 400 million years ago. Biochemical and fossil evidence suggests that freshwater green algae have a common ancestor with land plants. Some of these ancient green algae might have been able to survive periods of drought. Through natural selection, these drought-resistant green algae might have passed adaptations to future generations that helped them survive life on land.

Study Coach

Make Flash Cards Write a question about each paragraph on one side of a flash card. Then write the answer on the other side. Quiz yourself until you know the answers.

FOLDABLES™

Organize and Learn Make a three-tab Venn diagram Foldable, as shown below. Note what you learn about plants and algae, then determine what they have in common.

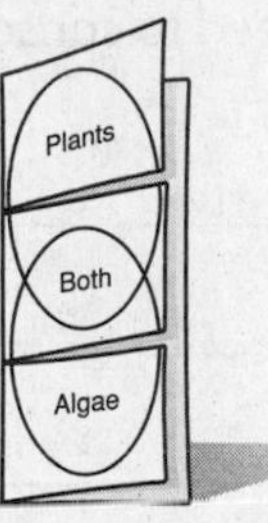

What do plants and algae have in common?

Scientists have compared present-day plants and algae. They have found the following common characteristics:

- cellulose cell walls
- the formation of a cell plate during cell division
- similar genes for ribosomal RNA
- food stored as starch
- the same types of enzymes in cellular vesicles.

The evolutionary tree below shows the relationship between ancient freshwater green algae and present-day plants.

Picture This

1. Circle the name of the ancestor of present-day land plants.

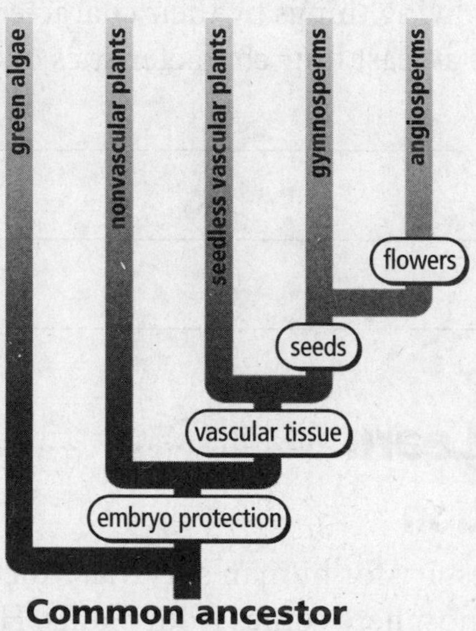

Plant Adaptations to Land Environments

Land organisms face many challenges that aquatic organisms do not face. Land organisms must survive with limited water resources. Over time, land plants developed adaptations that helped them survive when water was scarce. Land plants also developed adaptations to other environmental factors.

2. Conclude Why do land plants need to conserve water?

____

____

____

What purpose does cuticle serve in land plants?

Most plant parts that grow above ground have a coating called a cuticle on the outer surface of their cells. The cuticle is formed by wax and fats, which are lipids that do not dissolve in water. It is the wax in the cuticle that gives some plant leaves their gray appearance. The cuticle helps keep water from evaporating from the plant tissues. The cuticle also stops microorganisms from invading the plant.

What structure enables gas exchange?

Most plants carry on photosynthesis. Photosynthesis produces glucose and oxygen from carbon dioxide and water. For photosynthesis to occur, gases need to move between the plant and the environment. If the cuticle reduces water loss, it might prevent the movement of gases.

Stomata (STOH muh tuh) (singular, stoma) are adaptations that enable gas exchange. Stomata are openings in the outer cell layer of leaves and some stems. Most stomata are found in plant leaves, which are the site of most plant photosynthesis.

What are the functions of vascular tissues?

Vascular (VAS kyuh lur) **tissue** is a specialized transport tissue that is another plant adaptation to life on land. Plants with vascular tissues are called **vascular plants**. Plants that lack specialized transport structures are **nonvascular plants**. Water travels from cell to cell in nonvascular plants by osmosis and diffusion.

Vascular tissues provide support and structure to vascular plants. They help substances move faster than in nonvascular plants. Vascular tissues with thickened cell walls allow vascular plants to grow larger than nonvascular plants.

What are the features of a seed?

The evolution of the seed was another adaptation that helped vascular plants succeed on land. A **seed** is a plant structure that contains an embryo and nutrients for the embryo. The embryo and nutrients are protected by a seed coat. Seeds are adapted to survive in harsh environmental conditions and then sprout when favorable conditions exist.

Alternation of Generations

The life cycle of plants includes two stages, or alternating generations. One generation is the gametophyte generation. The other is the sporophyte generation. ☑

The gametophyte generation produces gametes—sperm and eggs. Sperm and eggs are both haploid cells. Some plants produce both sperm and eggs on one gametophyte. Other plants produce sperm and eggs on separate gametophytes. When the sperm fertilizes the egg, a diploid zygote forms. The zygote goes through mitosis repeatedly to form a multicellular sporophyte.

The sporophyte generation produces spores. The spores can grow to form the next gametophyte generation.

Think it Over

3. Explain how the cuticle and stomata function together.

____

____

____

____

Reading Check

4. Name the two plant generations.

____

____

How is the dominant generation identified?

Depending on the type of plant, one generation is dominant over the other. The dominant generation is larger and more easily seen. For example, the grass growing in a park is the sporophyte generation of the plant. Most plants you see are the sporophyte generation for those plants. The trend in plant evolution was from dominant gametophytes to dominant sporophytes that contain vascular tissue.

Think it Over

5. Predict Which generation is dominant in most nonvascular plants? (Circle your answer.)

a. gametophyte

b. sporophyte

Plant Classification

All plants belong to the plant kingdom. Over time, plant adaptations led to a diversity of plant characteristics. Botanists use these characteristics to classify plants into divisions. You will learn more about the characteristics of these divisions in the next three sections.

Division names end in an *a*. Botanists commonly drop the *a* from the division name and add *es*. Therefore, members of division Bryophyta are called bryophytes (BRI uh fites).

The 12 plant divisions are organized into two groups—nonvascular plants and vascular plants. Vascular plants are organized further into plants that produce seeds and plants that do not produce seeds. The basic organization of the plant kingdom is shown below.

Picture This

6. Label on the lines in the chart the ways in which water and other substances are transported in nonvascular and vascular plants.

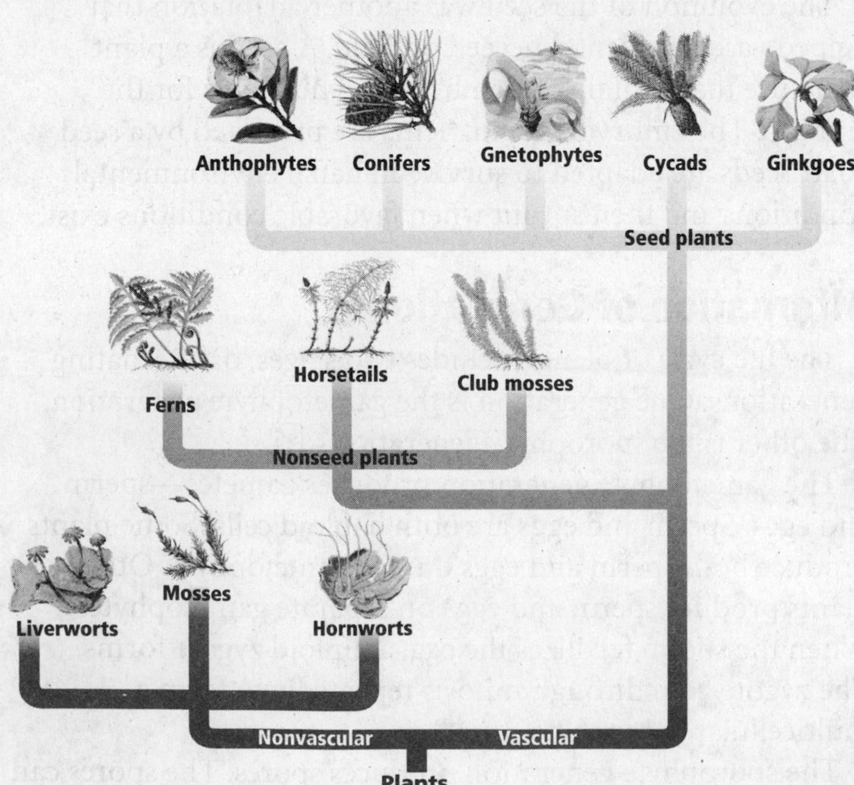

Introduction to Plants

section 2 Nonvascular Plants

Before You Read

One nonvascular plant often lives in a mutualistic relationship with another, where two organisms share their metabolism. On the lines below, explain this type of relationship.

MAIN Idea

Nonvascular plants are small and only grow in damp environments.

What You'll Learn

- the structures of nonvascular plants
- the differences among the nonvascular-plant divisions

Read to Learn

Diversity of Nonvascular Plants

Evidence suggests that four major groups of plants evolved along with green algae from a common ancestor. The figure below shows characteristics that separate each group.

Nonvascular plants are usually small, which enables materials such as water and nutrients to move easily within them. These plants are often found in damp, shady areas that provide the water needed for reproduction and the movement of nutrients.

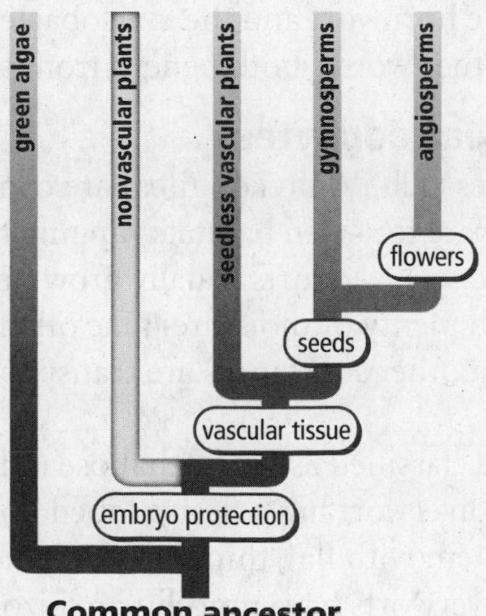

Mark the Text

Identify Plant Divisions Underline or highlight the name of each nonvascular plant division. Say the name aloud. Then circle the words or phrases that describe that plant division.

Picture This

1. **Highlight** the characteristic on the chart that separates nonvascular plants from all other plants.

What plants are bryophytes?

The most familiar bryophytes are mosses. These small, nonvascular plants often grow on a damp log or along the sides of a stream. They have structures that are similar to leaves. Photosynthesis occurs in these leaflike structures, which are usually only one cell thick. ☑

Mosses are anchored to the soil or another surface by multicellular rhizoids. Water and dissolved minerals can diffuse into a moss's multicellular rhizoids. Water and other substances move throughout a moss by osmosis or diffusion.

Some mosses have stems that grow upward from the plant. Others have stems that trail like a vine. Some mosses form mats that help hold soil in place on rocky slopes.

About 1 percent of Earth's surface might be covered with bryophytes. Mosses can survive climate changes. Many mosses freeze and thaw without damage. Some mosses can survive drought and begin growing again when water returns.

Reading Check

2. Describe where photosynthesis occurs in mosses.

______

______

______

What is the smallest division of nonvascular plants?

Anthocerophytes (an tho SAIR uh fites) are the smallest division of nonvascular plants. Members of division Anthocerophyta are commonly called hornworts because of their hornlike sporophytes.

These plants have one large chloroplast in each cell of the gametophyte and sporophyte. The sporophyte produces much of the food used by both generations of hornworts.

You can see the large chloroplast by looking at a hornwort under a microscope. You might also observe that the spaces around cells are filled with slime. Cyanobacteria often grow in this slime. The hornwort and the cyanobacteria exhibit mutualism. In other words, both benefit from the relationship.

3. Compare Name one characteristic that bryophytes and hepaticophytes share. Name one characteristic that differs.

______

______

______

______

______

What are hepaticophytes?

Hepaticophytes (hih PA tih koh fites) are commonly called liverworts. They are found in habitats ranging from the tropics to the arctic. Liverworts usually grow near the ground and in areas with plenty of moisture. Like other nonvascular plants, water and other substances are transported by osmosis and diffusion.

Liverworts are classified as either thallose (THAL lohs) or leafy. A **thallose** liverwort has a fleshy, lobed body. A leafy liverwort has a stem with flat, thin, leaflike structures arranged in three rows. Liverworts have unicellular rhizoids.

Introduction to Plants

section 3 Seedless Vascular Plants

Before You Read

The vascular tissues of plants serve as a type of plumbing for the plants. On the lines below, write what you think are the main purposes of plumbing.

____

____

MAIN Idea

Seedless vascular plants are adapted to drier environments.

What You'll Learn

- the characteristics of seedless vascular plants
- the differences between club mosses and ferns

Read to Learn

Diversity of Seedless Vascular Plants

Three plant groups have vascular tissues. The most diverse group in form and size is the seedless vascular plant group. Seedless vascular plants consist of club mosses and ferns. Club mosses are small plants. They are usually less than 30 cm tall. Tropical tree ferns can grow to a height of 25 m.

An adaptation seen in the sporophytes of many seedless vascular plants is the strobilus (STROH bih lus). A **strobilus** is a cluster of structures that produce spores.

Mark the Text

Highlight Main Ideas
Read the paragraphs under each question heading. Underline the part of the text that answers the question.

What is the dominant generation of lycophytes?

Fossil evidence suggests that lycophytes (LI kuh fites) were once the size of trees and were part of the early forest community. Modern members of division Lycophyta are much smaller and are called club mosses.

Unlike true mosses, the sporophyte generation is dominant in lycophytes. The gametophyte generation is small and grows from spores. Lycophytes have roots, stems, and small, scaly, leaflike structures. Their vascular tissue is found in a vein that runs down the middle of each leaflike structure.

Many tropical lycophyte species are known as epiphytes (EH puh fites). An **epiphyte** is a plant that lives anchored to another plant. In tropical forests, these lycophytes create a habitat in the forest canopy. ☑

Reading Check

1. **Define** What is an epiphyte?

____

____

____

✔ Reading Check

2. Identify the kind of environment in which most ferns grow.

____

____

____

Picture This

3. Highlight the name of the food storage structure. Color the structure where photosynthesis occurs green.

Where do ferns grow best?

Division Pterophyta includes ferns and horsetails. Approximately 350 million years ago, ferns were the most abundant land plants. They grow best in moist environments. However, some can survive in dry environments. ✔

How is the fern sporophyte produced?

The fern gametophyte is a tiny, thin structure. It grows from a spore and has male and female reproductive structures. After fertilization, the sporophyte grows from the gametophyte. Ferns that live in dry areas have an adaptation that allows them to produce sporophytes without fertilization.

The sporophyte produces roots and a thick underground stem called a **rhizome**. Food is stored in the rhizome. Photosynthesis occurs in the leafy structures known as fronds. The fronds shown below have branched vascular tissues and are part of the sporophyte generation.

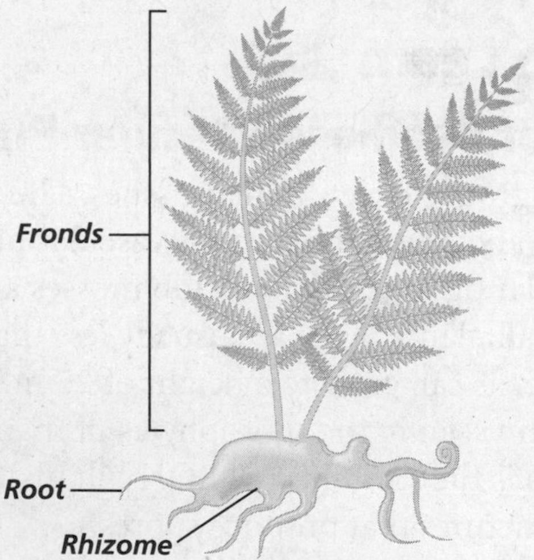

Where do fern spores form?

Fern spores form in a structure called a **sporangium** (plural, sporangia). Groups of sporangia form a **sorus** (plural, sori). Sori can be found on the underside of fronds.

What are the structures of horsetails?

Horsetails have hollow stems with circles of scalelike leaves. Spores are produced in strobili at the tips of reproductive stems. Horsetail spores develop into gametophytes when environmental conditions are right.

Present-day horsetails and ferns are much smaller than their ancestors. Like ferns, horsetails grow best in places where water is abundant. Horsetails are found most often in marshes, swamps, and along stream banks.

chapter 21 Introduction to Plants

section 4 Vascular Seed Plants

MAIN Idea

Seed plants are the most widely distributed plants on Earth.

What You'll Learn

- the characteristics of seed plants
- the divisions of gymnosperms
- the life spans of angiosperms

Before You Read

Think of the items on your kitchen shelves and in your refrigerator. Make a list on the lines below of products that come from seed plants. Examples might include grapes, corn, apples, and olives. You will read about seed plants in this section.

____

____

____

Read to Learn

Diversity of Seed Plants

Vascular seed plants produce seeds. The seed is actually a tiny sporophyte inside a protective coat. Seed plants do not require a film of water to reproduce. This is the most important difference between seed plants and the plants you have read about in the last two sections. Seeds are an adaptation that helps seed plants succeed in different environments, including places with little water.

Structures known as **cotyledons** (kah tuh LEE dunz) either store food or help take in food for the seed. Plants with seeds that are part of fruits are angiosperms (AN jee uh spurmz). Other seed plants with seeds that are not part of fruits are gymnosperms (JIHM nuh spurmz).

Why is seed dispersal important?

Seed plants are successful in diverse environments because they have adaptations that help spread, or disperse, seeds throughout their environments. Dispersal limits competition between the new plant and its parent and also between the new plant and other offspring. ☑

Study Coach

Read and Discuss the main topics of the section with a partner. Read a paragraph, and then take turns saying something about what you have learned. Continue until you both understand the main ideas.

Reading Check

1. **Explain** why seed dispersal helps make seed plants successful.

____

____

____

How are gametophytes formed?

The dominant generation in seed plants is the sporophyte. The sporophyte produces spores that divide by meiosis to form male gametophytes and female gametophytes. Male gametophytes are called pollen grains. Each female gametophyte consists of one or more eggs surrounded by protective tissue. Male gametophytes and female gametophytes depend on the sporophyte generation for survival.

Think it Over

2. **Compare** the dominant generation in seed plants with the dominant generation in seedless vascular plants.

____

____

____

What do cones contain?

Cycadophytes (si KAH duh fites) are gymnosperms, and they produce cones. **Cones** contain either male reproductive structures or female reproductive structures. As shown in the evolutionary tree below, members of division Cycadophyta and other plants with cones evolved before angiosperms.

Male cones produce clouds of pollen grains. The pollen grains produce male gametophytes. Female cones contain the female gametophytes. Male reproductive structures and female reproductive structures are on separate cycad plants.

What is the natural cycad habitat?

Cycads grew in abundance about 200 million years ago. Today, there are only about 250 species of cycads. They grow in the tropics and subtropics. The only native habitat for cycads in the United States is southern Florida.

Picture This

3. **Sequence** Write a number from one to five above each branch of the evolutionary tree to indicate the order in which the plant types evolved.

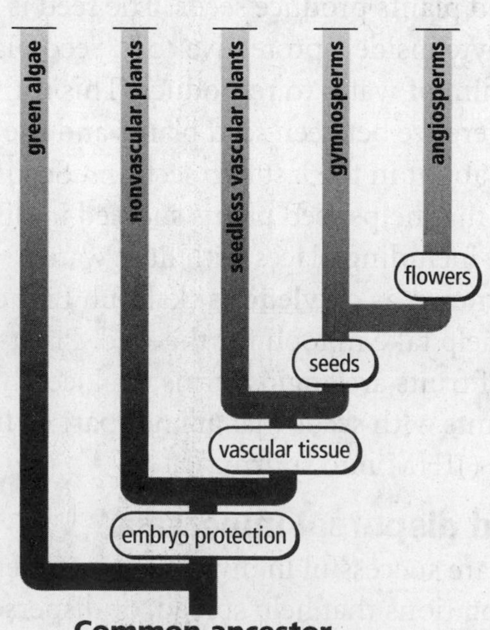

How long do gnetophytes live?

Gnetophytes (NEE tuh fites) are long-living seed plants. Some live as long as 1500 to 2000 years. The three genera of division Gnetophyta show unusual adaptations that help them survive in their environments. For example, one gnetophyte that is found only in the deserts of southwestern Africa has a large storage root and only two leaves. The leaves grow continuously and take in moisture from fog, dew, and rain.

What is the one living species of division Ginkgophyta?

Only one living species of ginkgophytes (GIHN koh fites) grows today—*Ginkgo biloba.* Like cycads, ginkgo male and female reproductive structures are on separate plants. The male ginkgo produces pollen grains from cones that grow from the bases of leaf clusters. When the cones on the female ginkgo are fertilized, they produce bad-smelling, fleshy seed coats. Male ginkgos are often planted in cities because they can survive smog and pollution. ☑

Reading Check

4. Explain why it would be better to plant a male ginkgo than a female ginkgo in a small yard.

____

____

____

How do conifers differ from ginkgophytes?

Conifers are part of forests throughout the world. They range in size from low-growing shrubs to towering trees. Plants in division Coniferophyta include pines, firs, cypresses, and redwoods. Most conifers produce male and female cones on different branches of the same tree or shrub.

Male cones are usually small and produce pollen grains. Female cones are larger than male cones and remain on the plant until the seeds inside the cone have matured. People often use the female cones to identify the tree or shrub.

Conifers are trees and shrubs with needlelike or scalelike leaves. The leaves are covered with cutin—a waxlike coating. Cutin is an adaptation that reduces water loss.

What is an evergreen plant?

Most evergreen plants in the northern temperate region are conifers. In tropical regions, evergreen plants include palm trees. Botanists refer to plants as evergreen when they have some green leaves all year long. Green leaves are an adaptation that allows evergreens to undergo photosynthesis whenever conditions are favorable.

A deciduous plant loses its leaves at the end of the growing season or when moisture is scarce. Many conifers are evergreen, but some are deciduous.

Think it Over

5. Evaluate You want to plant a row of trees to serve as a year-round windbreak. Would you choose evergreens or deciduous trees? Explain.

____

____

____

When did anthophytes first appear?

All flowering plants are anthophytes (AN thuh fites). Members of division Anthophyta are also known as angiosperms. They first appeared in the fossil record about 130 million years ago. Today, anthophytes make up more than 75 percent of the plant kingdom. Adaptations allow them to grow on land and in water environments.

Think it Over

6. Contrast What is the difference between angiosperms and gymnosperms?

____

____

____

What are monocots and eudicots?

At one time, botanists classified anthophytes as either monocots or dicots. Monocots and dicots were distinguished by the number of cotyledons in their seeds. Using this system, a plant with one cotyledon in its seeds was classified as a monocot, while a plant with two cotyledons in its seeds was classified as a dicot.

Botanists still classify plants with one cotyledon as monocots. But they now classify dicots as eudicots, based on the structure of the pollen.

About 75 percent of anthophytes are eudicots. Most trees, shrubs, and garden plants are eudicots. Monocots are the second largest group and include grass, onions, and palms. Oaks, maples, and sycamores are examples of eudicots.

What are the life spans of anthophytes?

The life-spans of anthophytes vary. The life-span is the time the plant sprouts from a seed, grows, produces new seeds, and dies. An **annual** plant completes its life-span in one growing season or less. Most weeds and many garden plants are annuals.

A **biennial** plant has a two-year life-span. During the first year, the plant produces leaves and a strong root system. The aboveground tissues die at the end of the first growing season. In the second year, stems, leaves, flowers, and seeds grow. The plant's life ends the second year. Some biennial plants, such as carrots, develop fleshy storage roots that are harvested after the first growing season.

Perennial plants can live for several years and usually produce flowers and seeds each year. In harsh conditions, their aboveground tissues die. They begin growing again when conditions are favorable. Fruit trees, irises, and roses are examples of perennial plants. ☑

7. Name the type of plants that have a life span of more than two years.

____

The life-span of a plant is genetically determined and reflects adaptations for surviving harsh conditions. A plant's life-span can be affected by environmental conditions.

Plant Structure and Function

section 1 Plant Cells and Tissues

Before You Read

You have already learned that cells contain many structures that have special functions. On the lines below, name three structures found in plant cells.

____

____

____

MAIN Idea

Different types of plant cells make up plant tissues.

What You'll Learn

- the major types of plant cells and plant tissues
- the functions of different types of plant cells and tissues

Read to Learn

Mark the Text

Key Concepts As you read, highlight the descriptions of the three types of plant cells. In a different color, highlight the descriptions of the four types of plant tissues.

Plant Cells

Every plant cell has a cell wall and a large central vacuole. Most plant cells also have chloroplasts. However, there are many types of plant cells that carry out specific functions. Plant cells have one or more adaptations that make it possible for these cells to perform their functions. For example, some plant cells lose their cytoplasm and organelles as they mature, leaving only the strong cell wall.

What functions do parenchyma cells perform?

Most cells in a plant have flexible, thin cell walls and are known as **parenchyma** (puh RENG kuh muh) **cells**. Mature parenchyma cells can undergo cell division. They help repair damage or wounds to a plant.

Functions performed in parenchyma cells include photosynthesis, storage, gas exchange, and protection. Their function determines the features of parenchyma cells. For example, parenchyma cells in leaves and green stems contain many chloroplasts. Remember, photosynthesis occurs in the chloroplast and produces glucose. The plant uses glucose for energy. Parenchyma cells found in roots and fruits have large central vacuoles for storing water, starch, or oils. ☑

Reading Check

1. **Conclude** Why do the internal structures of parenchyma cells vary?

____

____

____

Think it Over

2. Evaluate Why do plants need cells that allow them to bend without breaking?

What are collenchyma cells?

Collenchyma (coh LENG kuh muh) **cells** are elongated cells that occur in long strands or cylinders and support the surrounding cells. Collenchyma cells can expand and stretch. They allow plants to bend without breaking. Mature collenchyma cells can undergo cell division.

What distinguishes sclerenchyma cells?

Sclerenchyma (skle RENG kuh muh) cells differ from parenchyma and collenchyma cells. Mature **sclerenchyma cells** lack cytoplasm and other living components, but their rigid cell walls remain. Sclerenchyma cells support the plant and help transport materials within the plant.

Plants can have two types of sclerenchyma cells—sclereids and fibers. Sclereids (SKLER idz) can be found randomly throughout the plant. They give nuts and seed coats their toughness and they help transport materials within plants.

A fiber cell is needle-shaped with a thick wall and a small interior space. Fibers form a tough, flexible tissue. Humans use fibers to make things such as rope and canvas. The three types of plant cells are summarized in the table below.

Picture This

3. Compare What do collenchyma and sclerenchyma cells have in common?

Cell Type	Functions
Parenchyma	• storage • photosynthesis • gas exchange • tissue repair and replacement
Collenchyma	• support of surrounding tissues
Sclerenchyma	• transport of materials • support

Plant Tissues

A tissue is a group of cells that work together to perform a specific function. Four different tissues are found in plants: meristematic, dermal, vascular, and ground. Each tissue is composed of one or more types of cells.

Where are new plant cells produced?

Plants produce new cells in their meristematic tissues throughout their lifetime. Meristematic tissues make up **meristems** (MER uh stem), which are regions of rapidly dividing cells.

Cells in meristems have large nuclei and small or no vacuoles. Cells in meristems develop into many different kinds of plant cells.

Apical meristems Cells that result in an increase in length are apical (AY pih kul) meristems. This tissue is found at the tips of roots and stems. The growth in tissue is called primary growth.

Intercalary meristems Intercalary (in TUR kuh LAYR ee) meristems produce new cells that result in an increase in stem or leaf length. This tissue is found in one or more locations along the stems of most monocots.

Lateral meristems The two types of lateral meristems produce an increase in root and stem diameter. Nonflowering plants, eudicots, and a few monocots have secondary growth produced from this meristematic tissue.

Vascular cambium (VAS kyuh lur • KAM bee um) is a thin cylinder of meristematic tissue found in roots and stems. It produces two types of transport cells—xylem (ZI lum) and phloem (FLOH em)—in some roots and stems.

Cork cambium produces cells that develop tough cell walls. These cells form a protective outer layer on stems and roots.

What is dermal tissue?

The **epidermis** (eh puh DUR mus) is the layer of dermal cells that makes up the outer covering on a plant. Epidermal cells secrete a fatty substance that forms the cuticle. The cuticle prevents water loss and blocks disease-causing organisms from entering the plant.

What adaptations of the epidermis help plants survive?

Plants can have several adaptations of their epidermis. The stomata, found on most leaves and some green stems, are formed by two **guard cells**. As the guard cells swell and shrink, the stomata open and close. Hairlike projections from the epidermis of leaves and stems are called **trichomes** (TRI kohmz). They help protect the plant from predators and keep some plants cool by reflecting light. **Root hairs** extend from some root epidermal cells. They increase a root's surface area and help the root absorb water and nutrients. ☑

What are the two types of vascular tissue?

Water, food, and other dissolved substances need to move from place to place in plants. Schlerenchyma cells make up xylem and phloem, the two types of vascular tissue responsible for the transport function.

Think it Over

4. Summarize What are the three types of meristematic tissue?

______

______

______

✔ Reading Check

5. Describe the function of root hairs.

______

______

______

What is the function of xylem?

Xylem transports water and minerals from the root to other parts of the plant. It is composed of schlerenchyma cells called vessel elements and tracheids (TRA kihdz).

Each mature vessel element and tracheid consists of just its cell wall, as shown below. Water can flow freely through these cells because the cytoplasm and other living parts are gone.

Vessel elements are tubular cells that form strands of xylem. Mature vessel elements are open at each end, and water and dissolved substances move freely along the strands.

Tracheids appear as long cylinders with pitted ends. They form a tubelike strand. Mature tracheids have end walls that slow the transport of materials. They are less efficient in transporting substances than vessel elements.

Picture This

6. Circle the tracheid shown in the figure.

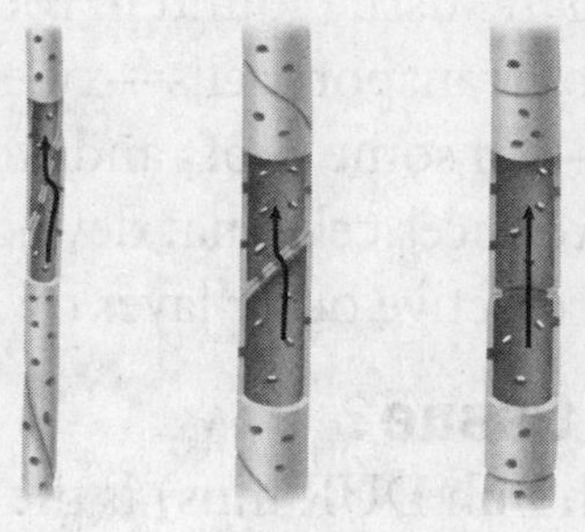

What does phloem transport?

Phloem is a vascular tissue that transports dissolved sugars and other organic compounds throughout a plant. It moves materials from the stems and leaves to the roots, and from the roots to the stems and leaves.

Phloem consists of sieve tube members and companion cells. Each mature **sieve tube member** contains cytoplasm but lacks a nucleus and ribosomes. **Companion cells** have a nucleus and are found next to each sieve tube member.

Plants produce and use glucose for energy. Glucose not needed by a plant is converted to carbohydrates and stored in regions of the plant called sinks. The process by which phloem transports carbohydrates and other substances to sinks is known as translocation (trans loh KAY shun).

FOLDABLES™

Organize and Learn

Make a three-tab Venn diagram Foldable, as shown below. As you read, note what you have learned about the xylem and phloem of a plant, then use this information to determine what they have in common.

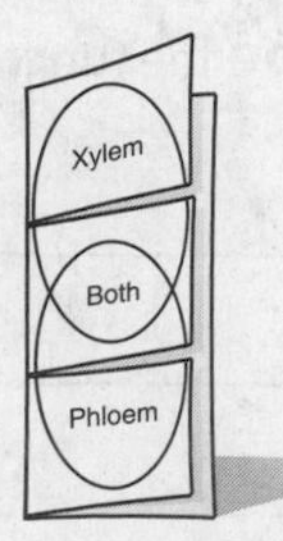

What are the functions of ground tissue?

Ground tissue consists of parenchyma, collenchyma, and sclerenchyma cells and has many functions, including photosynthesis, storage, and support. Most of a plant consists of ground tissue. The specific type of cells found in each ground tissue depends on the function of the tissue.

Plant Structure and Function

section 2 Roots, Stems, and Leaves

Before You Read

On the lines below, name the parts of a plant you have seen recently. Describe the function of one of the parts. In this section you will read more about the functions of plant parts.

____

____

MAIN Idea

The structures of roots, stems, and leaves relate to their functions.

What You'll Learn

the similarities and differences in the structure and function of roots, stems, and leaves

Read to Learn

Roots

For most plants, roots take in water and dissolved minerals, anchor plants in place, and support the plant against the effects of gravity, extreme wind, and moving water. The roots of most plants grow 0.5 to 5 m down into the soil. Plants that live in areas with little available water have root system adaptations that allow them to survive. For example, mesquite (mes KEET) trees have roots that grow as deep as 50 m toward available water.

Study Coach

Identify Answers After reading the section, reread each question heading. Answer the questions based on what you have read.

What protects roots as they grow?

The root is usually the first structure to grow out of the seed. The tip of the root is covered by the **root cap**. The root cap consists of parenchyma cells that help protect the root tissues. The root's apical meristem produces cells that increase the root's length and replace cells rubbed off as the root grows. Various root tissues develop from these cells.

An epidermal layer covers the root. The **cortex** (KOR teks) is the layer below the epidermal layer. It is composed of ground tissues made up of parenchyma cells. As water and minerals are taken in by the epidermis, they move through the cortex to the vascular tissues. The vascular tissues transport the substances to other parts of the plant. ☑

Reading Check

1. List the root tissues in the order that water reaches them.

____

____

____

What is the function of the endodermis?

The **endodermis** (en duh DUR mus) is the layer of cells at the inner boundary of the cortex. This layer of cells creates a waterproof seal around the root's vascular tissue. Water and dissolved minerals are forced to pass through the cells of the endodermis to reach the vascular tissue.

Next to the endodermis is the **pericycle** (PER ih si kul). It is the tissue that produces lateral roots. In some plants, the vascular cambium develops from part of the pericycle. The structures of a plant's roots are shown in the figure below.

Picture This

2. Identify Highlight the name of the structure that produces lateral roots.

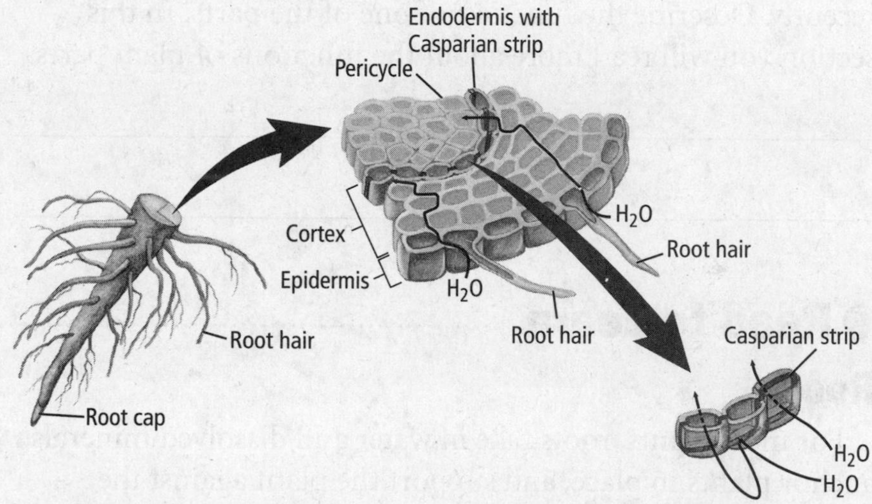

What are the two major types of root systems?

Taproots and fibrous roots are the two major types of root systems. Taproots are single, thick structures with smaller branching roots. Some taproots, such as carrots and radishes, store food in their parenchyma cells. Others grow deep in the soil in search of water.

Fibrous roots have many small branching roots that grow from a central point. Plants, such as sweet potatoes, also can store food in fibrous roots. Other root types are the result of adaptations to diverse environments.

Think it Over

3. Compare If you are pulling weeds from moist soil, which type of root system will likely be more difficult to remove without using any tools?

____

Stems

Herbaceous (hur BAY shus) stems are soft, flexible, and green. The chloroplasts in these stems perform photosynthesis. Most annual plants have herbaceous stems. Trees, shrubs, and many perennial plants have woody stems that do not perform photosynthesis. Some plant stems are covered with bark that protects them from damage and keeps insects out.

What is the structure and function of stems?

A stem's main function is to support the leaves and reproductive structures of the plant. Vascular tissue in the stem provides support and transports water and other substances. The tissues are arranged in bundles or groups and are surrounded by parenchyma cells. ☑

Primary growth from the apical meristems lengthens the stem. The stem's diameter widens as the plant grows taller. This provides additional support for the plant. Most of the increase in diameter is due to an increase in cell size. Cells produced in the vascular cambium of some plants cause the increase in stem diameter. Xylem and phloem, produced throughout the year, also increase stem diameter. They produce the annual growth rings found in tree trunks.

All stems have adaptations that help plants survive. These adaptations allow stems to store food and to survive weather extremes. Some stems do not look like typical stems. For example, the white potato is a stem called a tuber. Tulip bulbs are thick, short stems surrounded by leaves.

Reading Check

4. Identify the main function of a plant's stem.

____

____

____

Leaves

The main function of leaves is photosynthesis. The flattened portion of most leaves is called the blade. The blade provides a large surface area that receives sunlight. In some plant species, the blade is attached to the stem by a stalk called a **petiole** (PET ee ohl). In plants that don't have petioles, the blades attach directly to the stem.

Leaf structures, shown below, are well-adapted for photosynthesis. Most photosynthesis occurs in the cells directly below the leaf's upper epidermis. These column-shaped cells contain many chloroplasts and make up the **palisade mesophyll** (pa luh SAYD • MEHZ uh fihl) tissue.

Picture This

5. Highlight the name of the structure where photosynthesis occurs.

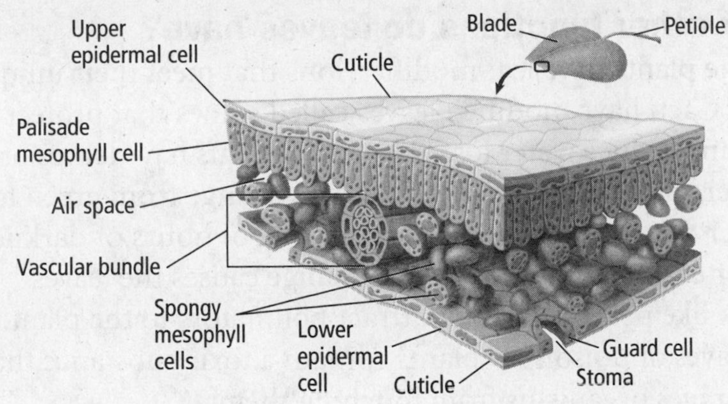

What is the shape of spongy mesophyll cells?

The spongy mesophyll lies below the palisade mesophyll. The **spongy mesophyll** consists of irregularly shaped cells that are loosely packed with spaces around them. Oxygen, carbon dioxide, and water vapor move through the spaces of this structure.

What is transpiration?

Most plant leaves contain stomata in their epidermis. In the last section, you read that the stomata is formed from two guard cells. The shapes of the guard cells change to allow water and gases to diffuse in and out of the leaf.

Water travels from the plant's roots through its stem to the leaves. Some water is used in photosynthesis. Some water evaporates from the inside of the leaf to the outside through the stomata in a process called **transpiration**.

Think it Over

6. Synthesize what regulates transpiration.

____

____

____

What differences in leaves help people identify plants?

Some people identify plants by the differences in size, shape, color, and texture of leaves. The arrangement of leaves on the stem and the arrangement of veins in a leaf can also be used to identify plants. Unique characteristics of leaves are shown in the figure below.

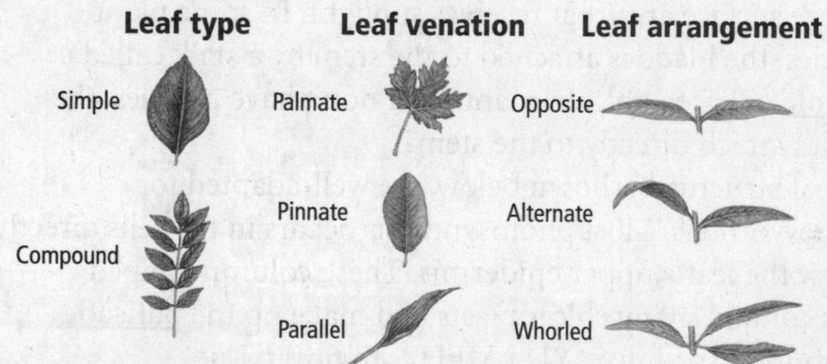

Picture This

7. Apply Circle the name of the leaf type that is most common in the plants around your school.

What other functions do leaves have?

Some plants have leaf modifications that meet their unique needs. Cacti have modified leaves called spines that protect the plant and reduce water loss. Some succulents have leaves used as water storage sites. Poinsettia leaves change from green to another color in response to the number of hours of darkness in their environment. This color change causes the leaves to look like petals and helps attract pollinators to the plant. The leaves of poison ivy plants contain a toxic substance that discourages organisms from touching them.

Plant Structure and Function

section ❸ Plant Hormones and Responses

Before You Read

On the lines below, describe how your body reacts when you enter a cold room. Your body is responding to a stimulus—the cold air. Plants also respond to their environment.

____

____

MAIN Idea

Hormones can affect a plant's responses to its environment.

What You'll Learn

- the major types of plant hormones
- different types of plant responses

Read to Learn

Mark the Text

Identify Details Highlight each hormone or group of hormones. Underline the action of each hormone.

Plant Hormones

Hormones are organic compounds that cause changes in organisms. They are made in one part of an organism and are transported to another part of the organism where they have an effect. Plant hormones control growth and development. Only tiny amounts of hormones are needed to cause a change.

What change does auxin cause?

Auxin (AWK sun) was one of the first plant hormones to be identified. It usually causes cells to get longer. The auxin, indoleacetic (IHN doh luh see tihk) acid, is produced in apical meristems, buds, and other tissues that grow rapidly. This hormone moves from one parenchyma cell to another, always moving away from where it was produced.

The effect of auxin varies widely depending on the concentration and location of the hormone. Low concentrations of auxin can cause cells to get longer. Higher concentrations can have the opposite effect.

Auxin produced in the apical meristem keeps side branches from developing. Auxin also delays fruit formation and keeps fruit from dropping off the plant. When the concentration of auxin decreases, ripe fruit drops to the ground and trees begin to shed their leaves.

Think it Over

1. **Explain** the error in the following reasoning: To make a house plant grow more quickly, simply apply more growth hormone.

____

____

____

Reading Check

2. Identify three effects of gibberellins.

____

____

____

____

How do gibberellins affect plants?

Gibberellins (jih buh REH lunz) are a group of plant hormones that cause cell elongation, increase enzyme production, and affect seed germination. Gibberellins are transported in vascular tissue. ☑

What is the main effect of ethylene?

Ethylene (EH thuh leen) is a gaseous hormone composed of two carbon and four hydrogen atoms. It primarily affects the ripening of fruits by causing the cell walls of unripe fruit to weaken and complex carbohydrates to break down into simple sugars.

Where are cytokinins produced?

Growth-inducing **cytokinins** (si tuh KI nihnz) are produced in rapidly dividing cells. They travel to other parts of the plant within xylem. Cytokinins promote cell division by stimulating protein production. The plant uses the proteins for mitosis and cytokinesis.

Plant Responses

Events such as vines climbing a pole and trees dropping their leaves are plant responses to their environments. A response of a plant that causes movement independent of the direction of the stimulus is a **nastic response**. A Venus flytrap plant exhibits a nastic response. Its leaves snap shut when it senses movement on the surface of the leaves.

Picture This

3. Draw on a separate sheet of paper a plant displaying each of the tropisms listed below. Label your drawing with the type of tropism and the type of response (positive or negative).

What are tropisms?

A plant's growth response to an external stimulus is called a **tropism** (TROH pih zum). Common tropisms are listed in the table below. Growth responses toward the stimulus are called positive tropisms. In contrast, negative tropisms result in plant growth away from a stimulus.

Tropism	External Stimulus	Plant Response	Positive or Negative Tropism
Phototropism	light	growth toward light source	positive
Gravitropism	gravity	growth downward	positive for roots negative for stems
Thigmotropism	mechanical	growth toward point of contact, such as a vine climbing a fence	positive

Reproduction in Plants

section 1 Introduction to Plant Reproduction

Before You Read

On the lines below, describe the characteristics you use to recognize friends and family members in old photos. In this section, you will learn that the way a young plant looks can be quite different from the way it will look when it is mature.

____

____

MAIN Idea

The life cycle of mosses, ferns, and conifers includes alternation of generations.

What You'll Learn

- forms of vegetative reproduction
- stages of alternation of generations
- reproduction of mosses, ferns, and conifers

Read to Learn

Vegetative Reproduction

<u>**Vegetative reproduction**</u> is a form of asexual reproduction in which new plants grow from parts of an existing plant. The new plants are genetically identical to the original plant.

It is often faster to grow plants by vegetative reproduction than to grow plants from seeds. Plants grown by vegetative reproduction are more similar than plants grown from seeds. Vegetative reproduction is the only way to reproduce fruits that do not produce seeds.

Does natural vegetative reproduction occur?

When there is little water, some mosses dry out and are easily broken and scattered by animals and wind. When it rains, some of these pieces will grow. New strawberry plants can grow at the end of horizontal stems, called stolons. If a stolon is cut, the new plant continues to grow.

How do humans use vegetative reproduction?

Farmers and scientists have used leaves, roots, and stems cut from some plants to grow new plants. A potato can be cut into pieces. Each piece with an eye (bud) can be planted and grown into a new potato plant. A few cells of plant tissue can also be used to grow some plants, using a technique called tissue culture.

Study Coach

Create a Quiz After you read this section, create a five-question quiz from what you have learned. Then, exchange quizzes with another student. After taking the quizzes, review your answers together.

Think it Over

1. **Draw Conclusions** Would a piece of potato without an eye grow into a new potato plant? Explain.

____

____

____

Alternation of Generations

Recall that the life cycle of a plant includes an alternation of generations. One generation is a diploid ($2n$) sporophyte stage. The other generation is a haploid (n) gametophyte stage.

As shown below, the sporophyte stage produces haploid spores. The spores divide by mitosis to form the gametophyte generation. The size of the gametophyte generation depends on the plant species. As plants have become more complex, smaller gametophytes have evolved. ☑

The gametophyte stage produces gametes—eggs and sperm. Nonvascular plants and some vascular plants need water for sperm to reach an egg. Flowering plants do not need water for sperm to reach an egg.

Fertilization of an egg by a sperm forms a zygote. A zygote is the first cell of the sporophyte stage. As plants have become more complex, larger sporophytes have evolved. The growth pattern of the sporophytes has also evolved. Most nonvascular plants have sporophytes that depend on the gametophyte for support and food. Flowering plants and other vascular plants have sporophytes that do not depend on the gametophyte for support and food. In these types of plants, the sporophytes live apart from the gametophyte.

2. Identify Which stage produces haploid spores?

____

Picture This

3. Explain to a partner the alternation of generations using the figure.

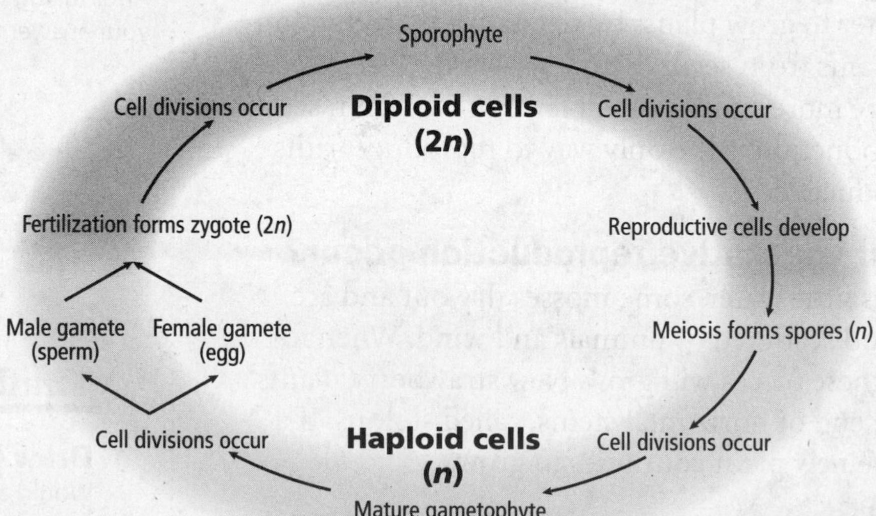

Moss Reproduction and Life Cycle

The life cycle of mosses includes an alternation of generations. As shown below, the gametophyte stage is the dominant generation. The gametophyte grows in damp shady places or on rocks along a stream. Gametophytes can produce archegonia and antheridia. These structures can be on the same moss plant but often are on separate moss plants. ☑

The archegonium of some moss species produces one egg. The archegonium of other moss species produces many eggs. The tissues of the archegonium surround the egg or eggs with a protective layer.

Antheridia produce sperm that have flagella. These sperm need water to move to the archegonium. Sperm exhibit **chemotaxis** (kee moh TAK sus) because they respond to chemicals produced by archegonia. The fertilized egg forms the first cell of the sporophyte stage, called the zygote. Tissues of the archegonium protect the zygote. The new sporophyte gets nutrients from the archegonium as it grows and matures. The sporophyte does not undergo photosynthesis. Instead, it depends on the gametophyte for nutrition.

A mature sporophyte consists of a stalk with a capsule at its tip. Some cells within the capsule undergo meiosis and produce spores. If a spore lands in a suitable place, mitosis begins. A small, threadlike structure called a **protonema** forms. A protonema can develop into the gametophyte plant, and the cycle repeats.

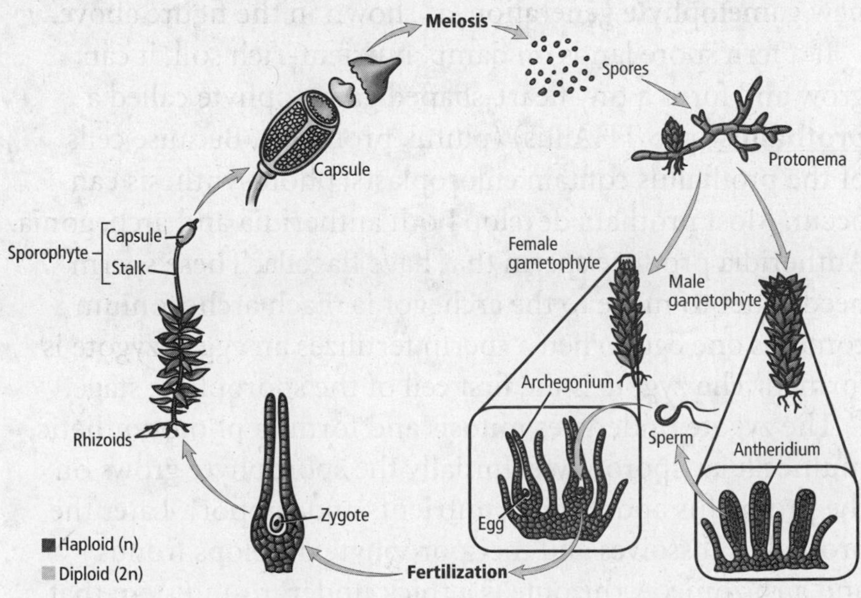

Reading Check

4. Name the dominant stage in the life cycle of a moss plant.

____

____

Picture This

5. Highlight the area where eggs are produced. Circle the area where sperm are produced.

Picture This

6. Identify Circle the name of the first cell of the sporophyte stage.

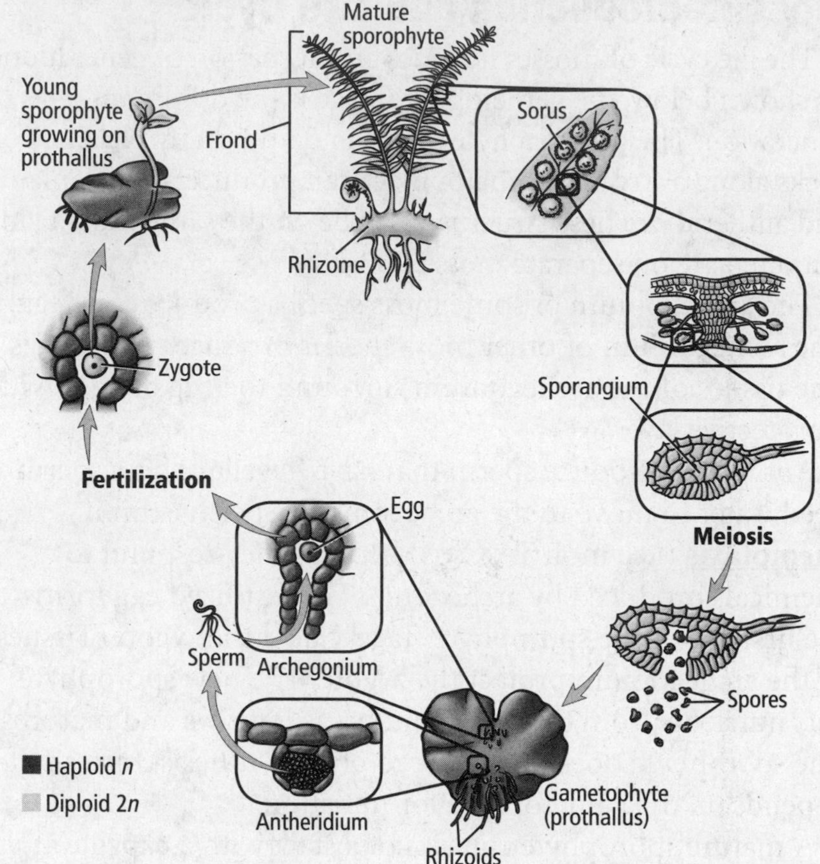

Think it Over

7. Predict What would happen if a fern spore landed on dry, nutrient-poor soil?

____

____

____

Fern Reproduction and Life Cycle

Spore-producing structures called sori are found on a fern's lacy fronds. Each sorus consists of sporangia. Cells in the sporangium undergo meiosis. The resulting spores begin a new gametophyte generation, as shown in the figure above.

If a fern spore lands on damp, nutrient-rich soil, it can grow and form a tiny heart-shaped gametophyte called a **prothallus** (pro THA lus) (plural, prothalli). Because cells of the prothallus contain chloroplasts, photosynthesis can occur. Most prothalli develop both antheridia and archegonia. Antheridia produce sperm that have flagella. These sperm need water to move to the archegonia. Each archegonium contains one egg. When a sperm fertilizes an egg, a zygote is formed. The zygote is the first cell of the sporophyte stage.

The zygote undergoes mitosis and forms a photosynthetic, multicellular sporophyte. Initially the sporophyte grows on the prothallus and receives nutrients and support. Later the prothallus dissolves and the sporophyte develops fronds and a rhizome. A rhizome is a thick underground stem that produces roots and supports the fronds. Photosynthesis occurs in the fronds.

Conifer Reproduction and Life Cycle

Pine trees are examples of the sporophyte generation of conifer plants. Conifers are heterosporous. **Heterosporous** (he tuh roh SPOR us) plants make two types of spores that develop into male and female gametophytes.

8. Compare What is the male gametophyte of a conifer?

What are the differences in conifer cones?

Each scale of a female cone has two ovules at its base. Within each ovule, meiosis produces four **megaspores**. Three megaspores dissolve. One megaspore undergoes mitosis and becomes the female gametophyte. A fully developed female gametophyte consists of hundreds of cells and contains two to six archegonia. Each archegonium contains one egg.

Male cones have many small scales. The scales have hundreds of sporangia. Cells in these sporangia undergo meiosis and form **microspores**. A pollen grain is made of four cells and develops from microspores. Pollen grains are carried by air.

How does pollination occur?

Pollination occurs when a pollen grain from one species of seed plant lands on the female reproductive structure of a plant of the same species. The opening of a female cone's ovule is called the **micropyle**. A conifer pollen grain that lands near the micropyle is trapped in a sticky substance called a pollen drop. As the pollen drop is absorbed into the ovule, the pollen grain is pulled closer to the micropyle. The pollen grain will continue to develop. ☑

Reading Check

9. Summarize pollination in conifers.

How do seeds develop?

Seed development can take as long as three years. After pollination, the pollen grain produces a pollen tube. The pollen tube grows through the micropyle and into the ovule. One of the four cells in the pollen grain undergoes mitosis, forming two sperm. The sperm travel through the pollen tube to an egg. When one sperm fertilizes an egg, a zygote is formed. The other sperm and the pollen tube dissolve.

The zygote depends on the female gametophyte for nutrition as it undergoes mitosis. After mitosis, an embryo with one or more cotyledons has formed. Cotyledons undergo photosynthesis and provide nutrition for the embryo when the seed sprouts.

The outside layer of the ovule forms a seed coat. The female cone opens and releases mature seeds.

chapter 23

Reproduction in Plants

section 2 Flowers

MAIN Idea

Flowers are the reproductive structures of anthophytes.

What You'll Learn

- about complete, incomplete, perfect, and imperfect flowers
- differences between monocots and eudicots
- about photoperiodism

Before You Read

On the lines below, describe the parts of your favorite flower. In this section, you will learn about the most important role of flowers—anthophyte reproduction.

____

____

Mark the Text

Identify Main Ideas As you read, underline or highlight the main ideas in each paragraph.

Read to Learn

Flower Organs

The colors, shapes, and sizes of flowers vary from species to species and are determined by each species' genetic makeup. All flowers have organs. Some organs provide support or protection, while other organs are involved in reproduction. Flowers can vary in structure and form from species to species.

Flowers usually have four organs—sepals, petals, stamens, and one or more pistils, as shown below. **Sepals** protect the flower bud and can look like small leaves or flower petals. **Petals** are usually colorful structures that can attract animals to pollinate the flower. Petals provide animals with a landing platform. Sepals and petals are attached to a flower stalk, called a peduncle.

Picture This

1. **Label** the structure to which sepals and petals are attached.

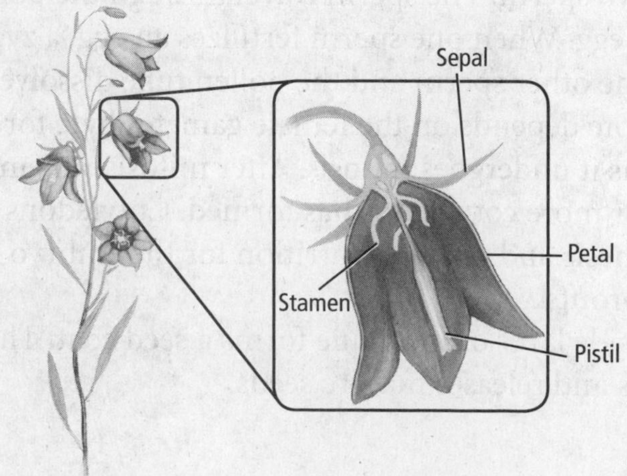

What are the reproductive flower structures?

Most flowers have several **stamens** which are the male reproductive organs. A stamen has two parts—a filament and an anther. The filament, also called the stalk, supports the anther. Inside the anther are cells that undergo meiosis and then mitosis to form pollen grains. Two sperm eventually form inside each pollen grain.

The female reproductive organ of a flower is the **pistil**. In the center of a flower is one or more pistils. A pistil usually has three parts—a stigma, a style, and an ovary. The style connects the stigma to the ovary. The ovary contains one or more ovules. A female gametophyte develops in each ovule, and one egg forms inside each female gametophyte.

Think it Over

2. Name What are the male and female reproductive organs?

Male: ________________

Female: ________________

Flower Adaptations

Many flowers can have modifications to one or more organs. Scientists categorize flowers using these modifications.

What structural differences do flowers have?

Flowers that have sepals, petals, stamens, and one or more pistils are called complete flowers. Flowers that are missing one or more of these organs are called incomplete flowers.

Flowers that have both stamens and pistils are called perfect flowers. Flowers that have either stamens or pistils are called imperfect flowers.

Flowers that have stamens are male flowers. Male flowers release pollen grains. Flowers that contain pistils are female flowers. Female flowers form fruit after they are fertilized.

The number of each flower organ varies from species to species. When the petal number of a flower is a multiple of four or five, as shown in the figure below, the plant usually is a eudicot. The number of the other organs often is the same multiple of four or five. Monocots usually have flower organs in multiples of three.

Picture This

3. Identify Label each flower as either monocot or eudicot.

How do plants pollinate?

Each anthophyte species has flowers of different sizes, shapes, colors, and petal arrangements. Many of these adaptations relate to pollination. Self-pollinating flowers are able to pollinate themselves or another flower on the same plant. Cross-pollinated flowers receive pollen from another plant. Some flowers must be cross-pollinated. Reproduction of plants that must be cross-pollinated depends on pollinators such as animals and wind. ☑

How do animals help pollinate flowers?

Many flowers that are pollinated by animals are brightly colored, have strong scents, or produce sweet nectar. When insects and other small animals move from one flower to another flower looking for nectar, they can carry pollen from flower to flower. Other insects move from flower to flower collecting pollen for food.

What types of plants are pollinated by the wind?

Flowers that are not brightly colored or do not have strong scents are usually pollinated by the wind. These flowers produce a lot of lightweight pollen that can be easily carried by the wind. This helps ensure that some pollen grains will land on the stigma of a flower of the same species. Also, the stamens of flowers that are pollinated by the wind are large and hang below the petals, exposing them to the wind. Most trees and grasses are pollinated by the wind.

What is photoperiodism?

After noticing that some plants only flowered at certain times of the year, plant biologists performed experiments to explain this. Scientists first focused on the number of hours of daylight that plants received. Later, scientists discovered that the number of hours of uninterrupted darkness influences the flowering of some plants. This flowering response is known as **photoperiodism** (foh toh PEER ee uh dih zum). Scientists also learned that the beginning of flower development for each plant species was a response to a range in the number of hours of darkness. The range of hours is known as a plant's critical period.

Scientists group plants into four types—short-day plants, long-day plants, intermediate-day plants, and day-neutral plants. The names reflect scientists' original focus—the number of hours of daylight. A more accurate name for a short-day plant would be a long-night plant.

Reading Check

4. Name What type of plant receives pollen from itself or another flower on the same plant?

____

____

Think it Over

5. Predict Which type of plant is most likely to flower only in the summer?

____

____

What are short-day plants?

As shown in the table below, a **short-day plant** flowers when it is exposed daily to a number of hours of darkness that is greater than its critical period. Short-day plants flower during the winter, spring, or fall when the number of hours of darkness is greater than the number of hours of light. Examples include pansies, poinsettias, and tulips. ☑

What are long-day plants?

A **long-day plant** flowers when it receives fewer hours of darkness than its critical period. Long-day plants flower during the summer. Examples of long-day plants are lettuce, asters, coneflowers, spinach, and potatoes.

What are intermediate-day plants?

An **intermediate-day plant** flowers as long as the number of hours of darkness is neither too great nor too few. Examples of intermediate-day plants are sugarcane and some grasses.

What are day-neutral plants?

A **day-neutral plant** flowers regardless of the number of hours of darkness as long as it receives enough light for photosynthesis to support growth. Examples of day-neutral plants are buckwheat, corn, cotton, tomatoes, and roses.

Reading Check

6. List In what seasons do short-day plants bloom?

____

____

Picture This

7. Classify Write an example of each plant type on the chart.

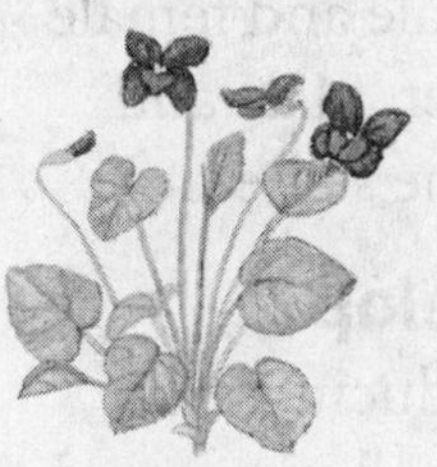
Longer than critical period

Shorter than critical period

Short-day plant

____

Shorter than critical period

Longer than critical period

Long-day plant

____

Longer or shorter than critical period

Intermediate critical period

Intermediate-day plant

____

Long night

Short night

Day-neutral plant

____

Reproduction in Plants

chapter 23

section 3 Flowering Plants

MAIN Idea

In anthophytes, seeds and fruits can develop from flowers following fertilization.

What You'll Learn

- the life cycle of a flowering plant
- about seed germination

Study Coach

Make Flash Cards Make a flash card for each underlined key term in this section. Write the term on one side of the card. Write the definition on the other side. Use the flash cards to review what you have learned.

Picture This

1. Calculate How many nuclei are present in the megaspore after the second round of mitosis?

____

Before You Read

Is a tomato a fruit or a vegetable? Explain your reasoning on the lines below. Then read the section to learn more about fruits.

____

____

Read to Learn

Life Cycle

Anthophytes are the most varied and widespread group of plants. Anthophytes have flowers. The sporophyte generation is dominant. The gametophyte generation depends on the sporophyte generation for support.

In anthophytes, the development of the male and female gametophyte begins in an undeveloped flower. Male and female gametophytes might not develop at the same time.

How do female gametophytes develop?

A cell in the ovary undergoes meiosis, producing four megaspores. Three megaspores break into small pieces and disappear. As shown in the figure below, the nucleus of the remaining megaspore undergoes mitosis three times. The result is one large cell with eight nuclei. There are three nuclei at each end and two nuclei in the center. The nuclei in the center are called <u>**polar nuclei**</u>. One of the three nuclei at the end closest to the micropyle becomes the egg. The cell that contains the egg and seven nuclei is the female gametophyte.

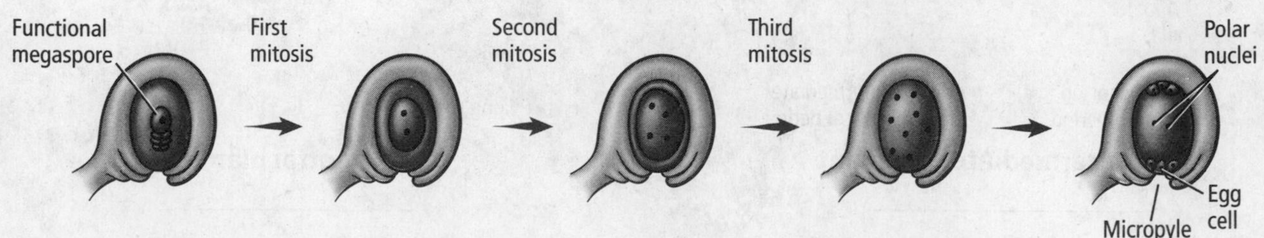

How do male gametophytes develop?

Specialized cells in the anther undergo meiosis and form microspores. The nucleus in each microspore undergoes mitosis and forms two nuclei called the tube nucleus and the generative nucleus. A thick protective cell wall forms around the microspore. The microspore is now an immature male gametophyte, or pollen grain.

What is exine?

Scientists can identify the family or genus of a pollen grain from the distinctive outer layer of its cell wall, called the exine. Paleontologists can trace the agricultural history of certain regions using pollen fossils. Forensic scientists use pollen evidence to help determine where and when some crimes were committed. ☑

How are flowering plants fertilized?

When pollen from the anther moves to the stigma of the pistil, the plant is pollinated. As shown in the figure below, after a plant is pollinated, the pollen grain forms a pollen tube. A pollen tube is an extension of the pollen grain that grows through the style to the ovary.

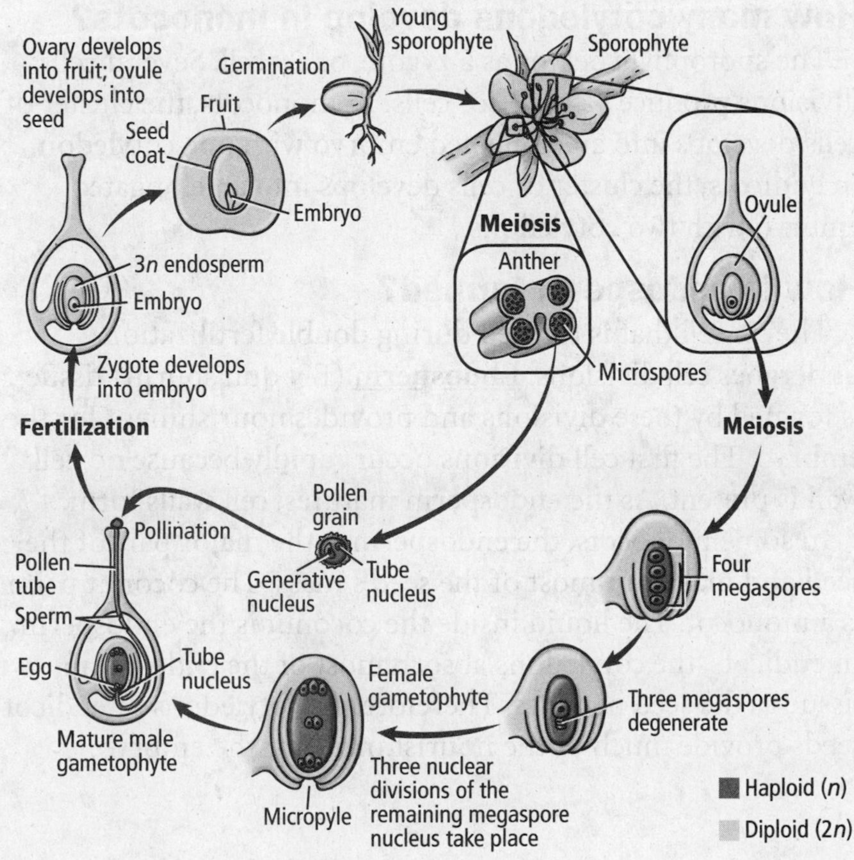

Reading Check

2. Identify What do paleontologists use pollen grains to determine?

______

______

______

Picture This

3. Highlight the area where the plant is pollinated. Circle the area where the embryo develops.

Think it Over

4. Explain Why does the length of the pollen tube depend on the pistil?

____

____

____

What determines the length of the pollen tube?

The length of the pollen tube depends on the length of the pistil. The pollen tube's length can vary from a few centimeters to more than 50 cm. As the pollen tube grows, the generative nucleus undergoes mitosis, forming two nonflagellated sperm nuclei. The pollen grain is now a mature gametophyte. The pollen tube grows through the micropyle and releases the two sperm nuclei. One sperm nucleus fuses with the egg, forming the zygote. The zygote is the new sporophyte. The other sperm nucleus and the two polar nuclei in the center of the ovule fuse, forming a triploid, or *3n* cell.

What is double fertilization?

Two fertilizations occur in an anthophyte egg. This is called double fertilization. Double fertilization occurs only in anthophytes. After fertilization, the ovule begins to develop into the seed and the ovary begins to develop into the fruit.

Results of Reproduction

Fertilization begins the long process of seed formation. Seeds of anthophytes are part of a fruit that develop from the ovary and sometimes other flower organs.

How many cotyledons develop in monocots?

The sporophyte begins as a zygote, or *2n* cell. Several cell divisions produce a cluster of cells. In monocots, the cluster of cells develops into an elongated embryo with one cotyledon. In eudicots, the cluster of cells develops into an elongated embryo with two cotyledons. ☑

Reading Check

5. Compare What is the difference between the seeds of monocots and eudicots?

____

____

____

____

How is endosperm formed?

The *3n* cell that is formed during double fertilization undergoes cell divisions. **Endosperm** (EN duh spurm) tissue is formed by these divisions and provides nourishment for the embryo. The first cell divisions occur rapidly because no cell wall is present. As the endosperm matures, cell walls form.

In some monocots, the endosperm is the major part of the seed and makes up most of the seed's mass. The coconut palm is a monocot. The liquid inside the coconut is the endosperm. In eudicots, the cotyledons absorb most of the endosperm tissue as the seed matures. Therefore the cotyledons of eudicot seeds provide much of the nourishment for the embryo.

What leads to the formation of fruit?

As the endosperm matures, the outside layers of the ovule harden and form protective tissue called the **seed coat**. Depending on the plant, the ovary can have one ovule or hundreds of ovules. As the ovule develops into a seed, the ovary changes. This change in the ovary leads to the formation of a fruit.

Fruits form from the ovary wall and other flower organs. For example, the seeds of the apple are within the core that develops from the ovary. The juicy tissue that we eat develops from other flower parts. As described in the table below, some fruits are fleshy, such as peaches and oranges, and some are dry and hard, such as walnuts and grains.

Fruit Type	Examples	Description
Simple fleshy fruits	peaches, apples, tomatoes, pumpkins	Fruits contain one or more seeds.
Aggregate fruits	strawberries, raspberries, blackberries	Fruits form from flowers with multiple female organs that fuse as the fruits ripen.
Multiple fruits	figs, pineapples, mulberries, osage oranges	Fruits form from many flowers that fuse as the fruit ripens.
Dry fruits	pods, nuts, grains	Mature fruits are dry.

How are seeds dispersed?

Fruits protect seeds. Fruits also help scatter, or disperse, seeds. Dispersal of seeds away from the parent plant increases the rate of survival of the offspring. When many plants grow in one area, they compete for light, water, and nutrients. If all seeds sprouted near the parent, the parent and all the offspring would compete for the same resources.

Fruits that are attractive to animals can be carried over long distances. Some animals gather and bury or store fruits. These animals might not recover all of the seeds, so some might sprout. Some seeds pass undamaged through the digestive tract of animals. These seeds are deposited on the ground with the animal's waste. Some of these seeds sprout. Seeds can be transported by water, animals, or wind. ☑

Picture This

6. Describe What is an aggregate fruit?

Reading Check

7. List three ways seeds are dispersed.

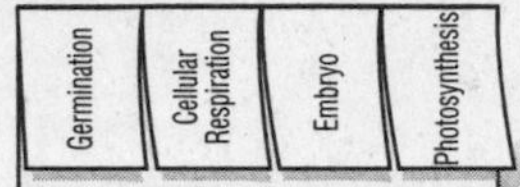

FOLDABLES

Take Notes Make a four-tab Foldable, as shown below. As you read, take notes and organize what you learn about the process of germination.

When does germination begin?

When the embryo in a seed starts to grow, the process is called **germination**. Many factors affect germination, including the presence of water, oxygen, and temperature range.

Germination begins when a seed absorbs water. As cells absorb water, the seed swells. This swelling breaks the seed coat. Water also transports materials to the growing regions of the seed.

Within the seed, digestive enzymes help break down the stored food. This broken-down food and oxygen are the raw materials needed for cellular respiration. Cellular respiration releases energy needed for growth.

As shown in the figure below, the **radicle** (RA dih kul) is the first part of the embryo that appears outside of the seed. The radicle starts absorbing water and nutrients from the environment. It will develop into the plant's root. The **hypocotyl** (HI puh kah tul) is the region of the stem nearest the seed. In many plants the hypocotyl is the first part of the seed to appear above the soil. In some eudicots, the hypocotyl pulls the cotyledons and the embryonic leaves out of the soil as it grows.

Photosynthesis begins as soon as the seedling's cells that contain chloroplasts are above ground and exposed to light. In monocots, seedling growth is slightly different. The cotyledons usually stay in the ground when the stem emerges from the soil.

Some seeds germinate soon after dispersal. Other seeds germinate after long periods of time. Most seeds that are produced at the end of the growing season enter dormancy. **Dormancy** is a period of little or no growth. Dormancy is an adaptation that increases the survival rate of seeds that are exposed to harsh environmental conditions. The length of dormancy varies from species to species.

Picture This

8. Explain How is germination different in monocots and eudicots?

____

____

____

____

____

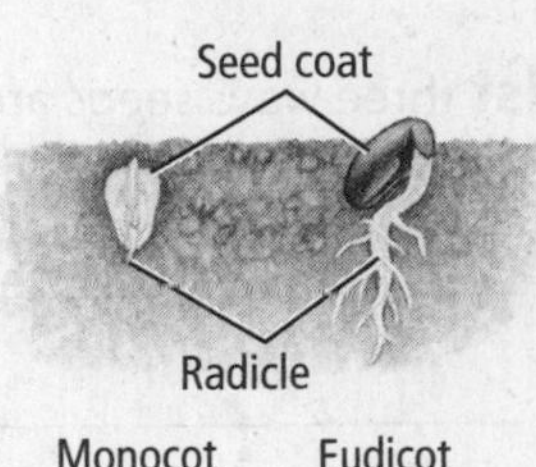

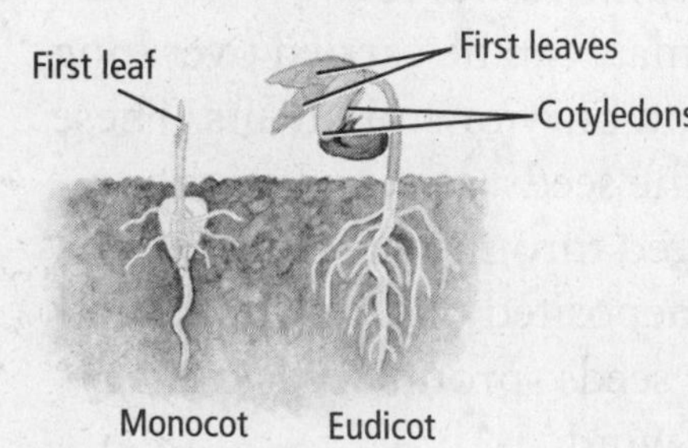

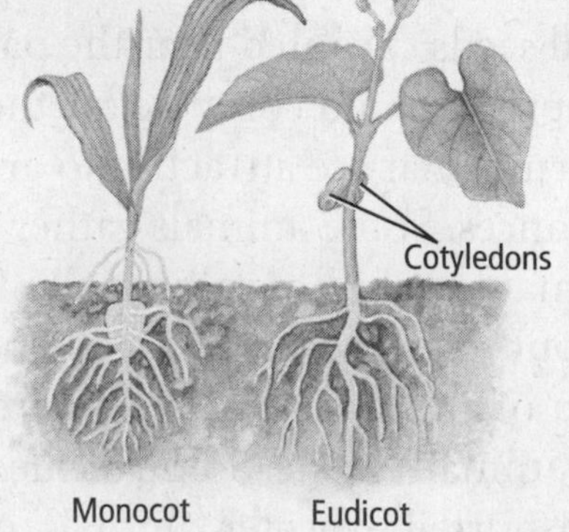

chapter 24 Introduction to Animals

section 1 Animal Characteristics

Before You Read

Animals have body coverings to help protect their bodies. On the lines below, list different types of body coverings you have observed among animals.

____

____

MAIN Idea

Adaptations enable animals to live in different habitats.

What You'll Learn

- the differences between animal structure and function
- the embryonic stages of animal development

Read to Learn

Study Coach

Make Flash Cards Make a flash card for each underlined term in this section. Write the term on one side of the card. Write the definition on the other side. Use the flash cards to review the key terms you have learned.

General Animal Features

All present-day animals might have evolved from choanoflagellates (KOH uh noh FLA juh layts), possibly the earliest true animals. Even the earliest animals were made up of many cells. As animals evolved, they developed adaptations that made it possible for them to survive in many habitats. These changes mark the branching points of the evolutionary tree.

Feeding and Digestion

Animals must feed on other living things. The structure of its mouth parts determines how an animal's mouth works. Some animals digest their food inside certain cells. Others digest their food in organs or cavities inside their bodies.

Support

Most animal species are **invertebrates**—animals without backbones. The bodies of many invertebrates have hard or tough outer coverings called **exoskeletons**. An exoskeleton supports and protects the animal's body. These animals shed their exoskeletons as they grow and make new ones.

Some invertebrates have internal skeletons called **endoskeletons**. An endoskeleton grows with the animal. Animals with both an endoskeleton and a backbone are **vertebrates**.

1. Draw Conclusions Which type of animal are you? (Circle your answer.)

a. invertebrate

b. vertebrate

Habitats

Animals have adaptations that make it possible for them to live in many habitats. Animals live in water and on land. They live in oceans, lakes, rivers, grasslands, forests, and deserts.

Animal Cell Structure

Plant cells have cell walls. Animal cells do not have cell walls. Animal cells, except in sponges, are organized into tissues that perform specific functions. ☑

Reading Check

2. Explain How are animal cells different from plant cells?

____

____

____

Movement

The evolution of nerve and muscle tissues is an important characteristic of the animal kingdom. Nerves and muscles enable animals to move in fast and complex ways.

Reproduction

Most animals reproduce sexually. Male animals produce sperm. Female animals produce eggs. Animals, such as earthworms, that produce both eggs and sperm are called **hermaphrodites** (hur MAF ruh dites). Hermaphrodites produce eggs and sperm at different times, so another individual of the same species is needed for reproduction.

In sexual reproduction, the sperm penetrates the egg. The result is a fertilized egg cell called a **zygote** (ZI goht). **Internal fertilization** occurs when the sperm and egg combine inside the animal's body. **External fertilization** occurs when the sperm and egg combine outside the animal's body. External fertilization requires a water environment for the sperm to swim to the egg.

Asexual reproduction means that a parent produces offspring that are genetically identical to itself. The table below summarizes some common methods of asexual reproduction.

Picture This

3. Highlight two types of asexual reproduction in which the offspring is the result of a lost body part.

Asexual Reproduction	
Budding	The offspring develops as a growth on the parent's body.
Fragmentation	The parent breaks into pieces, and each piece can develop into an adult animal.
Regeneration	A new organism can regrow from a lost body part.
Parthenogenesis	Eggs develop without fertilization.

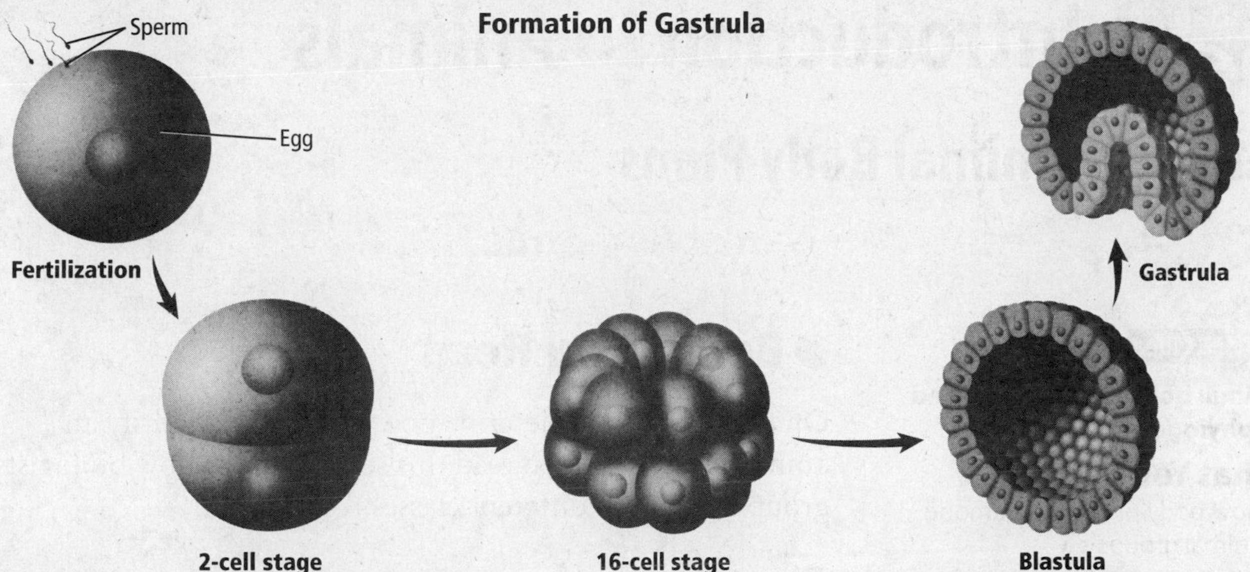

How does a zygote develop into a gastrula?

A zygote undergoes mitosis and cell division, as shown in the figure above. At the two-cell stage, the animal is called an embryo. The embryo's cells continue to divide and form a fluid-filled ball of cells called the **blastula** (BLAS chuh luh). As cell division continues, the blastula becomes indented as cells on one side move inward and form a gastrula. A **gastrula** (GAS truh luh) is a sac with two layers of cells and an opening at one end.

How do tissues develop?

The inner layer of cells in the gastrula is called the **endoderm**. The outer layer of cells in the gastrula is called the **ectoderm**. Some animals develop a layer of cells between the endoderm and ectoderm called the **mesoderm**. Each cell layer develops into certain types of tissues. The figure below identifies the tissues that form from the endoderm, ectoderm, and mesoderm in many animals.

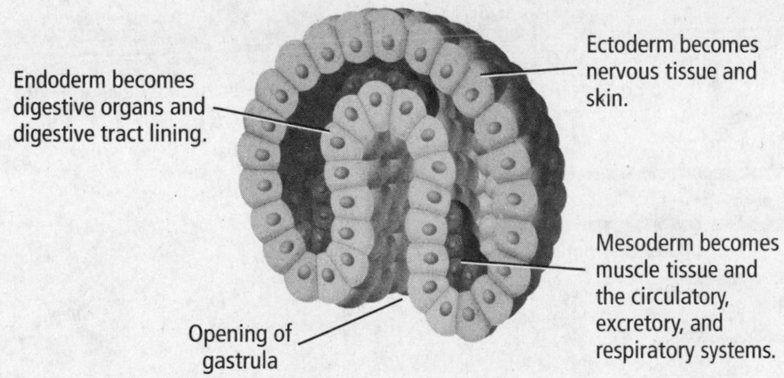

Picture This

4. Highlight the gastrula in the diagram. Label the opening of the gastrula.

Picture This

5. Identify What tissues develop from the gastrula's outer layer of cells?

____

____

Introduction to Animals

section 2 Animal Body Plans

MAIN Idea

Animal body plans are related to phylogeny.

What You'll Learn

- how body plans differ among animal groups
- the structure of body cavities
- how to distinguish between two types of coelomates

Before You Read

On the lines below, describe how you would identify an animal as a bird. Then read this section to see how biologists group animals in different classes.

Mark the Text

Identify Main Ideas As you read the section, highlight the main idea in each paragraph.

Read to Learn

Evolution of Animal Body Plans

On the evolutionary tree below, the branches show how major phyla of animals probably evolved from a common ancestor. Biologists map out relationships on the evolutionary tree by comparing the body features of organisms and the ways their embryos develop. Recently, biologists have begun to compare molecules. These studies have revealed that some animals might be more closely related to each other than their body features suggest.

Picture This

1. Determine What is distinguished at the first branching point on the tree?

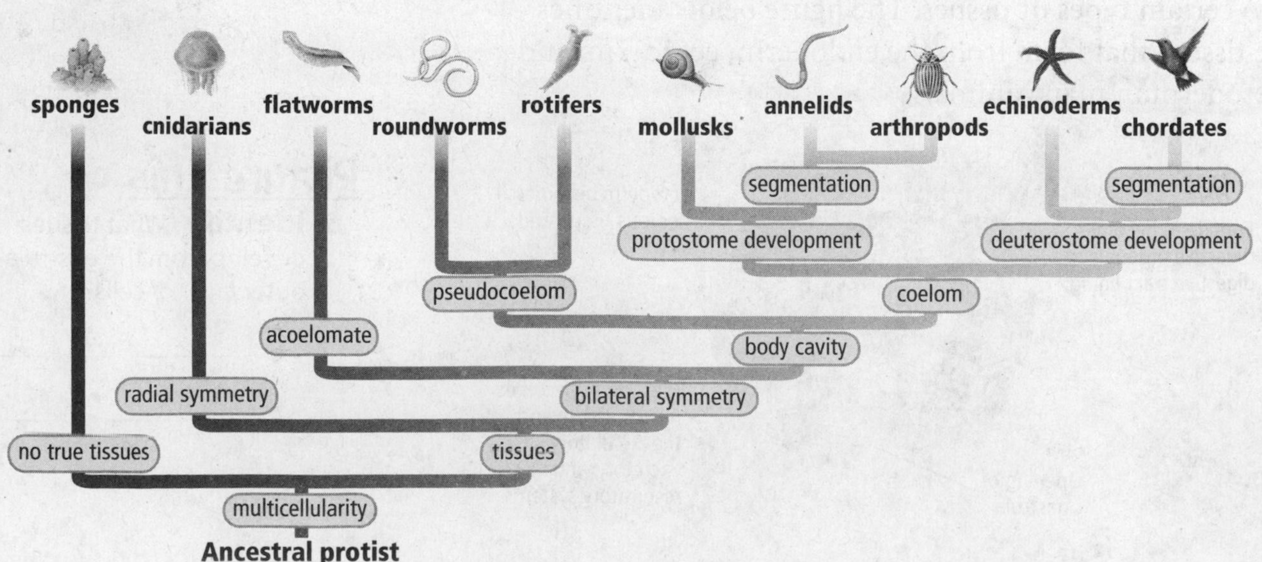

Development of Tissues

Tissues mark the first branching point on the evolutionary tree. Sponges are the only animals without tissues. They are on the no-true-tissues branch. All other phyla branch off from the tissues branch of the tree.

Symmetry

Symmetry (SIH muh tree) is the first branching point off the tissues branch. **Symmetry** describes the similarity, or balance, among body structures. The type of symmetry defines the kind of movements the animals can make.

What animal has an irregular shape?

Sponges are asymmetrical, which means their shapes are not regular. Animals with tissues have either radial or bilateral symmetry.

How do radial and bilateral symmetry differ?

Look at the figure below as you read the following discussion. Some animals have two halves that look almost the same when the animal is divided along any plane through its central line. This is **radial** (RAY dee ul) **symmetry**. A jellyfish is an animal with radial symmetry.

Animals with **bilateral** (bi LA tuh rul) **symmetry** have two halves that look like mirror images when the animal is divided along only one plane through its central axis. Birds and dogs are animals with bilateral symmetry.

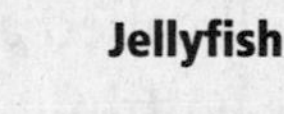

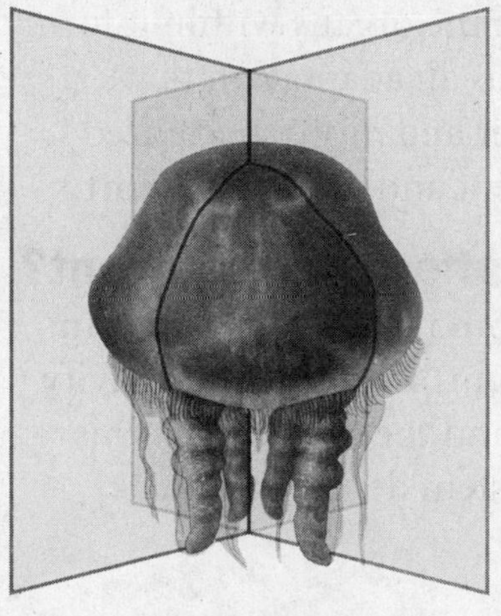

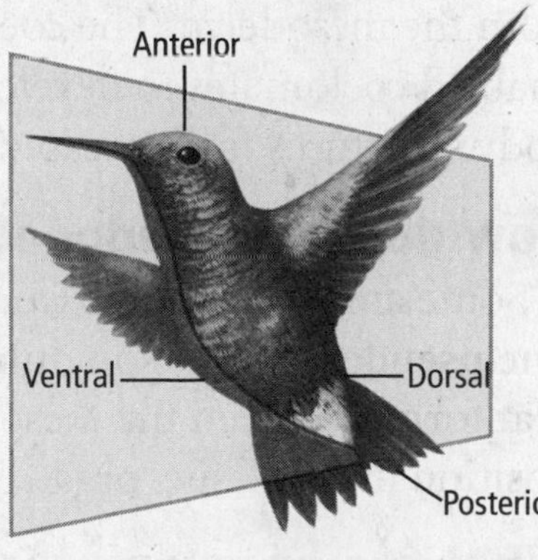

FOLDABLES™

Take Notes Make a three-column Foldables chart, as shown below. As you read, take notes and organize what you learn about three types of animal symmetry.

Symmetry		
Asymmetrical	Symmetrical	Bilateral

Picture This

2. Label Write the labels *bilateral symmetry* and *radial symmetry* below the appropriate animal in the figure.

Which cell layers are involved in development?

Most animals with radial symmetry develop from two cell layers—the ectoderm and the endoderm. All animals with bilateral symmetry develop from three cell layers—the ectoderm, the endoderm, and the mesoderm.

What is the body plan called cephalization?

An animal with bilateral symmetry has a head end and a tail end. The head end is called the **anterior** end. The tail end is called the **posterior** end. When the nervous tissue and sensory organs are located at the anterior end, the body plan is called **cephalization** (sef uh luh ZA shun). Most animals with cephalization move with their anterior ends first.

An animal with bilateral symmetry also has a backside and an underside. The backside is the **dorsal** (DOR sul) surface. The underside, or belly, is the **ventral** (VEN trul) surface.

Think it Over

3. Identify If an animal with bilateral symmetry has a brain, where is the brain located? (Circle your answer.)

a. posterior end
b. anterior end

Body Cavities

An animal with bilateral symmetry also has a gut where food is digested. The gut is either a sac inside the body or a tube that runs through the body. A saclike gut has one opening—the mouth. The mouth takes in food and disposes of wastes. A tubelike gut has an opening at each end. Food is taken in at the mouth and digested, nutrients are absorbed, and waste is then excreted through the anus.

Why was the coelom an important adaptation?

Most animals with bilateral symmetry have a fluid-filled cavity between the gut and the outside body wall called a **coelom** (SEE lum). The coelom and the organs within it form from the mesoderm. The coelom was an adaptation that enabled coelomates to develop larger and more specialized body structures for increased nutrient and waste transport.

How does a pseudocoelom affect development?

Some animals have a pseudocoelom rather than a coelom. The **pseudocoelom** (soo duh SEE lum) is a fluid-filled cavity that forms between the mesoderm and the endoderm. This position limits tissue, organ, and system development. ☑

Reading Check

4. Explain What is a disadvantage of a pseudocoelom?

____

____

____

What are acoelomates?

An **acoelomate** (ay SEE lum ate) is an animal that does not have a fluid-filled body cavity. This animal has a solid body with no circulatory system. Nutrients and wastes spread from one cell to another.

Development in Coelomate Animals

The figure below shows the coelomate part of the evolutionary tree. Notice that coelomates branch into two lines of development: protostomes (PROH tuh stohms) and deuterostomes (DEW tihr uh stohms). Biologists compare how embryos develop to decide if animals are closely related. ☑

How do protostomes develop?

The mouth of a **protostome** develops from the opening in the gastrula. The final outcome of each cell in the embryo cannot be changed during development. If one cell is removed, the embryo will not develop normally. Also, during development, the mesoderm splits down the middle, forming a coelom between the pieces.

How do deuterostomes develop?

In a **deuterostome**, the anus develops from the opening in the gastrula. The mouth forms from another part of the gastrula. The final outcome of each cell in the embryo can be changed during development. If any cell is removed, a new embryo can form from the cell. Also, during development, the coelom forms from two pouches of mesoderm.

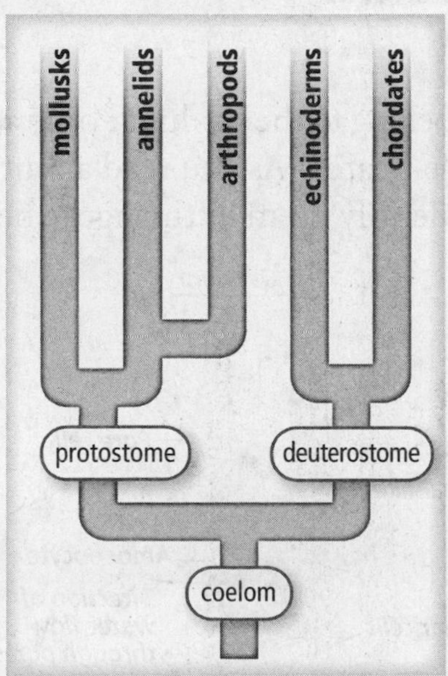

Segmentation

A segmented animal has a series of sections that are exactly alike. Segmentation has two advantages. First, other sections might be able to perform the function of a damaged section, enabling the animal to survive. Second, segments can move independently. This enables flexible and complex movement.

Reading Check

5. Determine What is one method biologists use to decide how to group animals into classes?

____

____

Picture This

6. Predict A snail is a mollusk. Based on the location of mollusks on the tree, how do you think a snail's coelom develops?

____

____

____

chapter 24 Introduction to Animals

section 3 Sponges and Cnidarians

MAIN Idea

Sponges and cnidarians were the first animals to evolve from a multicellular ancestor.

What You'll Learn

- the structures and functions of sponges and cnidarians
- how biologists classify sponges and cnidarians
- how cnidarians live with other animals in ways that benefit both species

Before You Read

On the lines below, describe ways that you contribute to your household, such as washing dishes or helping with yard work. In this section you will learn how cnidarians and other species benefit from living together.

____

____

____

Mark the Text

Identify Concepts

Highlight each question head in this section. Then use a different color to highlight the answers to the questions.

Read to Learn

Sponges

Living sponges belong to the phylum Porifera (po RIF uh ruh), which means "pore-bearer." As you read about the sponge's body structures, identify them in the figure below.

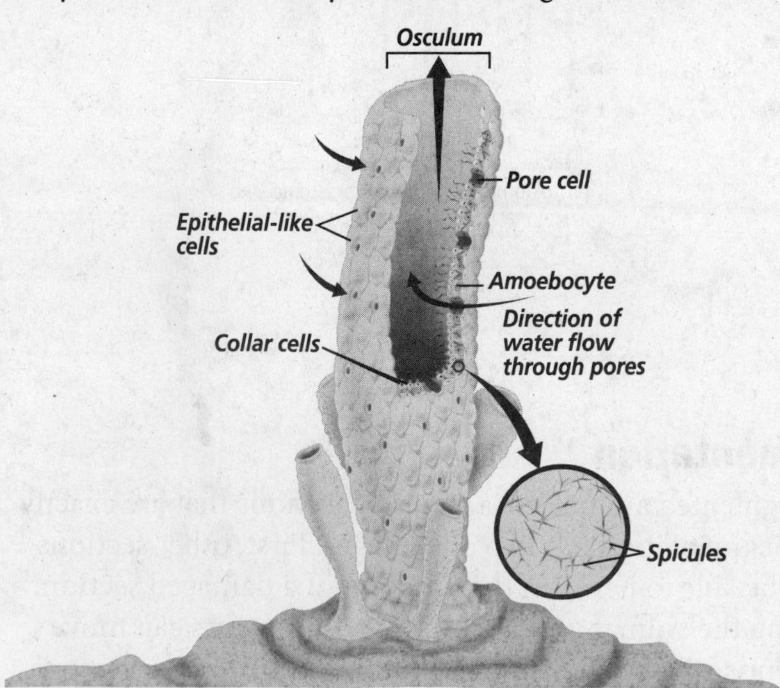

Picture This

1. Highlight the name of the structure through which water enters the body of the sponge. Circle the name of the structure through which water leaves the body of the sponge.

What body structures make up a sponge?

A sponge has no tissues. Its life functions take place in two layers of cells with a jellylike substance between the layers. An outer layer of epithelial-like cells protects the sponge. Collar cells with flagella line the inside. The flagella whip back and forth, drawing water into the sponge through pores. Water and waste leave the body through the mouthlike opening at the top called the osculum (AHS kyuh lum). ☑

How does a sponge take in and digest food?

The sponge is a filter feeder. **Filter feeders** get food by filtering small food pieces from water. The food pieces are digested within each cell. Adult sponges are **sessile** (SE sul)—they are attached to and stay in one place.

What structures support the sponge's body?

Amoebalike cells, called archaeocytes (ar kee OH sites), move within the jellylike substance between the cell layers. Archaeocytes help the sponge digest, reproduce, and excrete. Some archaeocytes also secrete spicules (SPIH kyuhls). Spicules are needlelike structures that support the sponge.

How do biologists classify sponges?

Biologists place sponges into three classes based on the sponges' support systems. Most sponges belong to the class Demospongiae (deh muh SPUN jee uh). These sponges have spicules made of spongin fibers, silica, or both. Sponges of the class Calcarea (kal KER ee uh) have spicules made of calcium carbonate and often feel rough. Sponges of the class Hexactinellida (heks AK tuh nuh LEE duh) have spicules made of silica, which make them look like spun glass.

How does a sponge respond to stimuli?

A sponge does not have a nervous system. When the outer cells detect stimuli, such as touch or chemical signals, the pores close to stop water flow.

Do sponges reproduce asexually?

Sponges reproduce asexually by fragmentation, by budding, or by producing gemmules (JEM yewlz). In fragmentation, a broken piece of sponge develops into a new adult. In budding, a small growth forms on a sponge, drops off, and settles in another spot to grow. In harsh conditions, some sponges form gemmules, which are seedlike particles that will grow again when conditions are favorable.

Reading Check

2. Explain Where do a sponge's life functions occur?

______

______

______

______

Think it Over

3. Draw Conclusions Why do you think a sponge reacts to stimuli by closing its pores?

______

______

______

How does a sponge reproduce sexually?

Most sponges are hermaphrodites, but some have separate sexes. As shown in the figure below, a sponge releases sperm into the water. Water currents carry the sperm to the collar cells of another sponge. The collar cells then change into cells that carry the sperm to an egg. After fertilization, the zygote develops into a free-swimming larva. The larva later attaches to a surface and develops into an adult.

Picture This

4. Apply Is a sponge's larva sessile? Explain.

____

____

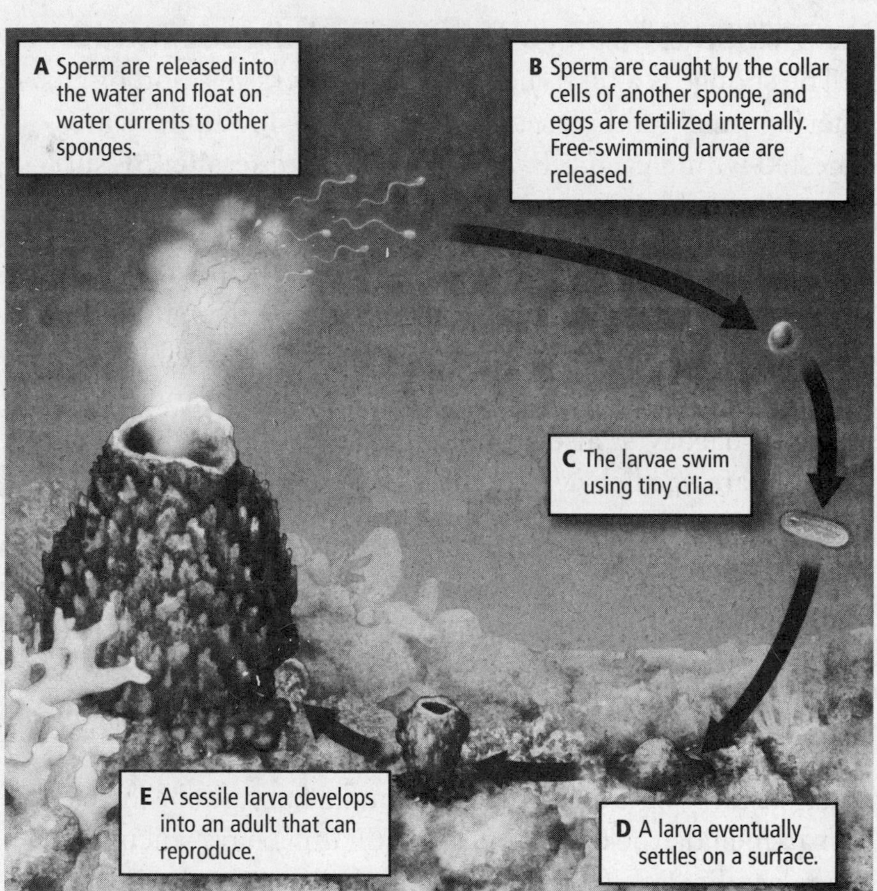

How might humans benefit from sponges?

Sponges produce chemicals that appear to discourage predators and prevent infection. Studies have shown that sponge chemicals might help fight infection, swelling, and tumors in humans. ☑

5. Name two ways sponges might benefit humans.

____

____

____

Cnidarians

Jellyfishes and sea anemones belong to the phylum Cnidaria (ni DARE ee uh). Most cnidarians live in the sea.

As you read about cnidarians, look at the table on the next page. The table compares the forms and functions of sponges and cnidarians.

What structures make up the cidarian body?

Like sponges, cnidarians have one body opening and two layers of cells. Unlike sponges, cnidarians have cell layers that are organized into tissues with specific functions. The outer layer protects the body. The inner layer digests food. Cnidarians have radial symmetry. Some are sessile, but others float.

Picture This

6. Highlight in one color ways that sponges and cnidarians are alike. Highlight in another color ways that sponges and cnidarians are different.

Comparing Sponges and Cnidarians

	Sponges	Cnidarians
Body plan	generally asymmetrical	radial symmetry
Feeding and digestion	filter feed; digest within individual cells	capture prey with nematocysts and tentacles; digest in gastrovascular cavity
Movement	sessile	sessile or float in the sea
Response to stimuli	no nervous system; cells react to stimuli	simple nervous system made up of a nerve net
Reproduction	most are hermaphrodites and reproduce sexually; asexual reproduction by fragmentation, budding, or gemmule production	separate sexes reproduce sexually; polyp stage can reproduce asexually by budding

How does a cnidarian capture food?

Cnidarians have tentacles armed with stinging cells called **cnidocytes** (NI duh sites). Cnidocytes contain nematocysts. A **nematocyst** (nih MA tuh sihst) is a capsule that holds a coiled tube containing poison and barbs. The nematocyst works like a tiny harpoon. In response to stimuli, the nematocyst releases a barb into its prey. After capture, the prey is brought to the cnidarian's mouth. ☑

Reading Check

7. Summarize What is the purpose of a nematocyst? (Circle your answer.)

a. to digest food
b. to capture food

How does digestion occur in a cnidarian?

The inner cell layer of a cnidarian surrounds its **gastrovascular** (gas troh VAS kyuh lur) **cavity**. The cells lining this cavity release digestive juices over the captured prey. Undigested food is disposed of through the mouth.

What is a nerve net?

Cnidarians have a nervous system made up of a nerve net. The **nerve net** conducts impulses to and from all parts of the body. These impulses cause musclelike cells to contract. Contractions of the musclelike cells in the tentacles help the animal capture prey. Cnidarians have no blood vessels, respiratory systems, or organs for excretion.

How do cnidarians reproduce?

Cnidarians have two body forms. The **polyp** (PAH lup) form has a tube-shaped body and a mouth surrounded by tentacles. The **medusa** (mih DEW suh) (plural, medusae) form has an umbrella-shaped body and tentacles that hang down. On its underside between the tentacles is a mouth. ☑

Jellyfishes use both body forms in their life cycles. In the medusa stage, jellyfishes release eggs and sperm into the water where fertilization occurs. Zygotes develop into free-swimming larvae. The larvae settle and grow into polyps. Polyps then reproduce asexually to form new medusae.

Reading Check

8. Identify What are the two main body forms of the cnidarian?

____

____

____

How do biologists classify cnidarians?

Biologists place cnidarians into four main classes: Hydrozoa (hydroids), Scyphozoa (jellyfishes), Cubozoa (box jellyfishes), and Anthozoa (sea anemones and corals).

What body forms do hydroids have?

Hydroids have both polyp and medusa stages in their life cycles. Most hydroids form colonies.

What is the main body form of jellyfishes?

Jellyfishes look like clear jelly floating near the water's surface. The saucer-shaped medusa is the main body form, although they do have a polyp stage. Box jellyfishes get their name from their boxlike medusae.

How are sea anemones and corals similar?

The polyp is the main body form of sea anemones and corals. The polyps extend their tentacles to feed. Sea anemones live as individual animals. Corals live in colonies of polyps. Corals secrete calcium carbonate shelters around their bodies. The living part of a coral reef grows on top of the shelters left by earlier generations.

Corals form a symbiotic relationship with a species of protists. The protists produce oxygen and food that the corals use. The protists use the carbon dioxide and waste materials the corals produce.

Think it Over

9. Apply A clown fish makes its home among the stinging tentacles of anemones. How would this benefit the fish?

____

____

____

What is mutualism?

Some anemones have a mutualistic relationship with hermit crabs. In a mutualistic relationship, both organisms benefit. The crabs carry sea shells with anemones attached, moving the anemones to better feeding sites. The crab is protected by the anemone's stinging cells.

Worms and Mollusks

section 1 Flatworms

Before You Read

Some animals produce mucus. On the lines below, suggest how animals use mucus. Then read the section to learn how flatworms use mucus to function.

MAIN Idea

Flatworms are acoelomates that can be free-living or parasitic.

What You'll Learn

- how flatworms maintain the proper internal balance
- the similarities and differences in the three classes of flatworms

Read to Learn

Mark the Text

Read for Understanding As you read this section, highlight any sentence that you do not understand. Reread the highlighted sentences to make certain that you understand their content. Ask your teacher to help you with anything that you still do not understand.

Body Structure

Flatworms belong to the phylum Platyhelminthes (pla tee HEL min theez). They are acoelomates—they have no coelom. A flatworm has a solid, thin, flat body with a head and body organs. Flatworms were among the first animals to develop bilateral symmetry.

Most flatworms are parasites. They live in the bodies of other animals. Other flatworms are free-living. They live in sea water, freshwater, or on moist land. The figure on the next page shows a free-living planarian. As you read about flatworms, locate the body structures in the figure.

What organs take in and digest food?

To feed, a free-living flatworm extends a tubelike organ out of its mouth. This organ is called the **pharynx** (FER ingks). The pharynx releases digestive juices that begin to digest the prey. Free-living flatworms then suck the food into their digestive tracts to finish digesting. Disposal of waste is through the mouth, which is the only body opening.

Parasitic flatworms attach to their hosts using hooks and suckers. Some have a reduced digestive system. They feed on blood and other body tissues. Others have no digestive system. They absorb food that has been digested by the host through their body walls. ☑

Reading Check

1. Explain What is the purpose of a flatworm's hooks and suckers?

How do respiration, circulation, and excretion occur?

Flatworms do not have organs for respiration or circulation. Oxygen and nutrients move through their body cells by diffusion. Cellular wastes are also disposed of by diffusion.

Flatworms have a system for excretion. As shown in the figure below, the system is made up of small tubes that branch throughout the body. **Flame cells** that are lined with cilia extend from the branches of these tubes. The cilia sweep water and waste into the small tubes. The water and waste then exit the body through pores. The mouth also helps the flatworm dispose of waste and maintain the proper balance of water.

Picture This

2. Highlight the body structures involved in feeding and digestion. Circle the names of the structures that are part of the excretory system. Which structure is involved in both feeding and waste disposal?

____

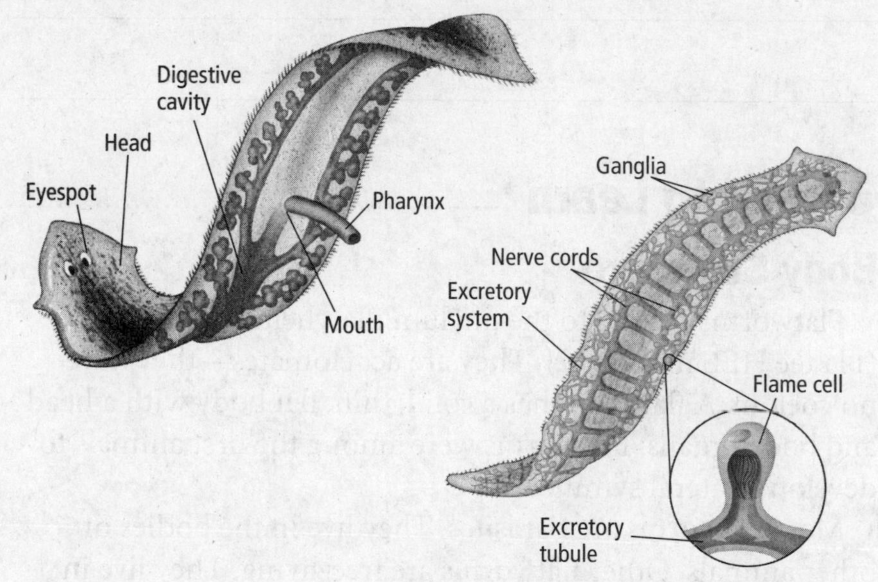

What structures make up the nervous system?

The nervous system controls the body's response to stimuli. Two nerve cords run the length of the flatworm's body. Nerve tissue connects the two nerve cords. A **ganglion** (plural ganglia) is a group of nerve cells that manages incoming and outgoing nerve signals. Ganglia at the head end of the nerve cords manage nerve signals in flatworms.

3. Determine Why is mucus an important adaptation for flatworms that live in streams?

____

____

____

How do flatworms move?

Muscle contractions in the body wall enable some flatworms to move. Most free-living flatworms use cilia on their undersides to propel themselves. Mucous lubrication makes movement easier. Mucus also helps flatworms attach themselves to rocks in swiftly moving streams. ☑

Do flatworms reproduce sexually and asexually?

Flatworms are hermaphrodites. They produce both eggs and sperm. During sexual reproduction, two flatworms exchange sperm. The eggs are fertilized inside their bodies.

Free-living flatworms can also reproduce asexually through regeneration. During **regeneration**, body parts that have been lost due to damage or predator attack can regrow, as shown in the figure below.

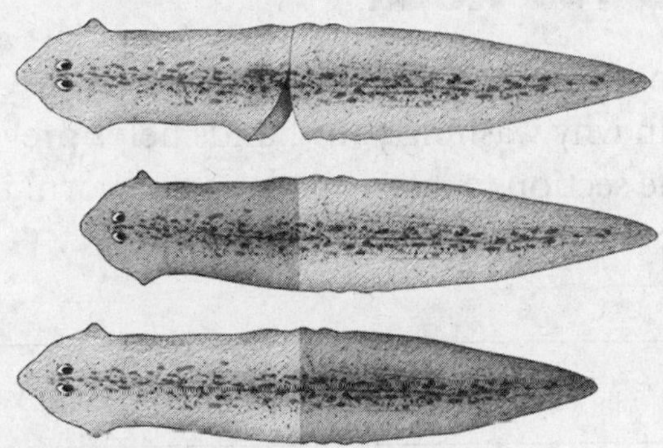

Picture This

4. Describe what happens when a planarian is cut in half across the middle.

Diversity of Flatworms

Biologists group flatworms in three main classes: Turbellaria (tur buh LER ee uh), Trematoda (trem uh TOH duh), and Cestoda (ses TOH duh). Turbellarians are free-living. Trematodes and cestodes are parasites.

How do sensory cells benefit turbellarians?

Turbellarians live in freshwater, salt water, and moist soils. A turbellarian has eyespots to detect light. At its anterior end, other sensory cells detect chemical signals given off by prey.

Think it Over

5. Explain How does the ability to detect chemical signals help turbellarians obtain food?

What trematode infects its host?

Flukes are one type of trematode. A fluke infects the blood and body organs of its hosts. Fluke eggs clog blood vessels. A fluke needs two hosts to complete its life cycle.

What is the life cycle of a cestode?

A tapeworm, a type of cestode, attaches to the lining of its host's intestines. It holds on using the hooks and suckers of its **scolex** (SKOH leks), a structure at its head end.

Behind the scolex are sections called **proglottids** (proh GLAH tihdz). As new proglottids form near the scolex, older ones move back. The last segments break off and pass out of the host. Other animals become infected by eating vegetation or drinking water that contains tapeworm proglottids.

chapter 25 Worms and Mollusks

section 2 Roundworms and Rotifers

MAIN Idea

Roundworms and rotifers have a more highly evolved gut than flatworms.

What You'll Learn

- how to compare the features of roundworms to the features of flatworms
- how different roundworms move
- how parasites infect their hosts

Before You Read

Frequent hand washing can help prevent disease. On the lines below, explain why washing your hands helps prevent disease. Then read the section to learn about roundworm infections and how to prevent them.

____

____

____

Study Coach

Create a Quiz After you read this section, create a five-question quiz from what you have learned. Then, exchange quizzes with another student. After taking the quizzes, review your answers together.

Read to Learn

Body Structure of Roundworms

Roundworms belong to the phylum Nematoda (ne muh TOH duh). Nematodes have a pseudocoelom and bilateral symmetry. They do not have segments. Their bodies become more narrow toward each end.

Roundworms come in many sizes and live in many habitats. Some are parasites.

How do roundworms feed and digest?

Most roundworms are free-living. Some capture tiny invertebrates. Others feed on decaying plants and animals. Food enters the digestive tract through the mouth. Waste is disposed of through the anus at the end of the digestive tract.

How do nutrients and gases move through the body?

Like flatworms, roundworms have no organs for circulation or respiration. Most nematodes exchange gases and get rid of cellular wastes by diffusion through their body coverings. Ganglia and nerve cords manage nematode responses to stimuli. Nematodes can detect touch and chemicals. Some free-living roundworms can detect light. ☑

Reading Check

1. Identify What process do roundworms use to move nutrients and gases through their bodies?

____

Why do roundworms thrash when they move?

A roundworm's muscles pull against the outside body wall and the pseudocoelom. The pseudocoelom acts as a **hydrostatic skeleton**—fluid within a closed space that provides rigid support for muscles to work against. As one muscle contracts, another relaxes. This causes a thrashing movement, as shown in the figure below.

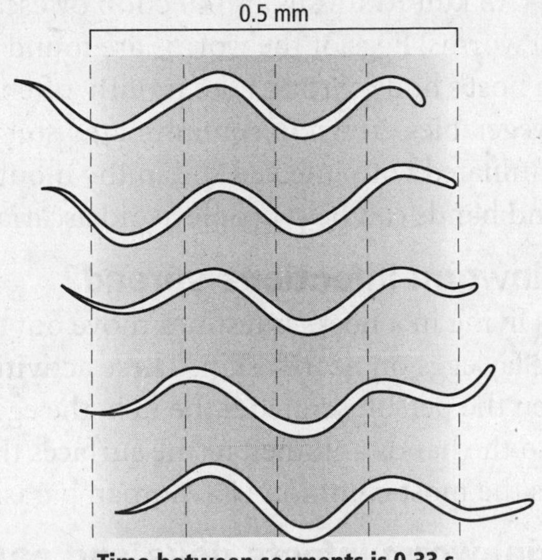

Time between segments is 0.33 s

How do roundworms reproduce?

Roundworms reproduce sexually. Fertilization occurs inside the body. The larvae of free-living roundworms hatch from fertilized eggs. Development of parasitic roundworms often requires more than one host or different locations within one host's body.

Diversity of Roundworms

Parasitic roundworms cause various diseases in plants and animals. Most roundworms do not harm plants, but some can kill trees and crops. Roundworm infection in humans is usually the result of poor personal hygiene or poor sanitation.

What is trichinosis?

The larvae of *Trichinella* worms cause trichinosis. **Trichinosis** (trih keh NOH sis) is a disease contracted by eating raw or undercooked pork or wild game infected with the larvae. Female worms burrow into the intestinal walls of the host animal. After their eggs hatch, the larvae burrow into organs and muscles, forming cysts.

Picture This

2. Calculate How long did it take the worm to move to its final location? Show your work.

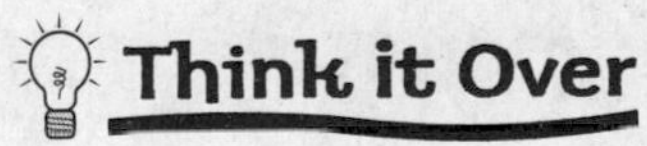

3. Draw Conclusions How can people avoid contracting trichinosis?

How do hookworms enter the host's body?

Hookworm infections occur when people walk barefoot on contaminated soil. A hookworm cuts through the bare skin. It travels through the bloodstream to the lungs and then to the windpipe. When it is coughed up and swallowed, it moves to the small intestine. There, it attaches and feeds.

How can people avoid ascariasis?

Ascariasis (AS kuh RI uh sus) is infection by ascarid (AS kuh rid) worms. Eggs of the worms are found in soil. They enter a host's body through the mouth, often by the host eating vegetables grown in contaminated soil or placing hands contaminated with infected soil in the mouth. Washing vegetables and hands can help people avoid ascariasis.

Think it Over

4. Infer Why do you think children are more likely than adults to become infected by roundworms?

____

____

____

____

____

How do pinworm infections spread?

Pinworms living in a host's intestines move out through the anus and lay eggs on nearby skin. These activities cause itching. When the person scratches the itch, the eggs are transferred to the hands and then to the surfaces that are touched. It is the most common U.S. human parasite.

What filarial worm infects dogs and cats?

Filarial (fuh LER ee uhl) worms cause the disease elephantiasis (el uh fun TI uh sus) in tropical areas. Mosquitoes transfer the infection from host to host. The worms block the flow of lymph fluid, causing legs to enlarge. One type of filarial worm—heartworm—infects dogs and cats in the United States.

Rotifers

Rotifers are tiny animals with wheel-like rings of cilia around their mouths. Most live in freshwater. They have bilateral symmetry and a pseudocoelom.

How do rotifers move?

Rotifers use their cilia to move through water. Glands at the tail end secrete a substance used to attach to a surface.

What organ systems do rotifers have?

Rotifers use their cilia to bring food into their digestive tract, which is open at both ends. Rotifers exchange gases and expel wastes through their body walls. Their heads include sensory bristles and eyespots. Some reproduce sexually. For others, reproduction involves diploid or haploid eggs. ☑

Reading Check

5. Identify Rotifers use their cilia for what purpose?

____

____

Worms and Mollusks

section 3 Mollusks

Before You Read

On the lines below, describe how it feels when you use your hands or swim fins to swim. In this section you will learn how some mollusks clap their shells together to propel themselves through water.

____

____

MAIN Idea

Mollusks are coelomates with a muscular foot, a mantle, and a digestive tract with two openings.

What You'll Learn

- the function of the mantle and its advantage to mollusks
- the importance of mucus and the muscular foot to mollusks

Read to Learn

Body Structure

Members of the phylum Mollusca include snails, slugs, clams, octopuses, and squids. Many mollusks live in the sea. Some live in freshwater and some in moist land. Mollusks might have been the first animals to develop coeloms. A coelom (SEE lum) enables more complex tissues and organs to develop.

A mollusk has bilateral symmetry. Its soft internal body holds a digestive tract with two openings. A mollusk's body also features a muscular foot and a mantle. The **mantle** (MAN tuhl) is a membrane that surrounds the internal organs. The mantle secretes calcium carbonate to form a shell around some mollusks. Other mollusks, such as slugs, have no hard covering.

Mark the Text

Restate the Main Point Highlight the main point in each paragraph. Then restate each main point in your own words.

How does a radula help mollusks obtain food?

Many mollusks have a rough structure called a radula in their mouths. The **radula** (RA juh luh) is a tonguelike organ with rows of teeth. A mollusk uses the radula to scrape food into its mouth. A mollusk that eats vegetation can use its radula to scrape algae off rocks. A mollusk that eats meat can use its radula to tear open prey. Filter feeders, such as clams, do not have a radula.

Think it Over

1. **Analyze** Why do you think filter feeders have no radulae?

____

____

____

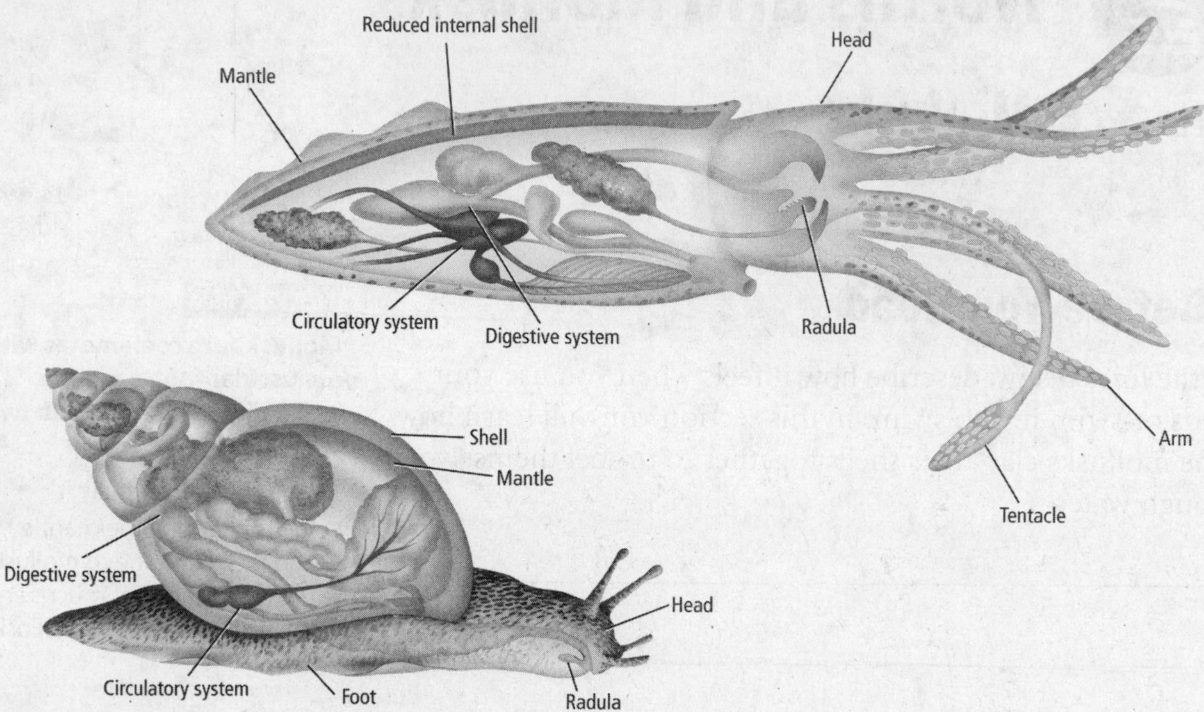

Picture This

2. Contrast How is the shell of the squid different from the shell of the snail?

____

____

____

____

What are the structures of snails and squids?

Compare the bodies of the snail and the squid in the figure above. Both have coelomate body plans. Their digestive and circulatory systems are highly developed.

What functions do gills perform?

Most mollusks use gills for respiration. **Gills** are a part of the mantle. Gills are made up of thin threads like the fringes of a blanket. As water moves over the gills, blood moving within the gills absorbs oxygen from the water and releases carbon dioxide from the blood. Some mollusks also use their gills for filter feeding.

How do mollusk circulatory systems work?

Most mollusks have an **open circulatory system**. Blood is pumped out of vessels into open spaces surrounding the body organs. The blood delivers oxygen and nutrients to the tissues while removing carbon dioxide. This slow delivery system works well for slow-moving mollusks.

A major evolutionary step in some mollusks was the development of a **closed circulatory system**. Blood is confined to vessels as it moves through the body. Fast-moving mollusks need energy quickly. A closed circulatory system quickly delivers nutrients and oxygen. ☑

Reading Check

3. Compare Which circulatory system delivers nutrients and oxygen more quickly? (Circle your answer.)

a. open circulatory system

b. closed circulatory system

How do mollusks dispose of wastes?

Most mollusks dispose of wastes using structures that filter the blood called **nephridia** (nih FRIH dee uh). Nephridia help mollusks maintain homeostasis, or balance, of their body fluids. ☑

Reading Check

4. State What is the primary purpose of nephridia?

Do mollusks have brains?

All mollusks have nervous systems. More highly evolved mollusks, such as octopuses, have brains. Octopuses even have eyes similar to those of a human.

How do mollusks move?

Clams use their muscular feet to burrow into wet sand. Some mollusks with two shells clap their shells together to produce short bursts of speed through water. Slugs and snails creep along on slime trails of mucus secreted by glands in their feet. Octopuses and squids produce bursts of speed by ejecting water forcefully through tubes called **siphons**.

What are the life stages of a mollusk?

All mollusks share a similar life cycle, illustrated in the figure below. Mollusks reproduce sexually. Mollusks that live in water release eggs and sperm into the water for fertilization. Many mollusks that live on land are hermaphrodites. For them, fertilization occurs internally.

Picture This

5. Identify the mollusk whose life cycle is shown in the figure.

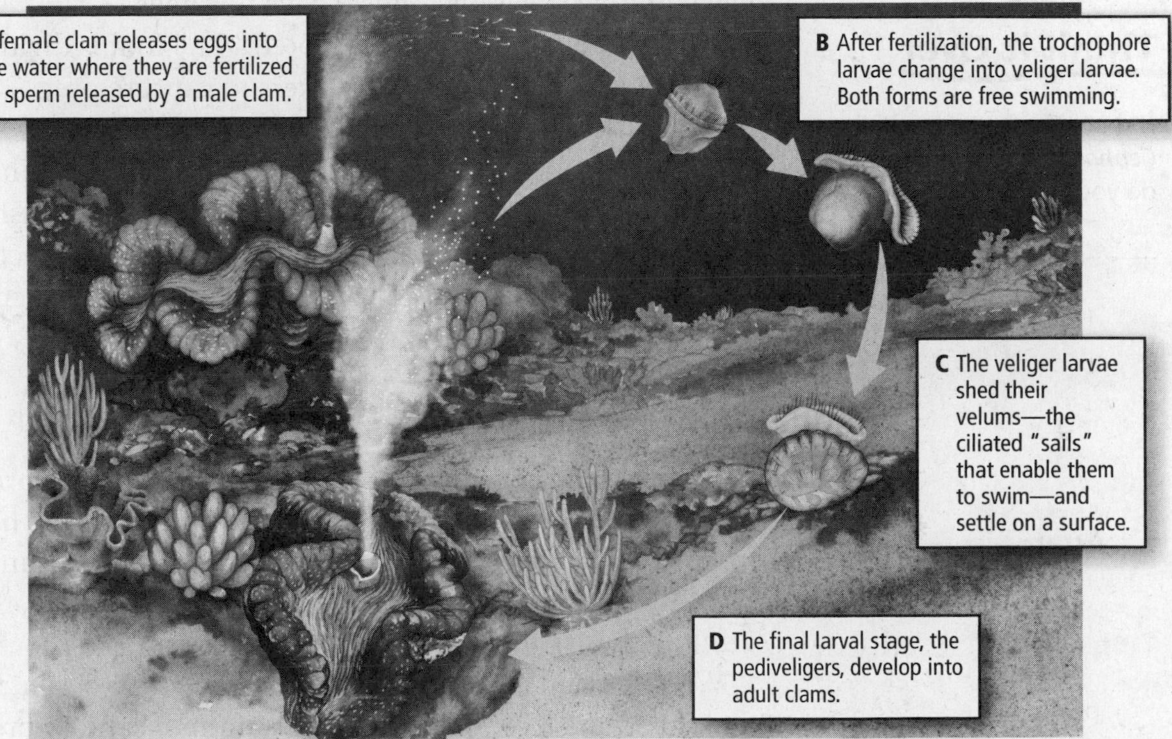

Diversity of Mollusks

Biologists group mollusks in three main classes based on the structure of the shell and foot. The three classes are gastropods, bivalves, and cephalopods.

What main feature distinguishes gastropods?

Gastropoda means "stomach-footed." The stomach is located on top of a long, muscular foot.

Most gastropods, such as snails and conches, have one shell. They retreat into their shells for protection. Slugs and nudibranchs have no shells. They secrete a mucous covering.

Slugs hide for protection. Nudibranchs have poisonous tissues. The poison comes from the poisonous nematocysts of the jellyfishes they eat. The nudibranchs' bright colors warn predators that they are poisonous.

What feature distinguishes bivalves?

Members of the class Bivalvia have two shells. A clam uses its muscular foot to burrow into wet sand. Mussels attach to rocks with a sticky substance. Scallops clap their shells together to produce short bursts of speed through water.

How do cephalopods protect themselves?

Members of the class Cephalopoda move quickly. Only the chambered nautilus has an outer shell. Squids and cuttlefishes have internal shells. Octopuses have no shells.

Cephalopoda means "head-footed." The foot is divided into arms and tentacles with suckers to capture prey.

Octopuses can change color to blend with their surroundings. When threatened, octopuses shoot an inky substance that confuses predators. Squids and cuttlefishes can also change color and use an inky substance for protection. The chambered nautilus can retreat into its shell, which blends with the ocean bottom.

Think it Over

6. Explain The bright colors of nudibranchs make them easy for predators to see. How is their bright color also a form of defense?

____

____

____

____

Think it Over

7. Infer *Gastro* means belly. *Cephalo* means head. What do you think *pod* means?

____

Ecology of Mollusks

Mollusks are considered keystone species in many areas. A keystone species is one whose health influences the health of the entire ecosystem. For example, some clams clean the ecosystem by filtering water. If their numbers decline, the unfiltered water harms the entire ecosystem.

When a grain of sand becomes trapped in an oyster, the oyster secretes a coating around the grain of sand to protect itself. Layers of this coating eventually become a pearl.

Worms and Mollusks

section 4 Segmented Worms

Before You Read

Do you sometimes wear a metal chain bracelet or necklace? On the lines below, explain why a string of links is a good structure for metal jewelry. Then read the section to learn how worms benefit from a body plan of links, or segments.

____

____

MAIN Idea

The segments of segmented worms allow for specialized tissues and efficient movement.

What You'll Learn

- how to distinguish annelids from flatworms and roundworms
- the features of the three main classes of annelids

Read to Learn

Body Structure

All members of the phylum Annelida have segments. Annelids live in many habitats. Most live in the sea. Others are earthworms. Leeches live as parasites.

Like flatworms and roundworms, annelids have bilateral symmetry and two body openings. Unlike flatworms and roundworms, annelids have segments and a coelom.

Walls of tissue separate the segments within the worms. Each segment contains structures for digestion, excretion, and movement. The fluid in the coelom of each segment provides rigid support. Segments move independently. If one segment is damaged, other segments can take over its functions. Groups of segments perform specific functions such as sensing or reproducing.

How do earthworms take in and digest food?

A digestive tract runs through all segments of an earthworm. Food and soil taken in by the mouth pass into the <u>**crop**</u>, where they are stored until passed on to the gizzard. The <u>**gizzard**</u> is a sac containing hard particles that help grind soil and food before passing them to the intestine. In the intestine, nutrients are absorbed. Waste is disposed of through the anus.

Study Coach

Make an Outline Make an outline of the information you learn in this section. Start with the headings. Include the boldface key terms.

FOLDABLES™

Organize and Learn Make a three-tab Venn diagram Foldable, as shown below. Note what you have learned about flatworms and roundworms. As you read, note what you learn about segmented worms. Then note what they have in common.

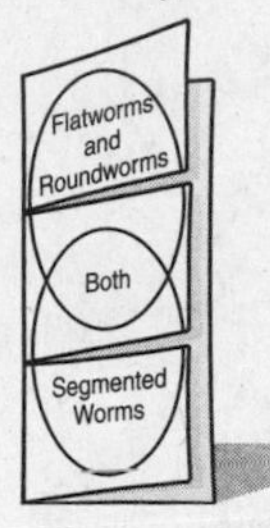

How do respiration, circulation, and excretion occur?

Most annelids have a closed circulatory system. Muscular vessels at the head serve as hearts that pump the blood. Blood moves toward the head in the dorsal vessel. Blood moves toward the tail in the ventral vessel.

Earthworms exchange gases through their skin. They absorb oxygen and release carbon dioxide. Almost every segment has two nephridia, as shown below. Wastes are collected in the nephridia and transported through the coelom and disposed of through the anus. The nephridia also maintain homeostasis, or balance, of body fluids.

Picture This

1. Highlight the structures involved in feeding, digesting, and excreting wastes. Highlight in another color the parts of the circulatory system. Circle the parts of the nervous system.

What stimuli can an earthworm detect?

Parts of the head end of most annelids are adapted to sense the environment. Annelids have brains and nerve cords composed of ganglia. Earthworms can detect both light and vibrations.

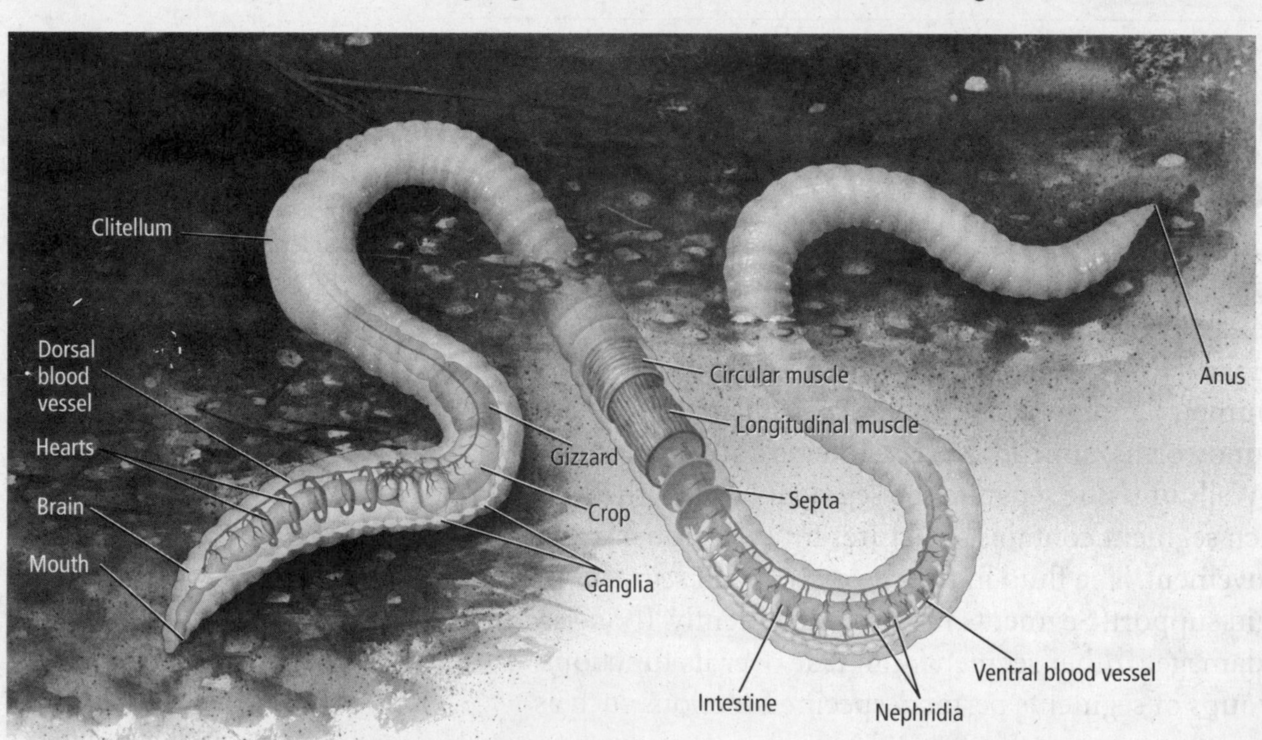

How do earthworms move?

To move, an earthworm contracts muscles that circle each segment. This contraction squeezes the fluid in each segment's coelom, causing each segment to get longer and thinner. Next, the worm contracts the muscles that run the length of its body, causing the segments to shorten. The result is movement. Tiny bristles called <u>**setae**</u> (SEE tee) on each segment push into the soil to anchor the worm as it moves.

How do annelids reproduce?

Most annelids have separate sexes. Earthworms are hermaphrodites. During sexual reproduction, sperm pass between two earthworms near the clitellum. The **clitellum** is a thickened band of segments that produces a cocoon. Sperm and eggs pass into the cocoon as it slips forward off of the body. The cocoon protects the young until they hatch. Some annelids reproduce asexually by fragmentation. ☑

Reading Check

2. Determine What is the advantage of a cocoon in reproduction?

Diversity of Annelids

Biologists group annelids in three classes: Oligochaeta (ohl ih goh KEE tuh), Polychaeta (pah lih KEE tuh), and Hirundinea (hur un DIN ee uh).

What worms belong to class Oligochaeta?

Earthworms are well-known oligochaetes (ah LEE goh keetz). They live in garden soil. As they eat the soil, they take nutrients from it. Tubifex worms live in polluted water. Lumbriculid (lum BRIH kyuh lid) worms live on the edges of lakes and ponds.

Where do the polychaetes live?

Polychaetes (PAH lee keetz) live mainly in the oceans. They have well-developed sense organs, including eyes. Each segment has a pair of parapodia—appendages for swimming and crawling. Fan worms are filter feeders. They trap food in the mucus of their fans.

How are leeches different from other annelids?

Leeches, in class Hirundinea, are parasites that live in freshwater. A leech attaches to the outside of its host's body using front and rear suckers. Chemicals in its saliva act as an anesthetic, reducing the pain of its bite. Other chemicals prevent its host's blood from clotting.

Think it Over

3. Infer Some chemicals in a leech's saliva prevent the host's blood from clotting. Why is this important to the leech?

Ecology of Annelids

Segmented worms play an important role in ecosystems. The table on the next page summarizes their benefits.

What role do earthworms play in ecosystems?

Earthworms are an important food source for many animals. Also, as earthworms move through the soil, they mix leaf litter into it and break it up, which allows air and water to move through the soil, helping plants grow.

Picture This

4. Highlight the characteristic that is an adaptation for a parasitic lifestyle.

What are benefits of polychaetes and leeches?

Polychaetes help change organic matter on the ocean floor into carbon dioxide. Plant plankton use the carbon dioxide for photosynthesis. Polychaetes also serve as an important food source for many marine predators. Today, leeches are sometimes used after minor surgery to prevent blood from building up in the surgical area.

Ecological Importance of Annelids			
Annelid	**Habitat**	**Characteristics**	**Benefit**
Earthworm	land	few setae on most body segments	breaks up soil, enabling air and water to get to plant roots; serve as food source for many animals
Polychaete	mainly marine	well-developed sense organs, many setae, parapodia	changes organic matter in oceans into carbon dioxide used by plankton for photosynthesis
Leech	mainly freshwater	usually no setae, front and rear suckers	maintains blood flow after minor surgery

Picture This

5. Identify Which mollusk has the most shell coiling? (Circle your answer.)

a. bivalve
b. cephalopod
c. gastropod

Evolution of Mollusks and Annelids

In the figure below, the cladogram on the left illustrates how mollusks might have evolved. Gastropods have more shell coiling than cephalopods. Bivalves lack a radula. For this reason, scientists think bivalves evolved earlier than gastropods and cephalopods.

The cladogram on the right shows how annelids might have evolved. The early segmented worms were the polychaetes. They developed parapodia. Later, annelids—the oligochaetes and leeches—developed clitella. Eventually, leeches developed suckers at the tail end.

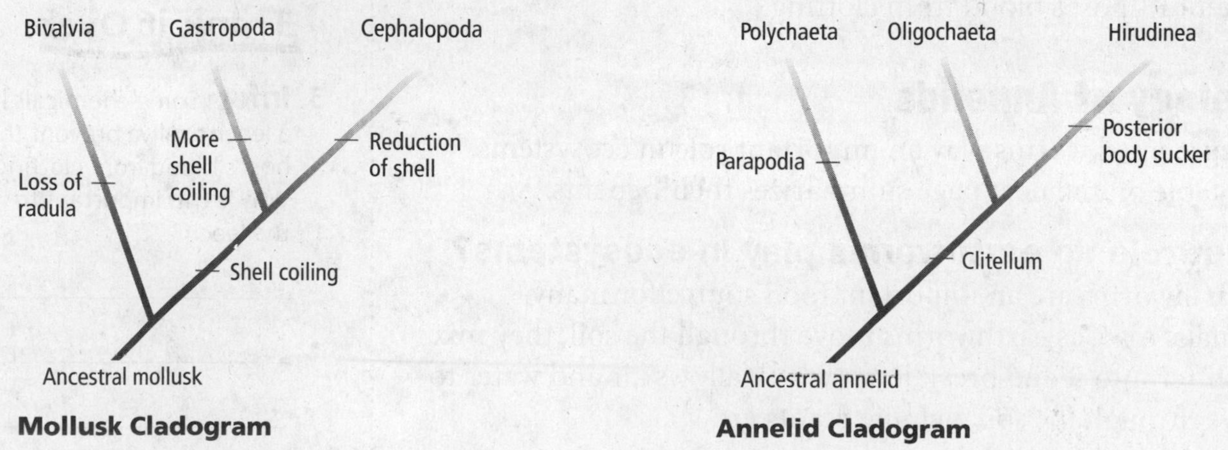

Arthropods

section 1 Arthropod Characteristics

Before You Read

Have you seen an illustration of a knight in armor? What was the purpose of the armor? On the lines below, list the advantages and disadvantages of wearing a heavy suit of armor. Then read the section to learn about an arthropod's armor.

____

____

MAIN Idea

Arthropods have segmented bodies and tough exoskeletons with jointed appendages.

What You'll Learn

- the importance of exoskeletons and jointed appendages
- how organ system adaptations differ among arthropods

Read to Learn

Mark the Text

Identify the Characteristics Highlight each feature and body structure of arthropods as you read about them. Underline the functions of each characteristic.

FOLDABLES

Define Make a vocabulary Foldable to list and define ten key terms in this section. Include underlined words or other words that are new to you.

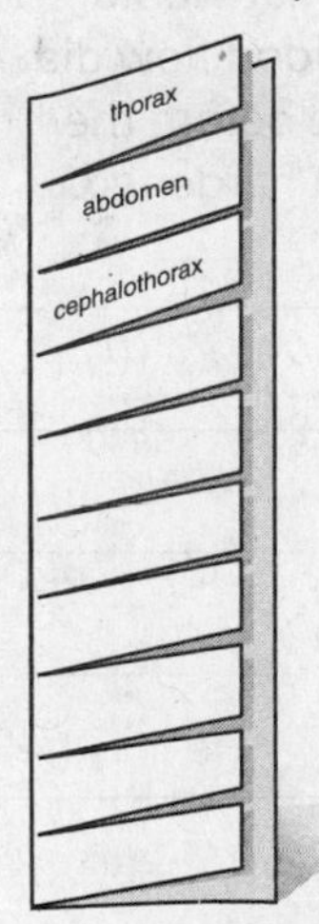

Arthropod Features

Between 70 and 85 percent of all animal species belong to phylum Arthropoda (ar THRAH puh duh). The majority of arthropods (AR thruh pahdz) are insects. Arthropods are segmented invertebrates. They are both coelomates and protostomes with bilateral symmetry. Unlike annelids, arthropods have exoskeletons with jointed appendages.

How are arthropods segmented?

Like annelids, arthropods are segmented, which enables complex movements. The three main body regions of an arthropod are a head, a thorax, and an abdomen. The head has mouthparts and eyes. Antennae on the heads of many arthropods sense smell and touch.

The **thorax** is the middle body region. It consists of three fused segments. In many arthropods, legs and wings are attached to the thorax.

The **abdomen** consists of fused segments at the posterior end of arthropods. The abdomen contains structures for digestion and reproduction. Additional legs are attached to the abdomen. In some arthropods, the thorax is fused with the head, forming a **cephalothorax** (sef uh luh THOR aks).

How does an exoskeleton benefit an arthropod?

Like a lightweight suit of armor, a hard exoskeleton covers the body of an arthropod. The exoskeleton provides support, protects soft body tissues, and slows water loss. Muscles attach to the exoskeleton. ☑

The exoskeleton is made of chitin. The exoskeleton of some arthropods is leathery. In others, the exoskeleton includes calcium that makes it hard. In some places, the exoskeleton is thin and flexible to allow joints to move.

The exoskeleton of small arthropods is thin to support the pull of tiny muscles. The exoskeleton in large arthropods is thick to support the weight and pull of larger muscles.

Reading Check

1. **Name** four functions of an exoskeleton.

______

______

______

Why are jointed appendages important?

<u>**Appendages**</u> (uh PEN dih juz) are structures that grow and extend from an animal's body. Legs and antennae are types of appendages. Arthropods have pairs of appendages with joints. Joints enable flexible movement. To understand the importance of joints, think of yourself without joints in your arms, hands, and legs. Without jointed appendages, you would be unable to do many of your normal activities.

Why do arthropods molt?

An arthropod's exoskeleton is made of nonliving material. The exoskeleton cannot grow. An arthropod must shed its outer covering in order to grow. The process of shedding the exoskeleton is called <u>**molting**</u>. Fluids from glands in the skin soften the old exoskeleton. As fluid builds up, the pressure cracks the old exoskeleton. Underneath, a new exoskeleton is forming. Before it hardens, blood flow increases throughout the body, causing the animal to puff up. As a result, the hardening exoskeleton will be a bit larger to make room for the arthropod to grow.

Think it Over

2. **Predict** what would happen if blood flow did not increase before the exoskeleton hardened.

______

______

______

______

______

Body Structure of Arthropods

Arthropods have complex organ systems that enable them to live in diverse habitats.

How are mouthparts adapted to feeding style?

Arthropods can be herbivores, carnivores, filter feeders, omnivores, or parasites. The mouthparts of most arthropods have <u>**mandibles**</u> (MAN duh bulz) that are adapted for biting and chewing. Others have mouthparts adapted for straining, stabbing, cutting, or sucking. An arthropod has a one-way digestive system. It includes a mouth, a gut, an anus, and glands that make digestive enzymes.

How do arthropods obtain oxygen?

Maintaining a homeostatic balance of oxygen in body tissues enables animals to have energy for a variety of functions. Arthropods get oxygen using one of three structures—gills, tracheal tubes, or book lungs. Like the crayfish in the figure below, most aquatic arthropods have gills. The feathery surface of gills creates a large surface area in a small space for gas exchange.

Land arthropods depend on respiratory systems rather than circulatory systems to carry oxygen to body cells. Most land arthropods have branching **tracheal** (TRAY kee ul) **tubes** as shown in the figure of the beetle. These tubes branch into smaller tubules that carry oxygen throughout the body.

Some arthropods have **book lungs**. These are saclike pockets with walls that have many folds for respiration. Notice in the figure that the folded walls look like pages of a book. The folded walls increase the surface area for exchanging gases. In both tracheae and book lungs, oxygen enters the body through openings called **spiracles** (SPIHR ih kulz).

Picture This

3. Underline the names of the structures that allow oxygen to enter the body of each arthropod.

Gills

Gill

Crayfish gills

Gills

Tracheal Tubes

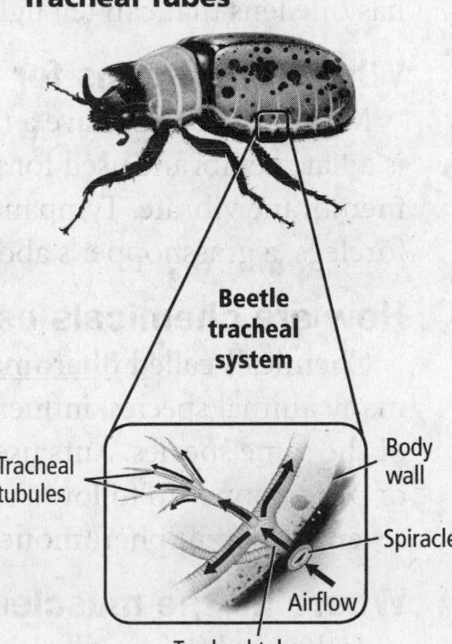

Book Lung

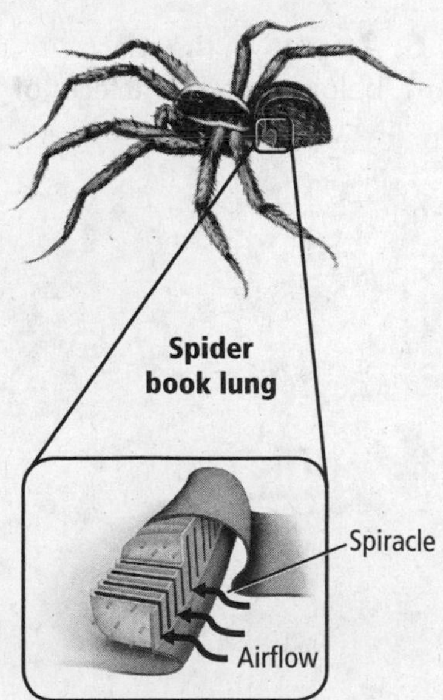

What is the function of the circulatory system?

The circulatory system maintains homeostasis in tissues by delivering nutrients and removing wastes. The heart pumps blood into vessels that carry the blood to body tissues. The blood bathes the tissues and returns to the heart through open body spaces. ☑

Reading Check

4. Describe the two main purposes of the circulatory system.

____

____

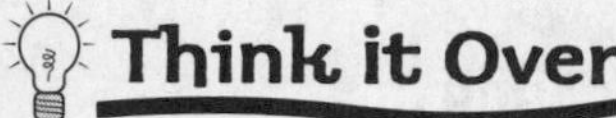

5. **Identify** two types of structures used by different arthropods for removing cellular wastes.

Think it Over

6. **Apply** In the space below, sketch one facet of a compound eye.

How are cellular wastes removed from blood?

Malpighian (mal PIH gee un) **tubules** remove cellular wastes from the blood. The tubules also help land arthropods maintain water balance for homeostasis. In insects, the tubules are located in the abdomen. The tubules empty into the gut, which contains undigested food wastes to be eliminated. Some arthropods have modified nephridia instead of Malpighian tubules to remove cellular wastes. ☑

What makes up an arthropod brain?

Most arthropods have double chains of ganglia throughout their bodies. Fused pairs of ganglia in the head make up the brain. The ganglia in each segment control most behaviors. However, the brain can prevent these actions.

How do compound eyes see an image?

Most arthropods have a pair of compound eyes. Each eye has many six-sided surfaces. Each surface sees part of an image. The brain puts the parts together into an image. Compound eyes can detect movement and color. Many arthropods also have three to eight simple eyes. A simple eye has one lens that can tell light from dark.

Where are organs for hearing located?

Many arthropods have a tympanum (tihm PA num), which is a flat membrane used for hearing. Sound waves make the membrane vibrate. Tympanums are located on a cricket's forelegs, a grasshopper's abdomen, and a moth's thorax.

How are chemicals used for communication?

Chemicals called **pheromones** (FER uh mohnz), secreted by many animal species, influence the behavior of other animals of the same species. Ants use their antennae to sense the odor of pheromones to follow a scent trail. Arthropods give off several different pheromones to signal mating or feeding.

Where do the muscles attach?

Arthropods have well-developed muscular systems. The muscles attach to the inner surface of the exoskeleton.

Do arthropods brood their eggs?

Most arthropods reproduce sexually. Most crustaceans have separate sexes, but some are hermaphrodites. Most crustaceans brood, or incubate, their eggs. Brooding might take place in a barnacle brood chamber or copepod brood sac.

chapter 26

Arthropods

section 2 Arthropod Diversity

Before You Read

Think about a spider web you have seen. On the lines below, explain how you think the spider made the web. What is the purpose of the web? Then read the section to learn how spiders build their silky homes.

____

____

MAIN Idea

Arthropods are classified based on the structures of their segments, types of appendages, and mouthparts.

What You'll Learn

- adaptations of the major groups of arthropods
- the similarities among crustaceans and among arachnids

Read to Learn

Mark the Text

Identify Main Ideas
As you read, highlight the main ideas in each paragraph.

Arthropod Groups

In this section, you will learn about two major groups of arthropods—the crustaceans (krus TAY shunz) and the spiders and their relatives. In the next section, you will learn about a third group—the insects and their relatives. The table below summarizes characteristics common to these groups.

Picture This

1. Circle the name of the arthropod group that lacks antennae.

Crustaceans	Spiders and Their Relatives	Insects and Their Relatives
two pairs of antennae, two compound eyes, mandibles, five pairs of legs (chelipeds and walking legs), swimmerets	no antennae, two body sections (cephalothorax and abdomen), six pairs of jointed appendages (chelicerae, pedipalps, and four pairs of walking legs)	antennae, compound eyes, simple eyes, three body areas (head, thorax, and abdomen), three pairs of legs, generally two pairs of wings on the thorax

Crustaceans

Most members of class Crustacea, such as lobsters and crabs, are aquatic. Most have two pairs of antennae, two compound eyes that can be on the tips of slender stalks, and mandibles. The mandibles close from side to side instead of up and down. Crustaceans have a free-swimming larval stage. A larva is an immature form of an animal that looks very different from the adult.

Most crustaceans have five pairs of legs. As shown in the figure below, the first pair are chelipeds. **Chelipeds** have large claws adapted to catch and crush food. Behind the chelipeds are four pairs of walking legs. Behind the walking legs are **swimmerets**, appendages used as flippers during swimming.

Sow bugs and pill bugs are crustaceans that live on land in damp places. They have seven pairs of legs.

Picture This

2. Describe the function of each labeled structure next to its label.

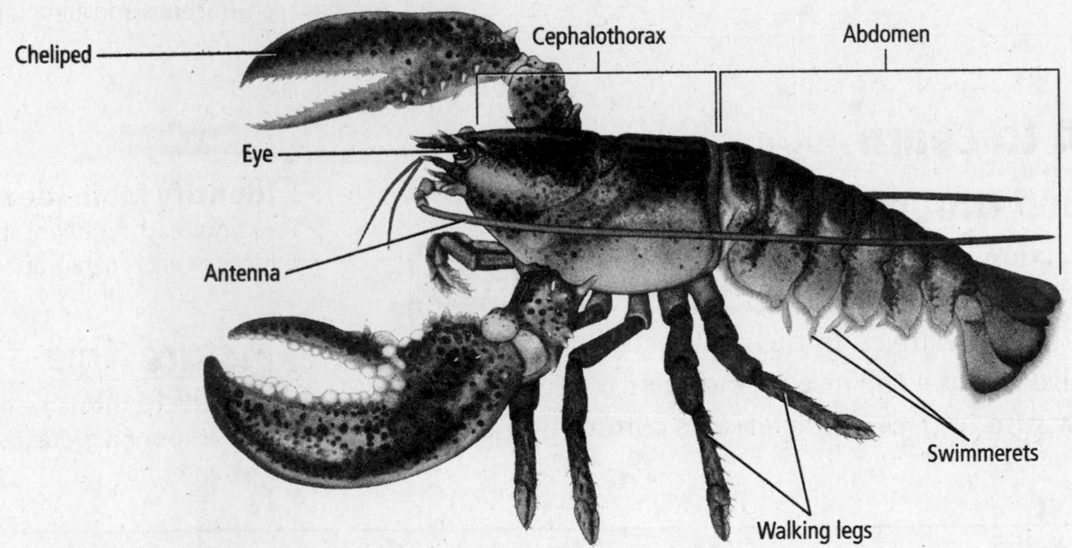

Spiders and Their Relatives

Spiders belong to class Arachnida (uh RAK nih duh). Ticks, mites, and scorpions are also arachnids.

Arachnids do not have antennae. Most arachnids have two body sections—a cephalothorax and an abdomen. Most have six pairs of jointed appendages. The most anterior pair is modified into mouthparts called **chelicerae** (kih LIH suh ree). Chelicerae act as fangs or pincers. Often, they are connected to a poison gland. The **pedipalps** are the second pair of appendages. They are used for sensing and holding prey. In scorpions, the pedipalps are the large pincers. The last four pairs of appendages are used for movement.

Think it Over

3. Draw Conclusions How would a poison gland attached to the chelicerae help a spider?

____

____

____

How do spiders make webs?

Some spiders hunt prey. Others catch prey in silk webs. Silk is made from a fluid protein secreted by glands. Structures called **spinnerets** on the spider's abdomen spin the fluid into silk.

Led by instinct, each spider can build only the kind of web common to its species. Because it builds only one kind of web, it can build the web skillfully time after time.

After catching an insect in the web, many spiders wrap the prey in a silken cocoon until they are ready to feed. Digestion begins when the spider secretes digestive enzymes onto its prey. After the prey becomes liquid, the spider ingests the food and absorbs the nutrients. Others inject the venom into the prey.

To reproduce, male spiders place sperm on a web, pick it up, and store it on the pedipalps. After courtship, the male inserts the sperm into the female. The femalc lays her eggs in a cocoon of spun silk. The young hatch after about two weeks. Before they reach adult size, young spiders molt between five and ten times.

How do ticks, mites, and scorpions feed?

Tiny mites have a cephalothorax and an abdomen that are fused into one oval-shaped body section. Mites can be parasites or predators. Ticks are parasites. They attach to their hosts and feed on blood. Ticks carry viruses, bacteria, and protozoa. They can spread these disease-causing agents to their host. Some diseases, such as Lyme disease and Rocky Mountain spotted fever, affect humans.

Scorpions feed on insects, spiders, and small vertebrates. They capture prey with their pedipalps and tear it apart with their chelicerae. Scorpions are active at night. A stinger at the end of the abdomen can cause a painful sting. ☑

How do horseshoe crabs reproduce?

Horseshoe crabs are an ancient group of marine animals related to arachnids. They have stayed about the same for more than 200 million years. Their heavy exoskeletons are shaped like horseshoes and have no segments. They use their chelicerae to capture annelids, small mollusks, and other invertebrates. Their posterior appendages are modified for digging or swimming.

Horseshoe crabs come to shore at high tide to reproduce. The female digs a hole in the sand and lays her eggs. A male adds sperm before the female covers the eggs with sand. After hatching, the young make their way to the sea.

Think it Over

4. Explain how spiders benefit from starting the process of digestion outside their bodies.

____

____

____

Reading Check

5. Identify three members of class Arachnida that are not spiders.

____

____

____

chapter 26

Arthropods

section 3 Insects and Their Relatives

MAIN Idea

Adaptations have enabled insects to become the most abundant and diverse group of arthropods.

What You'll Learn

- the characteristics and adaptations of insects
- how complete and incomplete metamorphosis are alike and how they are different

Study Coach

Create a Quiz After you read this section, create a five-question quiz from what you have learned. Then, exchange quizzes with another student. After taking the quizzes, review your answers together.

Before You Read

Think about the many kinds of insects in your area. On the lines below, list as many as you can recall. Then read the section to learn about the great diversity of insects and the adaptations that enable them to live in many habitats.

____

____

Read to Learn

Diversity of Insects

There are more insect species than all other animal classes together. Insects are the most abundant and widespread of all land animals. Some even live in Antarctica.

Insects have many adaptations that enable them to live in a variety of habitats. They can fly. They are small, so they can be moved easily by wind and water. Their hard exoskeletons protect them and keep them from drying out in dry habitats. They produce large numbers of offspring. Refer to the figure below as you read about insect adaptations.

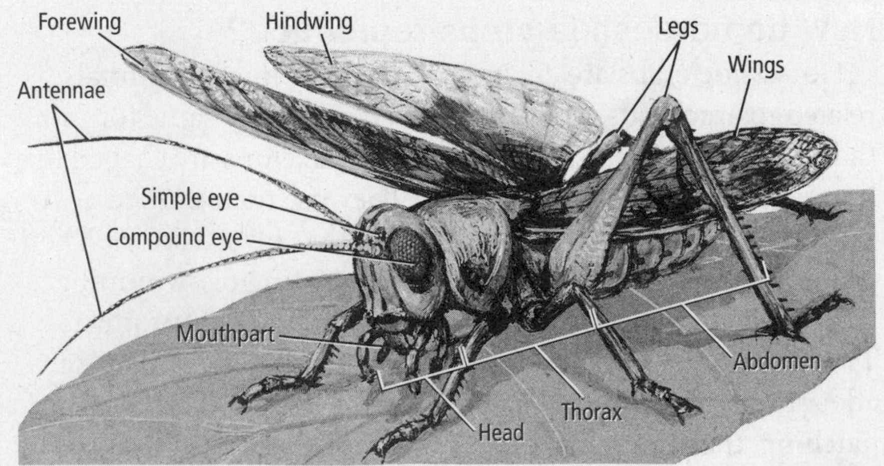

Picture This

1. Draw Conclusions Why do you think the grasshopper's hind legs are shaped differently than its forward legs?

____

____

____

____

External Features

Insects have three main body areas—the head, thorax, and abdomen. Head structures include antennae, compound eyes, simple eyes, and mouthparts. Attached to the thorax are three pairs of legs and usually two pairs of wings. Some insects have only one pair of wings. Others do not have wings.

Insect Adaptations

Adaptations in body structures enable insects to use all kinds of food and to live in many environments. Insects can be parasites, predators, or plant-sap suckers.

How have insect legs adapted?

Insect legs are adapted to a variety of functions. Claws on a beetle's legs enable it to dig in soil or crawl under bark. Sticky pads on a fly's legs enable it to walk upside down. A honeybee can collect pollen on its hind legs. The hind legs of grasshoppers and crickets are adapted for jumping.

How are mouthparts adapted to food sources?

The table below shows how an insect's mouthparts are adapted to the food it eats. Butterflies have a long tube for drawing nectar from flowers. Flies have mouthparts for sponging up liquids. The mouthparts of mosquitoes can pierce prey or plants to feed on juices. The mandibles of beetles can cut animal skin or plant tissue to reach the nutrients inside.

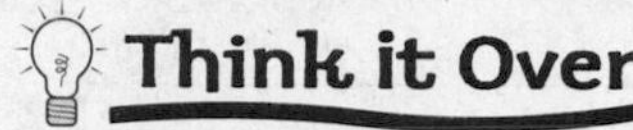

2. Summarize in one word why insects are so successful.

____

Picture This

3. Apply A sand fly feeds by drawing blood from humans and other animals. What type of mouthpart would you expect a sand fly to have?

____

Type of Mouthpart	Siphoning	Sponging	Piercing/Sucking	Chewing
Example				
Function	Feeding tube is uncoiled and extended to suck liquids into the mouth.	Fleshy end of mouthpart acts like a sponge to mop up food.	A thin, needlelike tube pierces the skin or plant wall to suck liquids into the mouth.	Mandible cuts animal or plant tissue, and other mouthparts take food into the mouth.
Insects	butterflies, moths	houseflies, fruit flies	mosquitoes, leafhoppers, fleas	grasshoppers, beetles, ants, bees

How do insect wings rotate?

Insects are the only invertebrates that can fly. Insect wings are made of two thin layers of chitin. Thick veins give wings strength. Most insects rotate their wings in a figure-eight pattern. This enables a variety of movements—forward thrust, upward lift, balance, and steering.

Think it Over

4. Name another arthropod structure that is made of chitin. (Review Section 1 if you need help.)

____

What sensory adaptations do insects have?

Insects use antennae and eyes to sense their environment. Insects also have hairlike structures that sense touch, vibration, and odor. Hairs that cover a fly's body sense changes in airflow, such as those caused by an approaching flyswatter.

Some insects have tympanic organs to detect sounds. Others have sensory cells on their legs that can detect ground vibrations. Chemical receptors are located on mouthparts, antennae, or legs. These receptors detect taste or smell. Pheromones enable insects to communicate with each other.

What is metamorphosis?

Insects that do not care for their young lay many eggs to ensure survival. They lay their eggs in habitats where the young can survive. For example, one kind of butterfly lays its eggs on a milkweed that will be food for the young. After hatching, most insects undergo metamorphosis, as shown in the figure below. **Metamorphosis** is a series of major changes from a larval form to an adult form. It can be complete or incomplete.

Picture This

5. Compare In which type of metamorphosis does the hatched young look most like the adult?

____

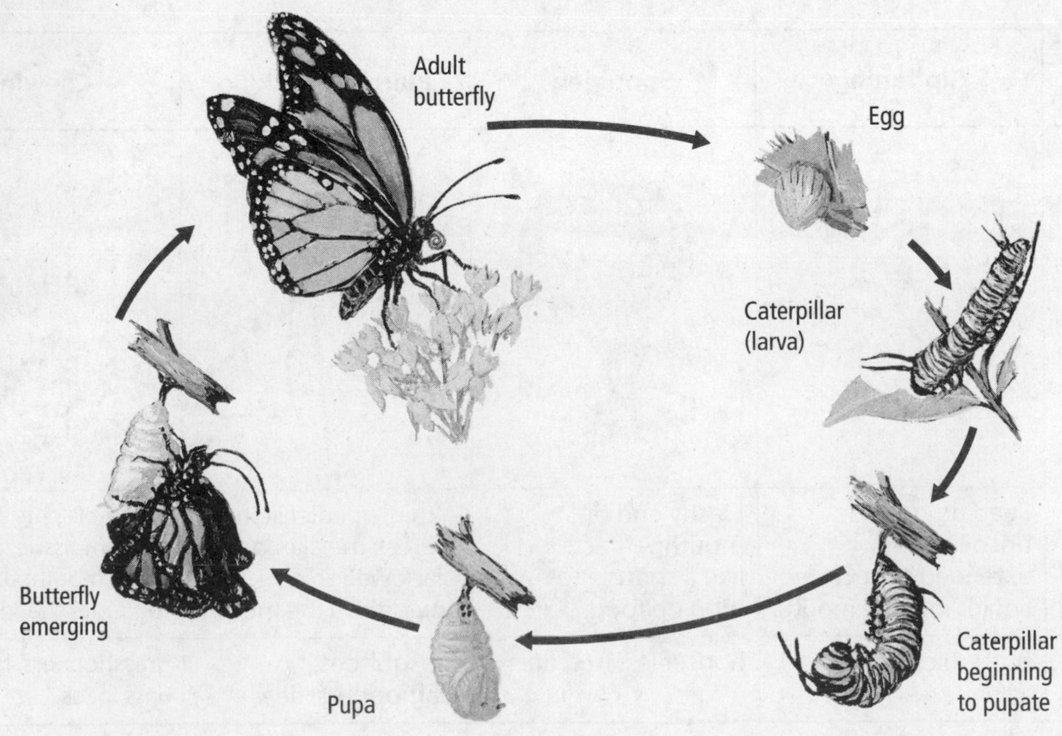

Complete Metamorphosis Most insects undergo complete metamorphosis. The four stages are egg, larva, pupa, and adult. When a butterfly egg hatches, the larva (caterpillar) appears. It has chewing mouthparts and eats constantly. A **pupa** (PYEW puh) is a nonfeeding stage of metamorphosis. In this stage, the animal changes from larval form to adult form. At the adult stage, the insect has structures adapted for reproducing and moving around the environment. Adults do not eat the same food as their larvae.

Incomplete Metamorphosis Insects that undergo incomplete metamorphosis hatch from eggs as **nymphs** (NIHMFS). A nymph is an immature form that looks like a small adult without fully developed wings. After several molts, nymphs become winged adults. ☑

6. Identify the process that nymphs undergo to become adults.

____

What are some insect societies?

Some insects, such as ants and honeybees, organize into cooperative societies. Members work together in activities important for survival.

Honeybees have a complex society, with as many as 70,000 bees in one hive. A honeybee society has three castes. A **caste** is a group of individuals within a society that perform specific tasks. The workers are females that do not reproduce. They gather food, care for the young, build the honeycomb, make honey, and guard the hive. Drones are reproductive males. The queen is the only female that reproduces.

How do honeybees communicate?

Honeybees use body movements to communicate the location and quantity of food sources. One movement is called the waggle dance. A bee performs this dance when it returns to the hive from a distant food source. First, the bee makes a circle. The bee then moves in a straight line while waggling its abdomen. This line indicates the direction to the food. Finally, the bee moves in a figure-eight pattern several times. If the food is close to the hive, a bee will perform a round dance instead of the waggle dance. ☑

Reading Check

7. Apply In the space above, sketch the three parts of the waggle dance. Number the parts in the proper sequence.

In what ways do insects benefit humans?

Most insect species are not harmful to humans. Insects pollinate most flowering plants, including human food crops. Insects make honey and silk that humans use. Insects are also food for many animals. An insect predator, such as a praying mantis, feeds on plant pests such as mites.

How can insects harm humans?

Some insects can harm humans. Lice and bloodsucking flies are human parasites. Fleas can carry plagues. House flies can carry typhoid fever. Mosquitoes can carry malaria and yellow fever. Ants and termites destroy property. Grasshoppers and boll weevils destroy crops. Certain types of beetles and moths can destroy parts of forests. ☑

Reading Check

8. Name two insects that are human parasites.

____

____

How are insect pests controlled?

In the past, large amounts of chemicals were used to kill insects. Overuse of chemicals disrupted food chains and killed many helpful insects. Also, many insects developed resistance to the insecticides. Today, many farmers use biological controls. They use pest-resistant plants, rotate crops, and time planting carefully. Some chemicals are used to control insect pests, but smaller amounts are applied.

Centipedes and Millipedes

Centipedes and millipedes are close relatives of insects. Centipedes of class Chilopoda move quickly, propelled by "hundreds" of legs. They live in moist places under logs, bark, and stones. Each segment has one pair of jointed legs. The first pair is modified to form poison claws for killing prey. Most species of centipedes do not harm humans.

Millipedes belong to class Diplopoda. They have two pairs of appendages on each abdominal segment and one pair on their thorax. Like centipedes, they live in moist places. Millipedes do not have poison claws because they are herbivores. They feed on decaying vegetation. Unlike centipedes, millipedes walk slowly.

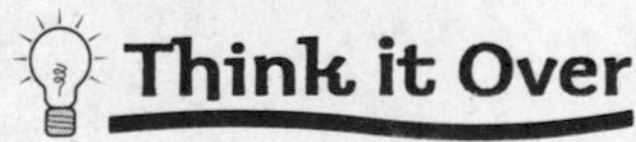

9. Compare and contrast centipedes and millipedes. State one way in which they are alike and one way in which they are different.

____

____

____

____

Evolution of Arthropods

Molecular evidence suggests that arthropods might share a common ancestor with trilobites and tardigrades. Trilobites, now extinct, were once abundant. Fossil records show that trilobites were early arthropods. Their oval bodies were divided into three sections like some present-day arthropods. Early arthropods had a large number of identical segments. Present-day arthropods have evolved fewer segments and more specialized appendages.

Tardigrades are related less closely to arthropods than are trilobites. Tardigrades are tiny and have four pairs of stubby legs. They live in water and land habitats, feeding on algae, decaying matter, and soil animals.

Echinoderms and Invertebrate Chordates

section 1 Echinoderm Characteristics

Before You Read

On the lines below, describe some ways that animals confuse predators to escape capture. In this section, you will learn about the unusual escape strategies of echinoderms.

____

____

MAIN Idea

Echinoderms are marine animals with water-vascular systems and tube feet.

What You'll Learn

- how the water-vascular system and tube feet enable echinoderms to live
- the differences in echinoderms

Read to Learn

Study Coach

Make Flash Cards Think of a quiz question for the material covered under each heading. Write the question on one side of a flash card. Write the answer on the other side. Use the flash cards to quiz yourself until you know all of the answers.

Echinoderms Are Deuterostomes

Echinoderms (ih KI nuh durmz) represent a major shift in evolutionary history. Echinoderms are the first animals to have deuterostome development. Recall that mollusks, annelids, and arthropods are protostomes. A protostome's mouth develops from the opening on the gastrula. A deuterostome's mouth develops from another place on the gastrula. Only echinoderms and chordates that evolved after echinoderms have this kind of development.

Echinoderms are marine animals, including sea stars, sea urchins, sand dollars, sea cucumbers, and sea lilies.

Body Structure

Echinoderms were the first animals to have endoskeletons. The endoskeleton of an echinoderm is made of calcium carbonate plates, often with spines. The endoskeleton is covered by a thin layer of skin. Small pincers on the skin called **pedicellariae** (PEH dih sih LAHR ee ee) aid in catching food and removing foreign materials from the skin. As adults, all echinoderms have radial symmetry. The larvae have bilateral symmetry. ☑

Reading Check

1. **Name** two evolutionary advances that began with echinoderms.

____

____

____

How does the water-vascular system work?

Refer to the sea star in the figure below as you read about the structures of echinoderms. The **water-vascular system** is a system of fluid-filled, closed tubes. The tubes work together to enable echinoderms to move and get food. ☑

✓ Reading Check

2. Identify the function of a water-vascular system.

Water enters the water-vascular system through the **madreporite** (MA druh pohr it), a strainerlike opening. The water then moves through the stone canal to the ring canal and on to the radial canals. Eventually, the water reaches the tube feet.

Tube feet are small, muscular, fluid-filled tubes that end in structures similar to suction cups. Tube feet are used for movement, food collection, and respiration.

At the opposite end of each tube foot is a muscular sac called an **ampulla** (AM pyew luh). Muscle contraction in the ampulla forces water into the tube foot, extending it. Contraction and relaxation of the muscle creates suction. The suction enables the echinoderm to attach to a surface and move. In some sea stars, the suction is strong enough to open a mollusk shell.

Picture This

3. Highlight each part of the water-vascular system.

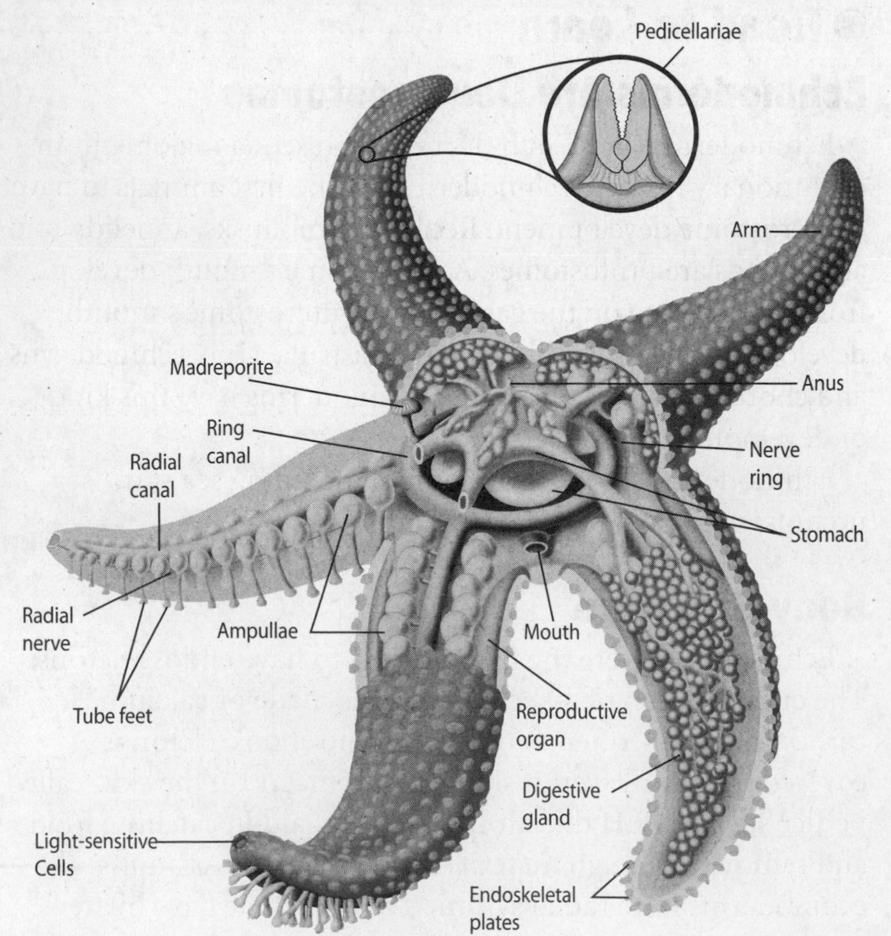

What feeding strategies do echinoderms use?

Echinoderms feed in various ways. Sea lilies extend their arms to trap food. Many sea stars push their stomachs out of their mouths and spread digestive juices over the food. Cilia bring digested materials into their mouths. Sea urchins scrape algae off surfaces with teethlike plates. Sea cucumbers trap floating food with their mucus-covered tentacles.

How does oxygen enter the body?

Oxygen diffuses from the water through the membranes of the tube feet. In some echinoderms, oxygen diffuses through all thin outer membranes. In others, pouches called skin gills extend from the body. Many sea cucumbers have branched tubes called respiratory trees through which oxygen diffuses.

Circulation takes place in the coelom and in the water-vascular system. Excretion occurs by diffusion through body membranes. Cilia move water and body fluids throughout these systems.

What kinds of stimuli can echinoderms detect?

A nerve ring surrounds the mouth. Branching neurons connect to other body areas. Sensory neurons respond to touch, chemicals in the water, water currents, and light. Sea stars have light-sensitive cells at the tips of their arms. Many echinoderms also sense the direction of gravity.

What structures enable echinoderms to move?

The movable bony plates in the endoskeleton enable echinoderms to move easily. Feather stars move by grabbing soft sediment along the ocean floor with their cirri—long, thin appendages on their ventral side. They also swim by moving their arms up and down. Brittle stars use tube feet and arms for snakelike movement. Sea stars use arms and tube feet to crawl. Sea urchins use their tube feet to move and their movable spines to burrow. ☑

How do echinoderms reproduce?

Most echinoderms reproduce sexually. Females shed eggs and males shed sperm into the water where fertilization takes place. The free-swimming larvae have bilateral symmetry. The larvae develop into adults with radial symmetry.

When attacked, many echinoderms drop off an arm. Others expel part of their internal organ system. These actions distract the predator, enabling the echinoderms to flee. The echinoderms can regenerate, or regrow, these lost body parts.

Think it Over

4. Draw Conclusions How does a sea star's digestive method help to defeat its prey's defenses?

____

____

____

____

____

____

Reading Check

5. Describe how the structure of the endoskeleton enables easy movement.

____

____

____

Picture This

6. Identify a distinguishing feature of brittle stars. (Circle your answer.)

a. five arms
b. radial symmetry
c. no suctions cups

Echinoderm Diversity

Echinoderms are marine animals. They have bilateral symmetry as larvae and radial symmetry as adults. They have a water-vascular system with tube feet and endoskeletons, often bearing spines. The table below compares echinoderms.

Class	Class Members	Distinctive Features
Asteroidea (AS tuh ROY dee uh)	sea stars	often five arms; tube feet used for feeding and movement
Ophiuroidea (OH fee uh ROY dee uh)	brittle stars	often five arms; arms break off easily and regenerate; move by arm movement; tube feet have no suction cups
Echinoidea (ih kihn OY dee uh)	sea urchins, sand dollars	body encased in a test with spines; sea urchins burrow in rocky areas; sand dollars burrow in sand
Crinoidea (kri NOY dee uh)	sea lilies, feather stars	sessile for some part of life; sea lilies have long stalks; feather stars have long branching arms
Holothuroidea (HOH loh thuh ROY dee uh)	sea cucumbers	cucumber shape; leathery outer body; tube feet modified to tentacles near mouth
Concentricycloidea (kahn sen tri sy CLOY dee uh)	sea daisies	less than 1 cm in diameter; no arms; tube feet located around a central disk

How do sea stars use their tube feet?

Most species of sea stars have five arms around a central disk. Some have more than five arms. The many tube feet of sea stars enable them to hold tightly to rocks, open mollusks for food, and crawl. Usually, sea stars are not food for other predators because of their spiny skin.

How do brittle stars move?

Most brittle stars have five thin, flexible arms. Brittle stars do not have suction cups, so they cannot use their tube feet to move. They row with snakelike movements of their arms. Brittle stars can release an arm to escape a predator and then regenerate the arm. They feed on food particles that stick in mucus between their spines. Some brittle stars can sense light.

How do sea urchins and sand dollars feed?

Both sea urchins and sand dollars burrow. Their bodies are enclosed in a hard endoskeleton, called a test, which looks like a shell. Sea urchins and sand dollars lack arms. Both have spines. Some sea urchin spines contain venom to paralyze prey and defend against predators. Sea urchins also scrape algae from rocks for food. Sand dollars filter organic materials from the sand. ☑

Reading Check

7. Explain how the endoskeleton of a sand dollar differs from the endoskeleton of other echinoderms.

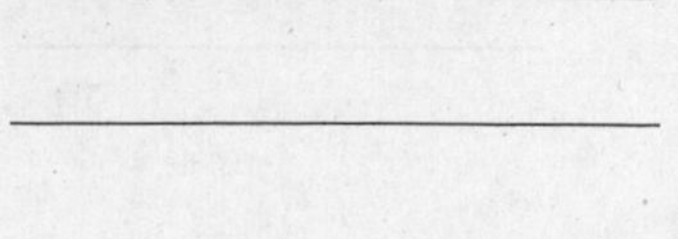

What distinguishes sea lilies and feather stars?

Fossil records show that sea lilies and feather stars are the most ancient echinoderms. Unlike other echinoderms, sea lilies and feather stars are sessile for part of their lives. Sea lilies have a flower-shaped body at the top of a long stalk. Feather stars have long, branched arms that extend up from a central area. Both sea lilies and feather stars catch food particles in the water by extending their tube feet and arms.

How do sea cucumbers trap food?

Sea cucumbers look like cucumbers creeping over the ocean floor with their tube feet. Unlike other echinoderms, their hard plates are smaller and unconnected. This makes their bodies look leathery. Some of their tube feet are adapted into mucous-covered tentacles around the mouth that trap food particles in the water.

Sea cucumbers are the only echinoderms with a respiratory tree. This many-branched tube system pumps water through the anus to obtain oxygen. The respiratory tree also excretes cellular wastes. ☑

When threatened, a sea cucumber can shed some internal organs through its anus. This action might confuse the predator. The sea cucumber can regenerate its lost parts.

What gives sea daisies their flowery look?

Few sea daisies have been found. Their tiny disc-shaped bodies have no arms. Tube feet around the edge of the disc give the animal a daisy-like appearance.

Ecology of Echinoderms

Some echinoderms live in commensal relationships with other marine animals. Recall that in a commensal relationship, one organism benefits, and the other is neither helped nor harmed. For example, some brittle stars live inside sponges. They gather food that settles on the sponge's body.

A decline in the population of echinoderms can warn of an upcoming change in the ecosystem. When disease reduced the sea urchin population in parts of the Caribbean, the algae population increased greatly, destroying many coral reefs.

Sea urchins and sea cucumbers stir up sediment on the ocean floor. This action benefits the marine ecosystem. It makes nutrients on the seafloor available to other organisms. When the sea star population increases, they eat so many coral polyps that they destroy reefs.

Reading Check

8. Contrast What respiratory structure distinguishes sea cucumbers from other echinoderms?

____

Think it Over

9. Predict Sea otters eat sea urchins. Sea urchins eat kelp. Kelp provides a habitat for snails. Predict what might happen to the snail population if sea otters declined in the area.

____

____

____

____

____

Echinoderms and Invertebrate Chordates

section 2 Invertebrate Chordates

MAIN Idea

Invertebrate chordates have features linking them to vertebrate chordates.

What You'll Learn

- features of invertebrate chordates that place them with invertebrates
- how the adaptations of lancelets compare with those of sea squirts

Before You Read

On the lines below, list the key feature that distinguishes invertebrates from vertebrates. If needed, review Chapter 24 to find the answer. Then read this section to learn the features of invertebrate chordates.

____

____

Mark the Text

Restate the Main Point As you read the section, highlight the main point in each paragraph. Then restate each main point in your own words.

Read to Learn

Invertebrate Chordate Features

Animals of phylum Chordata (kor DAH tuh) are chordates. At some point in their development, all **chordates** have four key features—a dorsal tubular nerve cord, a notochord, pharyngeal pouches, and a postanal tail. All chordates also might have some form of a thyroid gland. All have a coelom and segmentation. The figure below shows the main features of chordates. Most chordates are vertebrates—they have a backbone. Lancelets and tunicates are invertebrate chordates. **Invertebrate chordates** have the four key features of chordates, but they do not have backbones. They belong to two subphyla of chordates—Cephalochordata and Urochordata.

Picture This

1. **Highlight** each structure of chordates as you read about it.

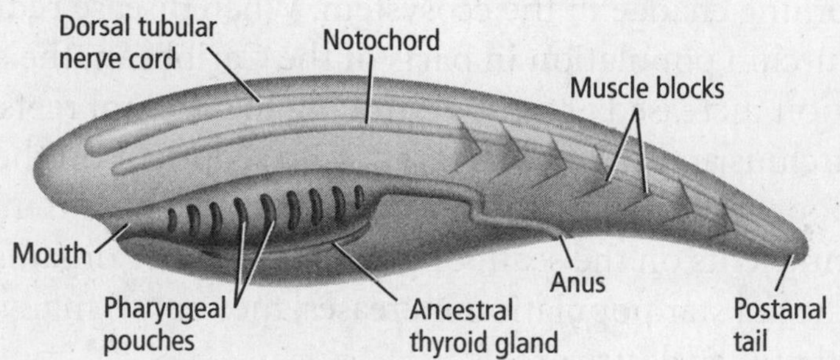

What movement does a notochord enable?

A **notochord** (NOH tuh kord) is a flexible, rodlike structure. It extends the length of the body below the dorsal tubular nerve cord. In most vertebrates, bone or cartilage eventually replaces the notochord. In invertebrates, the notochord remains. The flexibility of the notochord enables the body to bend from side to side rather than shorten. The notochord marks the start of fishlike swimming in evolutionary history.

How does a postanal tail benefit an animal?

A **postanal tail** is a structure used mainly for locomotion. It is located behind the digestive system and anus. In nonchordates, the anus is located at the end of the tail. A postanal tail enables more power movements than the body structure of invertebrates without a postanal tail.

Where is the dorsal tubular nerve cord located?

The nerve cords of nonchordates are located on the ventral side, or below, the digestive system and are solid. Chordates have a **dorsal tubular nerve cord** that is located on the dorsal side, or above, the digestive organs and is tube shaped. During development, the anterior end of this cord becomes the brain. The posterior end becomes the spinal cord.

What do pharyngeal pouches become?

All embryos have a pair of **pharyngeal pouches** that connect the mouth cavity to the esophagus. In aquatic chordates, the pouches contain slit openings, which were first used for filter feeding. Later, they evolved into gills for gas exchange in water. In land chordates, the pharyngeal pouches do not have slits and develop into other structures. These pouches are evidence of aquatic ancestry of all vertebrates.

How is an endostyle similar to a thyroid gland?

A thyroid gland regulates metabolism, growth, and development. This gland had its origin in early chordates. Only vertebrate chordates have a thyroid gland. Invertebrate chordates have an endostyle—cells in the same area as a thyroid gland that secrete similar proteins. ☑

Diversity of Invertebrate Chordates

Like echinoderms, all invertebrate chordates are marine animals. Lancelets belong to subphylum Cephalochordata. Tunicates belong to subphylum Urochordata.

FOLDABLES™

Take Notes Make a layered Foldable, as shown below. As you read, take notes and organize what you learn about invertebrate chordates.

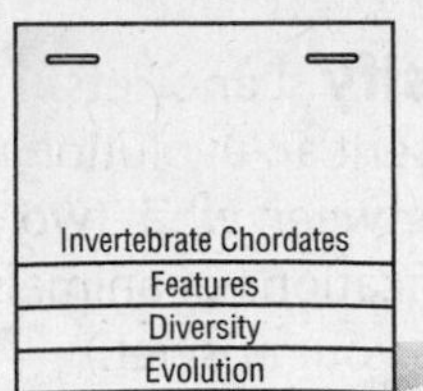

2. Identify the structure that evolved from the endostyle of invertebrate chordates.

____

Why are scientists excited about lancelets?

A lancelet is a small, fishlike animal without scales. Lancelets burrow their bodies into the sand. Fossil and molecular evidence shows that lancelets are one of the closest living relatives of vertebrates. You are more closely related to lancelets than to any other invertebrate. ☑

Lancelets do not have color in their skin. The skin is so thin that an observer can watch some inner body functions. As the lancelet filter feeds, water enters the mouth and passes through pharyngeal gill slits. Food passes to a stomachlike structure to be digested. Water exits through the gill slits.

Segmented muscle blocks enable lancelets to swim with a fishlike motion. Lancelets do not have heads. Their only sensory structures are light receptors and sensory tentacles near their mouths. Nerves branch from a simple brain. There is no true heart. Blood vessels pump blood through the body. Lancelets have separate sexes, and fertilization is external.

Reading Check

3. Classify Lancelets represent an evolutionary link between what two classifications of animals? (Circle your answer.)

a. chordates and nonchordates

b. vertebrates and invertebrates

Why are tunicates called sea squirts?

Tunicates (TEW nuh kayts), or sea squirts, are named for the thick outer covering, called a tunic, that covers their small, saclike bodies. They are sessile. Only the larvae show typical chordate features.

Beating cilia draw water into the adult's incurrent siphon. Food particles are trapped in a mucous net and moved to the stomach. Water leaves the body first through gill slits and then through the excurrent siphon. A heart and blood vessels distribute nutrients and oxygen. Neurons branch from a main nerve complex. When threatened, tunicates may eject a stream of water from the excurrent siphon.

Evolution of Echinoderms and Invertebrate Chordates

Echinoderm larvae have bilateral symmetry. Therefore, scientists think echinoderms evolved from ancestors with bilateral symmetry. The deuterostome development of echinoderms links them to chordates.

Both lancelets and tunicates have the main chordate features. These shared features show that they are closely related, even though tunicates have them only as larvae. The notochord provides support and a place for muscles to attach. This enables a side-to-side swimming motion, a development in chordate evolution that led to the first large animals. ☑

Reading Check

4. Explain the importance of notochord development in evolutionary history.

____

____

____

____

Fishes and Amphibians

section 1 Fishes

Before You Read

Have you ever watched a fish swim? On the lines below, describe how a fish's body moves as it swims. In this section you will learn how the bodies of fishes are well adapted for life in their watery world.

MAIN Idea

Fishes are vertebrates that live and reproduce in water.

What You'll Learn

- the differences between vertebrates and invertebrates
- the characteristics common to most fishes
- how fishes are adapted to life in water

Read to Learn

Characteristics of Vertebrates

Recall that chordates have four main characteristics: a dorsal nerve cord, a notochord, pharyngeal pouches, and a postanal tail. Vertebrates are chordates that also have a vertebral, or spinal, column and specialized cells that develop from the nerve cord. These animals belong to the subphylum Vertebrata. Classes of vertebrates include fishes, amphibians, reptiles, birds, and mammals.

Mark the Text

Identify the Characteristics

Highlight each characteristic of vertebrates and fishes as you read. Underline the functions of each characteristic.

What are the functions of the vertebral column?

The vertebral column replaces the notochord as a vertebrate embryo develops. The vertebral column surrounds and protects the dorsal nerve cord. It also functions as a strong, flexible rod that muscles can pull against during movement. Separate vertebrae allow an animal to move quickly and easily.

Vertebrate skeletons are made of cartilage or a combination of bone and cartilage. **Cartilage** (KAR tuh lihj) is a tough, flexible material that makes up part or all of the skeleton of a vertebrate. ☑

Reading Check

1. **Identify** the two main building materials in the skeletons of vertebrates.

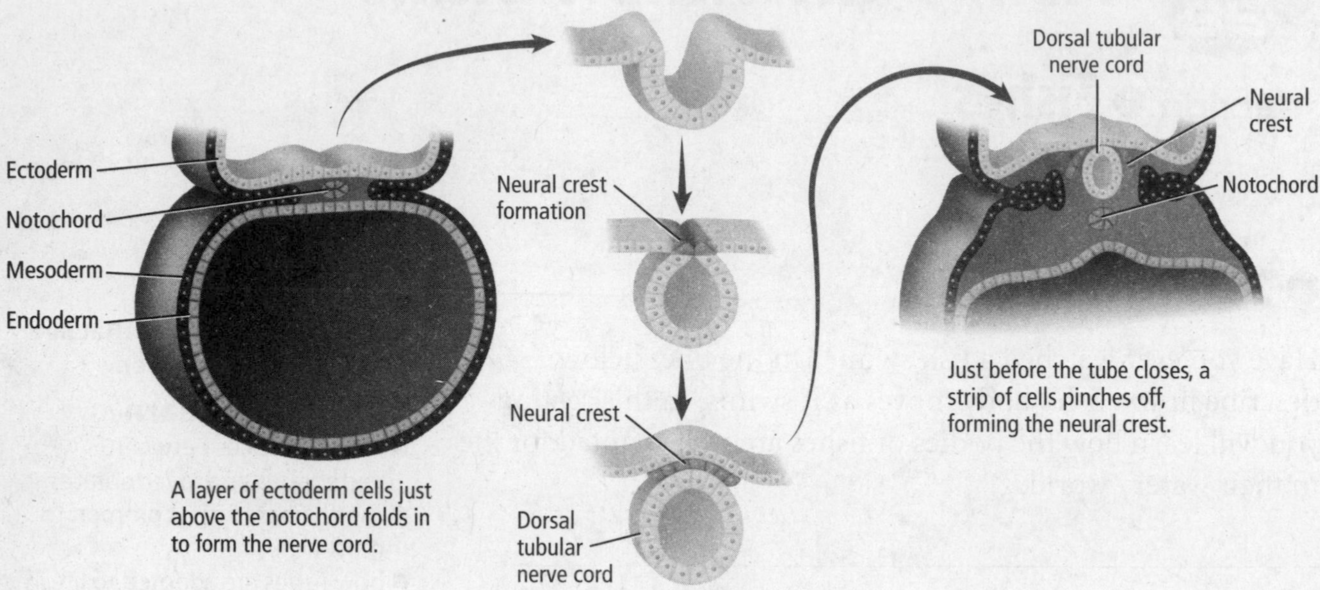

A layer of ectoderm cells just above the notochord folds in to form the nerve cord.

Just before the tube closes, a strip of cells pinches off, forming the neural crest.

Picture This

2. Describe the location of the neural crest

______

______

______

What is a neural crest?

As vertebrate embryos develop and the nerve cord forms, a neural crest also forms. A **neural** (NOOR ul) **crest** is a small group of cells that develop from the nerve cord in vertebrates. The neural crest is located just above the nerve cord.

The figure above shows how the neural crest forms. Many important vertebrate features develop from the neural crest. These features include parts of the brain and skull, some sense organs, parts of pharyngeal pouches, some nerve fibers, insulation of nerve fibers, and some gland cells. Other vertebrate features are a heart, a closed circulatory system, and internal organs, such as kidneys.

Characteristics of Fishes

Fishes live in most aquatic habitats on Earth, including seas, lakes, ponds, streams, and marshes. Some fishes live in complete darkness at the bottom of the deep ocean. Others live in cold polar waters and have special proteins in their blood to keep them from freezing. They can range from 18 m long to the size of a human fingernail. There are more species of fish than all other vertebrates combined.

Most fishes have vertebral columns, jaws, paired fins, scales, gills, and single-loop circulation. They also cannot make certain amino acids. Some characteristics of fishes, including jaws and, in some fishes, lungs, provided the structural starting place for the development of land animals during evolution.

Think it Over

3. Name three characteristics that you share with fish.

______

______

______

How did the development of jaws benefit fishes?

In ancient fishes, gill arches evolved to form jaws. Jaws enabled fishes to prey on more kinds of animals, including larger and more active fishes. Fishes grasp prey with their teeth and crush them with powerful jaw muscles. Fishes also use their jaws as a biting defense against predators.

What advantages do paired fins provide?

As jaws were evolving, paired fins—one on each side of the body—were also appearing in fishes. A **fin** is a paddle-shaped structure on a fish or other aquatic animal that is used to balance the body, control the direction of its movements, and move its body through the water.

A fish uses pelvic fins and pectoral fins, like those shown in the figure below, to keep its body steady as it moves. Paired fins, shown in the figure below, enable a fish to control the direction of its movements and keep its body from rolling to the side.

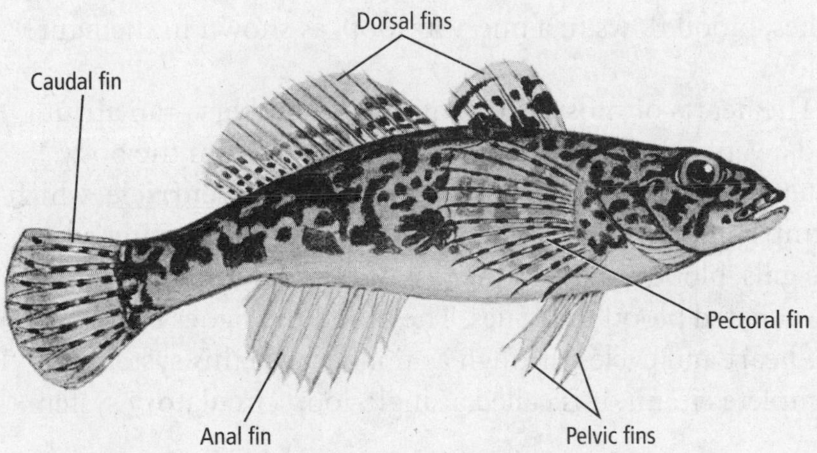

Picture This

4. Explain how a fish's movement might change if it had a pectoral fin on only one side.

What types of scales do fishes have?

A **scale** is a small, flat, platelike structure near the surface of the skin of most fishes. There are four types of fish scales. Two types, ctenoid (TEH noyd) scales and cycloid (SY kloyd) scales, are made only of bone. These scales are thin and flexible. Sharks have rough and heavy placoid (PLA koyd) scales that are made of toothlike materials. Thick ganoid (GAN oyd) scales are diamond-shaped and made of both enamel and bone.

How do gills help fishes live in water?

Gills give fishes the ability to get oxygen from water. As water enters the mouth and flows across the gills, oxygen from the water diffuses into the blood. Gills are made of thin filaments that are covered with platelike lamellae (luh MEH lee). The lamellae are highly folded and have many blood vessels to take in oxygen and give off carbon dioxide.

In the gill, blood flows in the opposite direction of the flow of water on the gill's surface. This opposing flow helps fishes get up to 85 percent of the oxygen from the water. Some fishes have a moveable flap called an **operculum** (oh PUR kyuh lum) that covers and protects the gills. An operculum also helps to pump water coming in the mouth and over the gills.

Some fishes, such as lungfishes, can live out of water for a short time using structures similar to lungs. An eel can breathe through its moist skin when it is out of water.

Think it Over

5. Describe a situation in which having structures similar to lungs would benefit a lungfish.

____

____

____

____

How does blood circulate in fishes?

Vertebrates have a closed circulatory system. This means that the heart pumps blood through blood vessels. In most fishes, blood flows in a one-way loop, as shown in the figure below.

The hearts of most fishes have two chambers—an atrium and a ventricle. The **atrium** receives blood from the body. After leaving the atrium, blood passes to the **ventricle**, which pumps blood from the heart to the gills. After flowing over the gills, blood travels to the rest of the body, delivering the oxygenated blood to tissues. The blood then goes back to the heart and cycles through again. Because this system is a complete circuit, it is called a single-loop circulatory system.

Picture This

6. Circle the names of the chambers of a fish's heart.

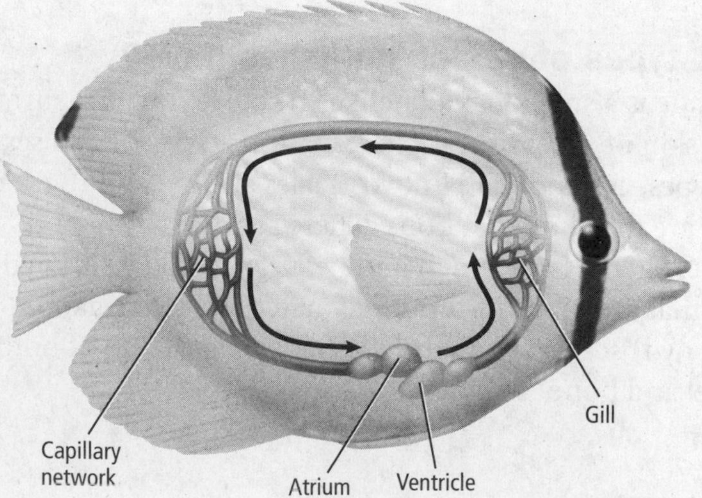

How has feeding changed from ancient times?

Ancient fishes most likely were filter feeders or scavengers, sucking up organic matter on the ocean floor. The evolution of jaws enabled fishes to become predators. Most fishes today swallow their prey whole.

What are the structures of the digestive tract?

The digestive tract of a fish is illustrated in the figure below. After a fish swallows its prey, the food passes through a tube called the esophagus (ih SAH fuh gus) to the stomach, where digestion begins. Food then passes to the intestine, where most digestion occurs. Some fishes have pyloric (pi LOR ihk) ceca (SEE kuh) (singular, cecum), which are small pouches where the stomach and intestine meet. The pyloric ceca secrete enzymes for digestion and absorb nutrients into the bloodstream. Digestive juices from the liver, pancreas, and gallbladder complete digestion.

One important thing fishes cannot do is make certain amino acids. Therefore, fishes and all vertebrates that evolved from fishes must get these amino acids from foods they eat.

Picture This

7. Highlight in one color the structures where digestion occurs. Highlight in a different color the structures that produce digestive fluids.

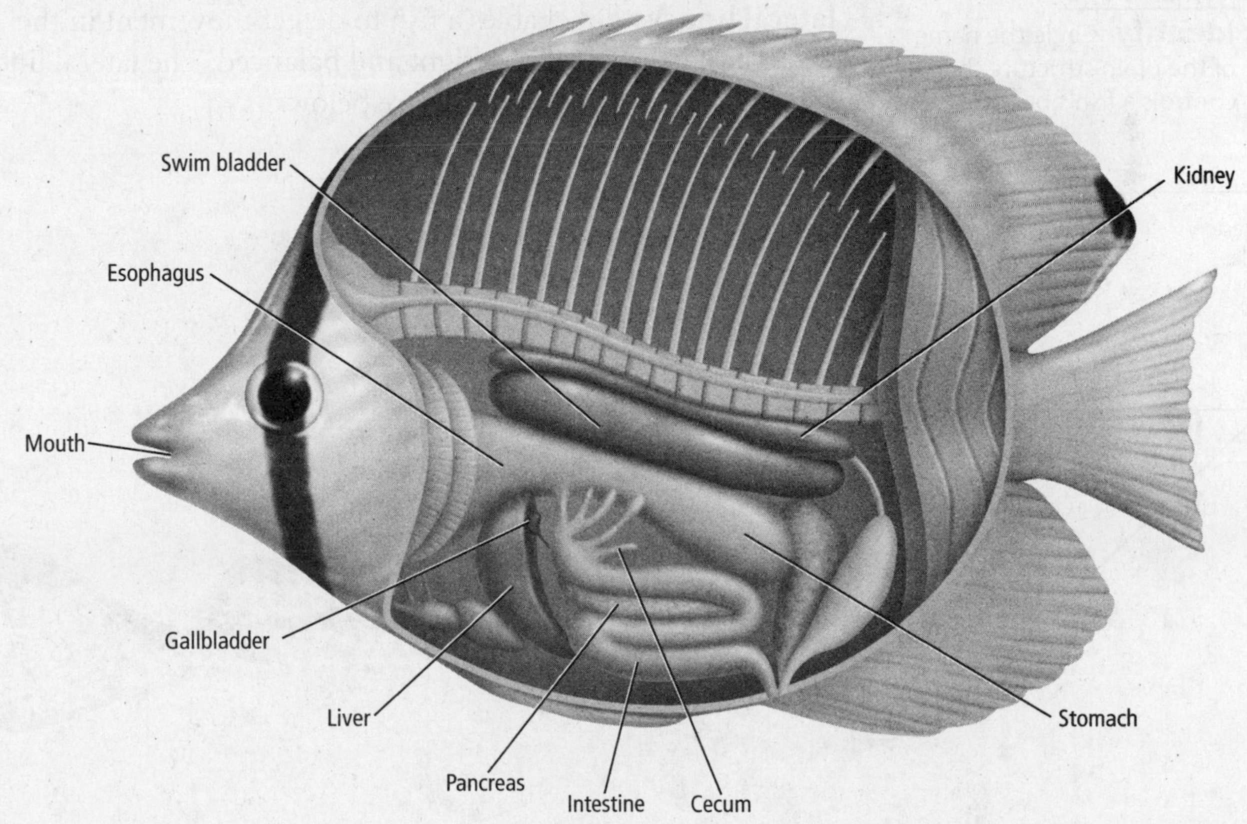

What are the functions of the excretory system?

Kidneys filter cellular wastes from a fish's blood. A **nephron** is a filtering unit within the kidney that removes cellular wastes from the blood. Nephrons also help keep a balance between salt and water in the body. Some cellular wastes are excreted by the gills. ☑

Reading Check

8. Name the excretory organ that contains filtering nephrons.

____

What can fishes sense?

The nervous system of all vertebrates consists of a spinal cord and a brain. A fish brain is shown in the figure below. The cerebellum controls movement and balance. Receptors for the sense of smell detect chemicals in the water. In the brain, olfactory (ohl FAK tree) bulbs receive and process this chemical input. Fishes also have color vision. The optic lobes in the brain are responsible for visual input. The cerebrum coordinates input from the rest of the brain. The medulla oblongata controls internal organs.

What is the purpose of the lateral line system?

Fishes also have receptors called a lateral line system. The **lateral line system** enables a fish to detect movement in the water and helps keep it upright and balanced. The lateral line system is also shown in the figure below.

Picture This

9. Identify Circle the name of the brain structure that controls a fish's balance.

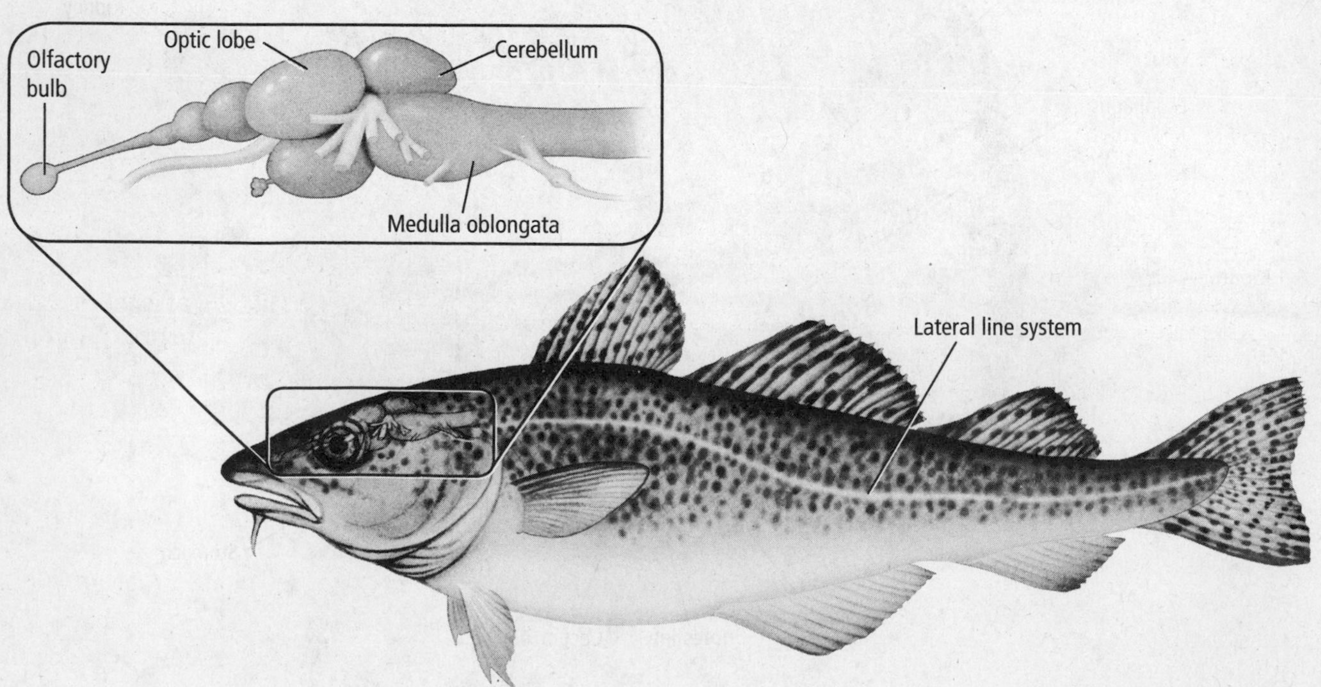

How do fishes reproduce?

For most fishes, fertilization occurs outside the body in a process called **spawning**. During spawning, male and female fishes release their gametes near each other in the water.

The yolk of the egg provides nutrition for the developing embryos. Some fishes, like sharks, reproduce through internal fertilization, with the offspring developing outside the body after the fertilized eggs are laid. Others reproduce through internal fertilization and internal development of the offspring. The developing offspring obtain nutrients from the female's body.

Why do most fishes produce large numbers of eggs?

Fishes that spawn can produce millions of eggs in a season. Most fishes do not protect or care for their eggs or their offspring. As a result, many eggs and young fishes become prey to other animals. The large number of eggs ensures that some will survive to develop and reproduce. ☑

Reading Check

10. Summarize the survival strategy of reproduction by spawning.

What features aid movement?

Most fishes have a streamlined shape. Most also have a mucous coating that reduces friction between the fish and the water for easier swimming. Fins enable fishes to control their movement through water. The buoyant force of water reduces the effect of gravity on fishes. Also, bony fishes have a **swim bladder**, which is a gas-filled space that allows a fish to control its depth in the water. When gases diffuse out of the swim bladder, the fish can sink. When gases from the blood diffuse into the swim bladder, the fish can rise.

A fish moves by contracting muscles on either side of its body. As muscles on one side contract, the fish bends, pushing against the water. The water pulses and the fish moves forward at an angle. Alternating contractions of muscles from side to side move the fish forward in an *s*-shaped pattern, as shown in the figure below.

Picture This

11. Name What is the shape of the pattern of a moving fish?

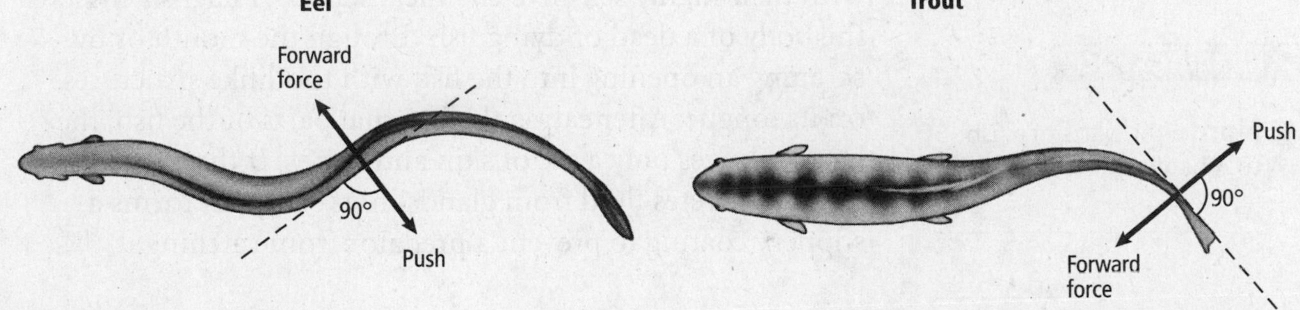

Fishes and Amphibians

section 2 Diversity of Today's Fishes

MAIN Idea

Fishes belong to one of three groups based on body structure.

What You'll Learn

- characteristics of different groups of fishes
- key features of various types of fishes
- how fishes evolved

Before You Read

When you hear the word *shark*, what image comes to mind? On the lines below, describe a shark. In this section you will learn the features of sharks that make them successful hunters.

____

____

____

Study Coach

Create a Quiz As you read this section, write quiz questions based on what you have learned. After you write the questions, answer them.

Read to Learn

Classes of Fishes

Fishes are grouped into three classes based on body structure. The classes are jawless fishes, bony fishes, and cartilaginous (kar tuh LAJ uh nus) fishes.

What are the characteristics of jawless fishes?

Hagfishes belong to class Myxini (mik SEE nee). They are jawless, eel-shaped fishes that do not have scales, paired fins, or a bony skeleton. Hagfishes have a notochord throughout life. They do not develop a vertebral column. However, they do have gills and many other characteristics of fishes. They live on the ocean floor and are scavengers, which means that they feed on dead or dying fishes.

Because they are almost blind, hagfishes locate their food with their highly sensitive chemical senses. A hagfish enters the body of a dead or dying fish through the mouth or by scraping an opening into the fish with toothlike structures on its tongue. After eating the internal parts of the fish, the hagfish leaves only a sac of skin and bones. If threatened, a hagfish secretes fluid from glands in its skin that forms a slippery coating to prevent a predator from catching it. ✓

✓ Reading Check

1. **Name** the parts of a fish that hagfish do not eat.

____

____

Are there other jawless fishes?

Class Cephalaspidomorphi (ceh fah las pe doh MOR fee) includes another jawless, eel-shaped fish—the lamprey. Like hagfishes, lampreys do not have scales, paired fins, or bony skeletons. Lampreys also keep a notochord through life. They have gills and other features of fishes. Adult lampreys are parasites. They attach to other fishes with a suckerlike mouth and scrape away scales and skin to feed on the blood and body fluids of their hosts. ☑

Reading Check

2. Contrast Lampreys differ from hagfishes in what main way?

______

______

______

______

What features do cartilaginous fishes share?

Sharks are members of class Chondrichthyes (kon DRIK thees) and have several rows of sharp teeth. As teeth are broken or lost, new ones move forward to replace them. The main feature that sets sharks apart from other fishes, however, is their skeleton. Sharks are cartilaginous fishes. All cartilaginous fishes have skeletons made of flexible cartilage and calcium carbonate for strength.

Most sharks have a streamlined shape with a pointed head and a tail that turns up at the end, as shown below. Their skin is covered with tough placoid scales. These features, along with strong swimming muscles and sharp teeth, make sharks effective predators.

A shark's chemical sensors enable it to detect prey from a distance of 1 km. As it moves closer, its lateral line system detects movement in the water. In the final chase, a shark uses vision and receptors that detect electricity given off by the prey.

The largest sharks—whale sharks—do not have rows of teeth. They are filter-feeders. Their mouth structures are adapted for straining food from the water.

Skates and rays are also cartilaginous fishes. Their flattened bodies are adapted for life on the bottom of the ocean. Their pectoral fins are enlarged and attached to their heads. These winglike fins flap slowly as they search for mollusks and crustaceans, which they crush with their teeth.

Picture This

3. Name one feature shown in the figure that helps make a shark an effective predator.

______

______

What are the characteristics of ray-finned bony fishes?

Class Osteichthyes (ahs tee IHK theez) contains the bony fishes, which are separated into two groups: ray-finned fishes and lobe-finned fishes. The fins of ray-finned fishes are thin membranes supported by thin, spinelike rays. These fishes also have a bony skeleton, ctenoid or cycloid scales, an operculum covering the gills, and a swim bladder. Most fishes alive today, such as salmon and trout, are ray-finned fishes.

What are the characteristics of lobe-finned bony fishes?

Only eight species of lobe-finned fishes are alive today. Their fins have muscular lobes and joints similar to those of land vertebrates. Most lobe-finned fishes, such as lungfishes, have structures similar to lungs. During droughts, lungfish burrow into the mud with their fins and breathe air. When rain returns, they come out of their burrows.

Coelacanths (SEE luh kanths), another small group of lobe-finned fishes, were thought to be extinct until one was caught in 1938. Others have been caught since then.

A third group of lobe-finned fishes is now extinct. It is thought to be the ancestor of tetrapods. A **tetrapod** is a four-footed animal with legs that have feet and toes with joints. Tetrapods walked on land. Their limbs might have evolved from the fins of lobe-finned fishes.

4. Draw Conclusions The first vertebrates to live on land probably evolved from which class of fishes? (Circle your answer.)

a. bony fishes
b. cartilaginous fishes
c. jawless fishes

Evolution of Fishes

The cladogram on the next page shows how fishes might have evolved. Notice the features that developed during the course of evolution.

What were the characteristics of the first fishes?

The first fishes appeared in the Cambrian period. These jawless, toothless fishes sucked up organic matter from the ocean floor. Ostracoderms (OS tra koh dermz) were the next group of fishes that appeared in the Ordovician period. Their bony head shield, bony outer covering, and paired fins marked a milestone in vertebrate evolution. Stronger movement was possible with muscle attached to bone. Scientists hypothesize that present-day fishes share a common ancestor with ostracoderms.

Cladogram of Fishes

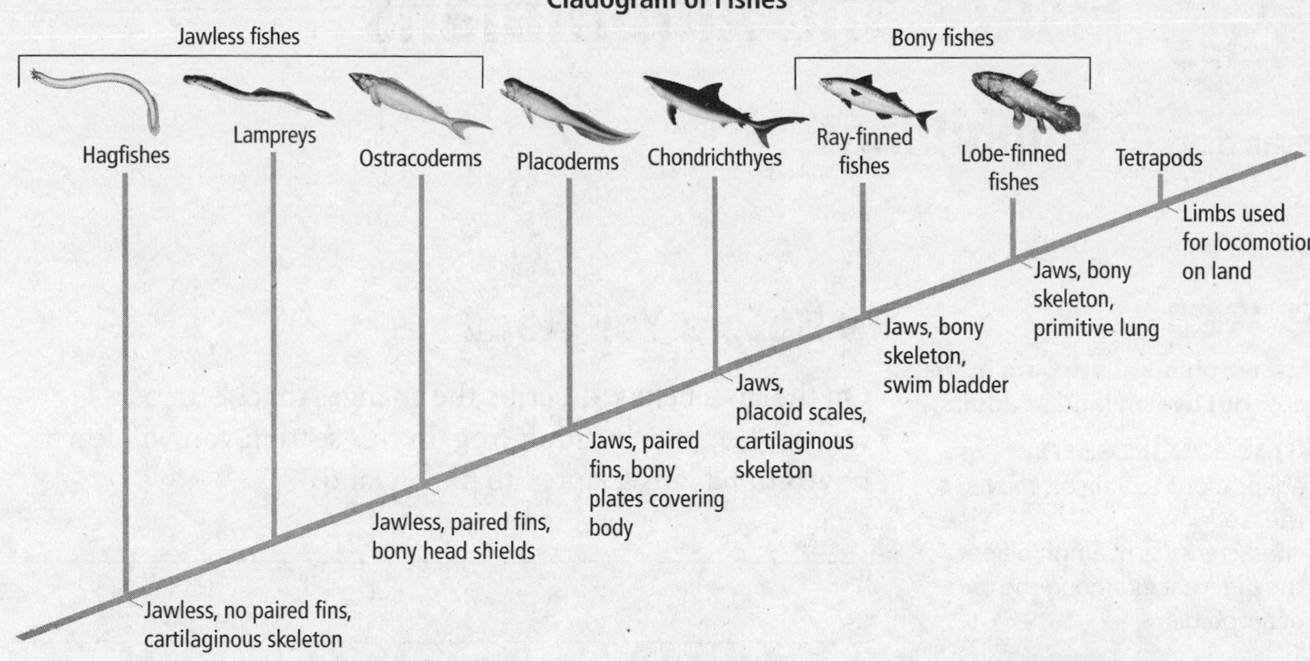

What period is called the Age of Fishes?

Many adaptations appeared during the Devonian period, also called the Age of Fishes. During this time period, now-extinct placoderms had three features of modern fishes: jaws, paired fins, and an internal skeleton.

Picture This

5. **Identify** which fishes did not have jaws.

Ecology of Fishes

Fishes are an important source of food in aquatic ecosystems. Human activities such as damming rivers and dumping pollutants in waterways are changing fishes' freshwater and saltwater habitats. When the numbers of fishes decline, people that make their living from fishing and related industries will be negatively affected. In addition, ecosystems can become unbalanced.

How have dams affected salmon?

Each year, salmon return to spawn in the freshwater stream where they hatched. Dams on rivers interfere with this migration. As a result, the salmon population is smaller in these areas.

How does pollution affect aquatic ecosystems?

Pollution of lakes and streams can reduce the quality of water in lakes, rivers, and streams. This can result in a decline of both number and diversity of fishes in an area. Sometimes fish return when the pollution stops. ☑

Reading Check

6. **List** two ways in which humans have altered ecosystems, causing the number and diversity of fishes to decline in the affected area.

chapter 28 Fishes and Amphibians

section 3 Amphibians

MAIN Idea

Most amphibians are born in the water but live on land as adults.

What You'll Learn

- adaptations as animals moved to the land
- characteristics of amphibians
- the differences among the orders of amphibians

Before You Read

On the lines below, describe the changes that occur as a tadpole becomes an adult frog. In this section you will learn how amphibians adapted to life on land.

____

____

Mark the Text

Identify Main Ideas As you read, highlight the main ideas in each paragraph.

Read to Learn

Evolution of Tetrapods

Tetrapods first appeared on Earth approximately 360 million years ago. The amphibians evolved as they adapted to life on land.

How did vertebrates adapt to life on land?

Conditions are much different on land than in water. The table below compares conditions in water and on land. It also lists some important adaptations that enabled vertebrates to live on land.

Picture This

1. Highlight the adaptation that occurred in response to more available oxygen.

Conditions in Water	Conditions on Land	Adaptations for Life on Land
Water has buoyancy. That is, it has an upward force that works against gravity.	• Air is less buoyant. • Movement is against gravity.	Limbs develop and skeletal and muscular systems become stronger.
Oxygen is dissolved in water and must be removed by gills.	• Oxygen is more available in air than in water.	Lungs enable animals to get oxygen from air more efficiently.
Water holds heat, so the temperature of water does not change quickly.	• Temperature changes occur more rapidly in air. • Temperatures vary between day and night.	Animals develop behaviors, such as migration, and physical adaptations that protect them from extreme temperatures.
Sound waves travel more quickly through water.	• A lateral line system cannot detect sound in air.	Animals develop ears that detect sound waves in the air.

What land habitats do animals occupy?

Land provides many habitats for animals. With proper adaptations, animals can occupy tropical rain forests, temperate forests, grasslands, deserts, taiga, and tundra.

Characteristics of Amphibians

Most amphibians begin life as aquatic organisms. After undergoing metamorphosis (me tuh MOR fuh sihs), they are equipped to live on land.

A frog begins life as a limbless, gill-breathing tadpole. The tadpole undergoes metamorphosis daily. Hind legs form and grow longer, and forelimbs sprout. The tail shortens. Lungs replace gills. Soon the tadpole becomes an adult frog.

Most amphibians have four legs, moist skin with no scales, a double-loop circulatory system, and aquatic larvae. They exchange gases through both their skin and lungs.

FOLDABLES™

Take Notes Make a vocabulary Foldable and label as shown. As you read, record information about the characteristics of amphibians under the tabs.

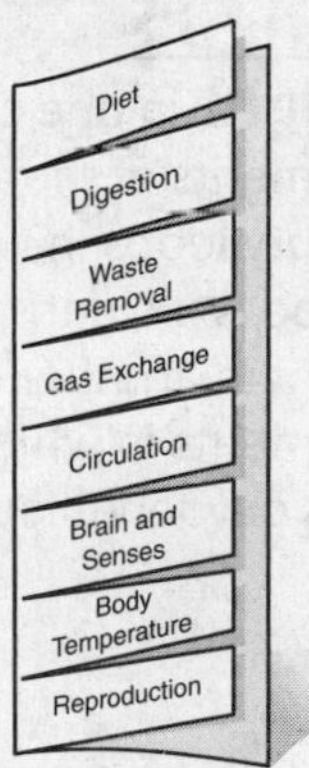

How do amphibians feed and digest?

Frog larvae are herbivores. Salamander larvae are carnivores. As adults, both frogs and salamanders are predators. They feed on a variety of invertebrates and small vertebrates. Some salamanders and legless amphibians catch prey in their jaws. Frogs flick out their long, sticky tongues to catch flying prey.

Food moves from the mouth through the esophagus to the stomach, where digestion begins. The food moves to the small intestine, which receives enzymes from the pancreas to digest food. From the intestine, food is also absorbed into the bloodstream. From the small intestine, food moves to the large intestine before waste is eliminated.

At the end of the intestine is a chamber called the cloaca. (kloh AY kuh). The **cloaca** receives digestive and urinary wastes as well as eggs or sperm before they leave the body.

How are wastes removed from the body?

The kidneys filter wastes from the blood. Amphibians that live in water excrete waste as ammonia. Amphibians that live on land excrete urea. Urea is made from ammonia in the liver. Unlike ammonia, urea is stored in the urinary bladder until it leaves the body through the cloaca.

What structures are used for gas exchange?

As larvae, most amphibians exchange gases through their skin and gills. As adults, most amphibians breathe through their lungs, mouths, and thin, moist skin.

Think it Over

2. Explain how the method of gas exchange reflects the difference in habitat between larvae and adult amphibians.

____

____

____

Amphibian Circulatory System

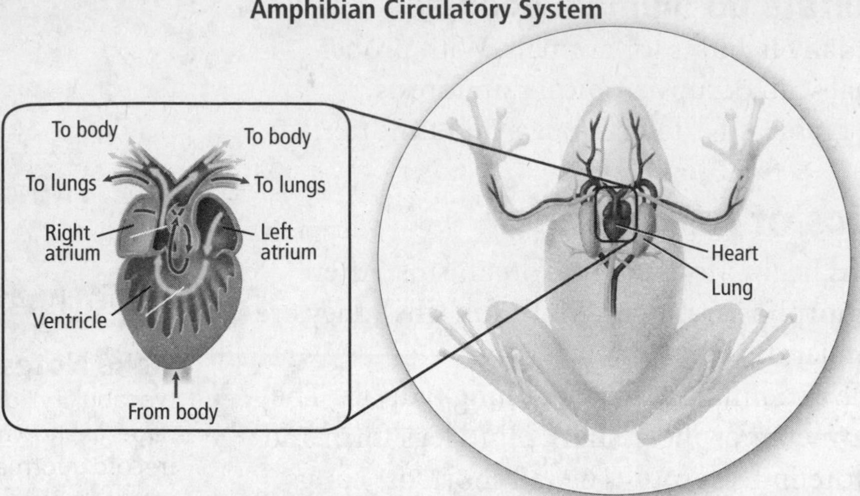

Picture This

3. Highlight in one color the segments of the circulatory loops in which the blood is oxygen-poor. Highlight in another color the segments in which the blood is oxygen-filled.

How does blood circulate through the body?

Amphibians have a double-loop circulatory system, as shown in the figure above. In the first loop, oxygen-poor blood moves from the heart to pick up oxygen in the lungs and skin. Oxygen-filled blood then moves back to the heart. In the second loop, oxygen-filled blood moves from the heart through vessels to the body, where oxygen diffuses into cells.

The amphibian heart has three chambers. The right atrium receives oxygen-poor blood from the body. The left atrium receives oxygen-filled blood from the lungs. The ventricle remains undivided.

How have the brain and senses adapted?

The amphibian brain is adapted for life on land. A frog's forebrain can detect odors in air. The cerebellum, important for balance to fishes, is not well developed in amphibians that live on land.

Vision is important to amphibians for catching prey and escaping predators. A **nictitating** (NIK tuh tayt ing) **membrane** is a transparent eyelid that covers a frog's eye. It protects the eye underwater and keeps it from drying out on land.

Amphibians have developed a **tympanic** (tihm PA nihk) **membrane**, or eardrum. In frogs, it is a thin external membrane on the side of the head. It is used to hear high-pitched sounds, such as mating calls.

Think it Over

4. Apply Why might a salamander lie on a rock on a sunny morning?

____

____

____

How do amphibians control body temperature?

Amphibians are ectotherms. **Ectotherms** are animals that get their body heat from the external environment. Because they cannot control their body temperature internally, they must be able to sense where to go to get warmer or cooler.

How do amphibians reproduce and develop?

A frog's reproduction cycle is typical of many amphibians. Female frogs lay eggs in the water to be fertilized by males. The eggs do not have shells and can dry out if not kept in water. The eggs are covered in a jelly-like substance that helps them stick to plants in the water. After fertilization, the embryo uses the egg yolk for nutrition until it hatches into a tadpole. Chemicals in the tadpole's body control its metamorphosis into an adult frog. During metamorphosis, the tadpole changes from gill-breathing to lung-breathing. The legless herbivore becomes a four-legged carnivore. Its two-chambered heart changes to a heart with three chambers. ☑

Reading Check

5. Explain Why do frog eggs need to be laid in water?

______

______

______

Amphibian Diversity

Biologists group amphibians into three orders. Frogs and toads belong to order Anura (a NOOR ah). Salamanders and newts belong to order Caudata (KAW day tah). Caecilians make up order Gymnophiona (JIHM noh fee oh nah).

What features distinguish frogs from toads?

Frogs have longer and more powerful legs and can make more powerful jumps than toads. Frogs have moist, smooth skin, while toads have bumpy, dry skin. Both need to be near water for reproduction. However, toads generally live farther from water than do frogs. Unlike frogs, toads have glands near the back of their heads. These glands release a foul-tasting poison to discourage predators from eating them.

What are features of salamanders and newts?

Salamanders and newts have long, slim bodies with necks and tails. Most salamanders have four legs; thin, moist skin; and lay their eggs in water. Their larvae look like small salamanders with gills. Salamanders must live near water. They live in moist areas, such as under logs or in leaf litter. They feed on worms, frog eggs, and insects. Newts are aquatic throughout their lives. ☑

Reading Check

6. Contrast the habitat of adult salamanders with the habitat of newts.

______

______

______

How do caecilians differ from other amphibians?

Unlike other amphibians, caecilians (si SILH yenz) have no legs. They are wormlike. They burrow in the soil and feed on worms. Skin covers their eyes, so they might be nearly blind. They have internal fertilization and lay their eggs in moist soil near water. Caecilians live in tropical forests.

Evolution of Amphibians

Fossils show that the first tetrapods evolved limbs in water before they moved to land. Many adaptations that are useful on land first evolved in water. The cladogram below shows one interpretation of amphibian evolution. Many scientists believe that early tetrapods are more closely related to extinct lobe-finned fishes called rhipidistians (RI pih dihs tee unz). They share similar skull and limb bone structure, nostril-like openings in the tops of their mouths, and tooth structure.

Early tetrapods had legs with feet, but the legs were too weak for walking on land. Ichthyostegans had stronger shoulders, heavier leg bones, and more muscular features. These features enabled their movement on land. Tetrapods branched out to produce the three major groups of amphibians alive today, as well as reptiles, birds, and mammals.

Picture This

7. Highlight the groups of amphibians alive today that branched out from tetrapods.

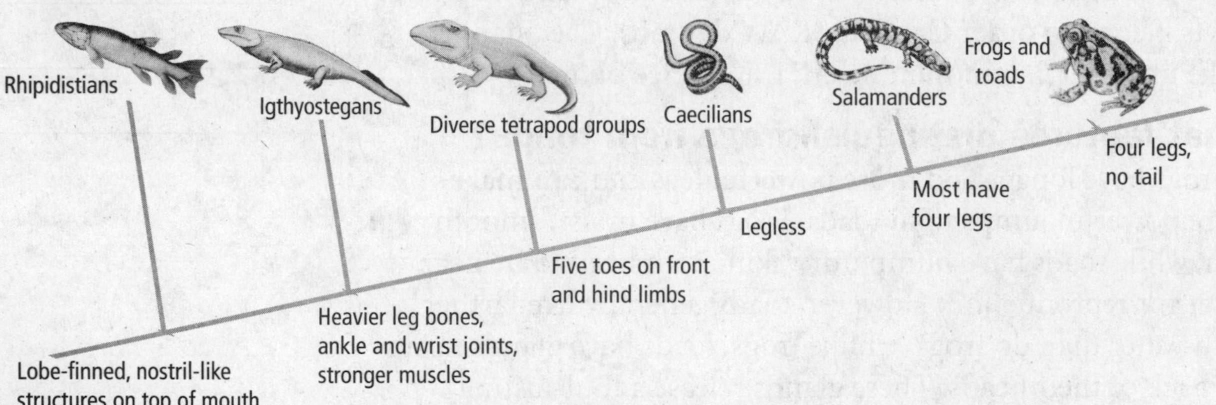

Ecology of Amphibians

Amphibian populations have been declining worldwide. Scientists are collecting data to find possible causes.

What local factors contribute to their decline?

Habitat destruction is a cause of the decline of some amphibians. Wetlands are drained to provide land for new buildings. Without water, amphibians cannot reproduce. The introduction of species not naturally found in an area also adds to the decline. These species might prey on amphibians or compete with them for food and space. ☑

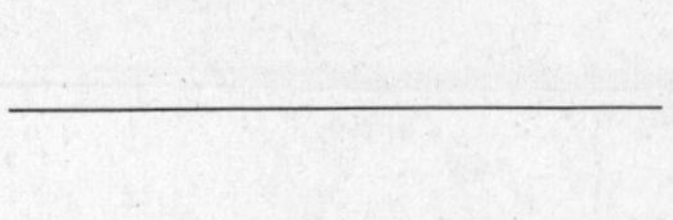

8. Explain why draining wetlands might result in the decline of amphibians.

____

____

____

What global factors contribute to their decline?

Global climate change, such as warmer temperatures and less rainfall, can cause death or disease among amphibians. Increased exposure to UV light might increase the risk of fungal infection that damages amphibian eggs.

Reptiles and Birds

section 1 Reptiles

Before You Read

Think about a movie you saw that featured dinosaurs. On the lines below, describe the characteristics of the dinosaurs portrayed in the movie. Then read the section to see if the movie gave an accurate picture of what dinosaurs were like.

____

____

MAIN Idea

Reptiles are fully adapted to life on land.

What You'll Learn

- how the amniotic egg made the move to land complete
- the characteristics of reptiles
- the differences among orders of reptiles

Read to Learn

Characteristics of Reptiles

Reptiles were the first completely terrestrial vertebrates. Shelled eggs, scaly skin, and more efficient circulatory and respiratory systems enabled reptiles to live on land.

How does an amniotic egg protect the embryo?

Reptiles, birds, and mammals are all amniotes. As the embryos develop, they are surrounded by a membrane called the **amnion** (AM nee ahn). An **amniotic egg** is covered with a protective shell, as shown in the figure below. Within the shell are several membranes with fluids between them. The shell and fluids protect the embryo as it develops.

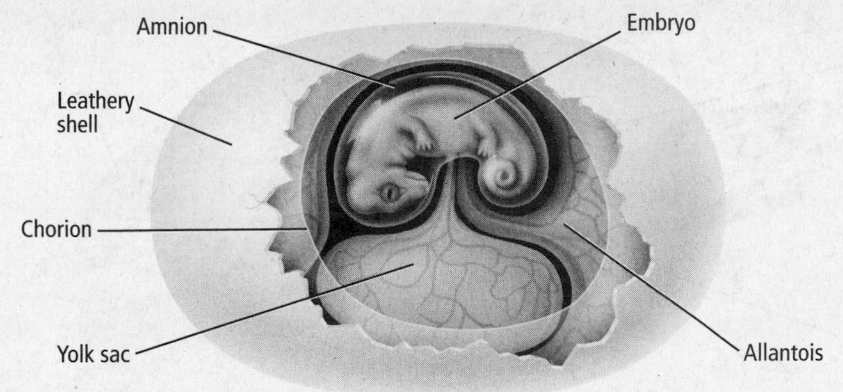

Study Coach

Make Flash Cards
Make a flash card for each question heading in this section. On the back of the flash card, write the answer to the question. Use the flash cards to review what you have learned.

Picture This

1. **Label** Read about how embryos develop on the next page. Then label each membrane in the figure with a brief description of its functions.

How does the embryo develop inside the egg?

The embryo gets nutrition from food in the egg's yolk sac. Amniotic fluid within the amnion provides a watery environment for the embryo. The allantois (uh LAN tuh wus) is a membrane that contains the embryo's waste products. The chorion (KOR ee ahn) is the outside membrane of the egg. It allows oxygen to enter and keeps fluids inside the egg. Reptile eggs have a leathery shell. Bird eggs have a hard shell.

FOLDABLES™

Take Notes Make a three-pocket Foldable from an 11″ × 17″ sheet of copy paper, as shown below. As you read, use quarter sheets of paper or note cards to take notes and organize what you learn about reptiles.

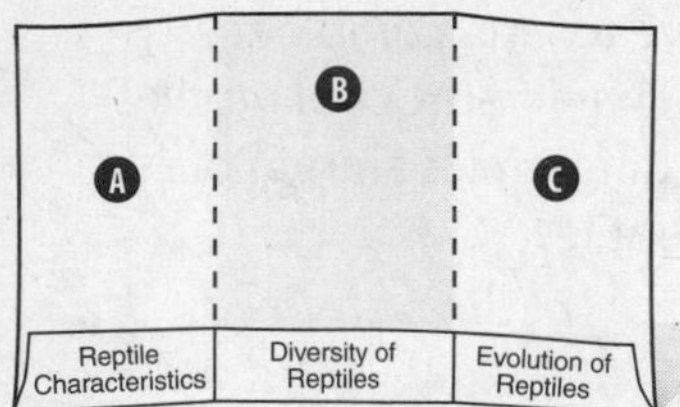

What advantage does dry, scaly skin offer?

Dry skin keeps reptiles from losing internal fluids to the air. Outside scales also help prevent the reptile from drying out. One problem with having a tough outer covering is that an organism could have a hard time growing larger. To grow, some reptiles must shed their skins in a process called molting.

How do reptiles take in air?

Amphibians must squeeze their throats to force air into the lungs. Reptiles are able suck air into their lungs, or inhale, by contracting muscles in the chest. These muscle contractions expand the upper part of the body around the lungs. Reptiles exhale by relaxing these same muscles. Reptiles' lungs have larger surface areas than the lungs of amphibians. With more oxygen, reptiles have more energy for more complex movements.

How many chambers are in a reptile's heart?

Most reptiles have two atria and one ventricle. As shown in the figure below, the ventricle is partly divided by an incomplete septum. In crocodiles, the ventricle is completely divided by the septum, creating a four-chambered heart. The two ventricles keep oxygen-rich blood separate from oxygen-poor blood. Because they are larger than amphibians, reptiles need a higher blood pressure.

Picture This

2. Highlight the structure that is complete in crocodiles but not in other reptiles.

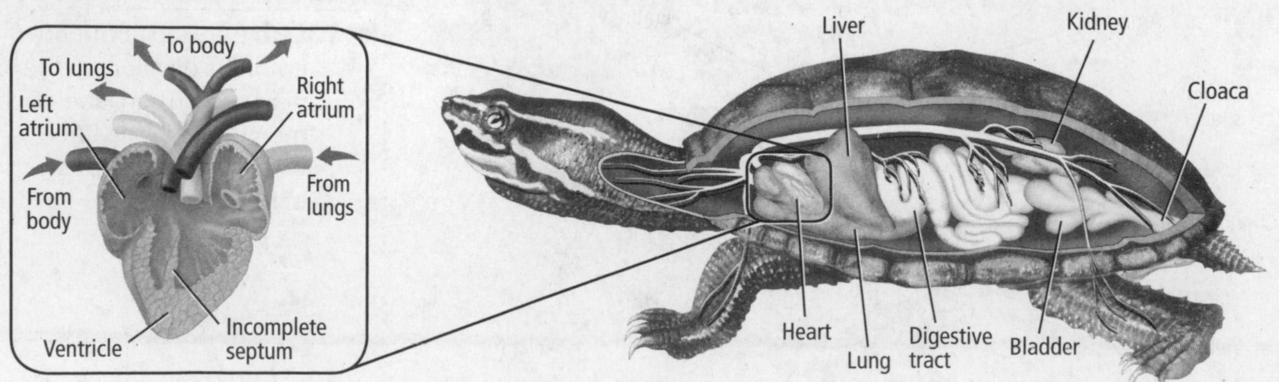

How do reptiles feed and digest?

The digestive system of a reptile, shown in the previous figure, is similar to those of fish and amphibians. Most reptiles are carnivores, but some are herbivores.

Snakes can swallow prey much larger than themselves. The bones of the skull and jaws are joined loosely so they can spread apart to swallow large prey. Some snakes have venom that can paralyze the prey and start the digestion process.

Why does reabsorption occur before excretion?

A reptile's kidneys remove waste from the blood. When urine enters the cloaca, water is reabsorbed to form uric acid. This semisolid waste is then excreted. Reabsorption enables reptiles to conserve water and maintain homeostasis of water and minerals in their bodies. ☑

Reading Check

3. Explain how reabsorption relates to homeostasis.

____

____

____

What senses are highly evolved in reptiles?

Because vision and muscle function are more complex, reptiles have larger optic lobes and cerebellum than those of amphibians. Vision is the main sense for most reptiles. Some reptiles have tympanic membranes for hearing. Others such as snakes detect vibrations through their jaw bones.

Reptiles have a highly evolved sense of smell. A snake flicks its forked tongue to smell odors. Odor molecules stick to its tongue. When the snake pulls in its tongue, odor molecules are transferred to a pair of saclike structures called **Jacobson's organs**. These structures, on the roof of the mouth, sense odors. Snakes use their sense of smell to find prey and mates.

Think it Over

4. Draw Conclusions A salamander drags its body along the ground with legs extended from its sides. The legs of a crocodile are positioned farther under its body. How does leg position enable the crocodile to carry more weight than the salamander?

____

____

____

____

____

How do reptiles control body heat?

Reptiles are ectotherms. They cannot create their own body heat. They adjust their body temperature through behaviors. A turtle might bask in the sun to get warm and move to the shade to cool. Some reptiles survive winter by burrowing or going into a state of inactivity. Some snakes huddle in masses during winter to reduce heat loss.

How are reptile bodies adapted for movement?

Like amphibians, some reptiles move with limbs extended from their sides. Other reptiles such as crocodiles have limbs farther under their bodies which allow them to support more weight and move faster. A stronger, heavier bone structure also helps reptiles support more body weight on land. Reptiles have claws that aid in digging, climbing, and gripping the ground.

Do reptiles care for their young?

Reptiles have internal fertilization. A female reptile usually digs a hole and lays eggs in the ground or in plant debris. Most female reptiles then leave the eggs alone to hatch. Alligators and crocodiles build nests in which to lay eggs and tend to their young after they hatch. Some snakes and lizards keep the eggs in their bodies until they hatch. ☑

Reading Check

5. Identify reptiles that care for their young.

____

____

Diversity of Modern Reptiles

After dinosaurs disappeared, four living orders of reptiles remained. They are order Squamata (skwuh MAHD uh) (snakes and lizards), order Crocodilia (crocodiles and alligators), order Testudinata (turtles and tortoises), and order Sphenodonta (sfee nuh DAHN tuh) (tuataras).

What are the features of lizards and snakes?

Most lizards have legs with clawed toes. Most lizards also have movable eyelids, a lower jaw with a movable hinge joint for flexible jaw movement, and tympanic membranes.

Snakes do not have legs, movable eyelids, or tympanic membranes. Snakes have shorter tails than lizards. Some snakes have venom that can slow or kill prey. Other snakes are constrictors. They wrap their bodies around their prey, tightening their hold until the prey can no longer breathe and dies.

How are turtles different from tortoises?

Unlike other reptiles, turtles have a shell. The dorsal (upper) part is the **carapace** (KAR ah pays). The ventral part (underside) is the **plastron** (PLAS trahn). The vertebrae and ribs are fused to the inside of the carapace. When threatened, many turtles can pull their heads and legs inside their shells for protection. Some turtles are aquatic. Others live on land. Turtles that live on land are called tortoises.

Think it Over

6. Compare If you wanted to determine whether an animal was an alligator or a crocodile, what characteristics would you look for?

____

____

____

____

How do crocodiles and alligators differ?

All members of order Crocodilia have a four-chambered heart. Crocodiles have long snouts, sharp teeth, and powerful jaws. A crocodile's upper and lower jaws are about the same width. When the mouth is closed, some teeth in the lower jaw can be seen. Alligators have broad snouts. An alligator's upper jaw is wider than its lower jaw. As a result, the upper jaw overlaps the lower jaw and the teeth are almost completely covered when its mouth is closed.

What is the function of a tuatara's third eye?

Tuataras (tyew ah TAR ahz) resemble large lizards. A tuatara has a spiny crest down the back. It has a scaly structure known as a "third eye" on top of its head. The eye senses sunlight and might keep tuataras from overheating in the Sun. Their upper jaws have two rows of teeth and their lower jaws have one row.

Evolution of Reptiles

The cladogram below shows that early amniotes might have separated into three lines, each with different skull structure. Skulls of anapsids, which might have given rise to turtles, had no openings behind the eye sockets. Skulls of diapsids, which gave rise to crocodiles, dinosaurs, modern birds, tuataras, snakes, and lizards, had two pairs of openings behind each eye socket. Skulls of synapsids, which gave rise to modern mammals, had one opening behind each eye socket.

Dinosaurs were a diverse group. Some were huge and predatory. Others had large horns and ate plants. Dinosaurs are divided into two groups, based on hip structure. Hip bones of one group extended out from the center of the hip area. Hip bones of the other group extended back toward the tail.

Dinosaurs and many other species went extinct during the Cretaceous Period. Some scientists think this extinction resulted from a meteorite that crashed to Earth. Dust clouds might have blocked the Sun, causing a cooler climate. When dinosaurs disappeared, other vertebrates were able to evolve.

Picture This

7. Draw Conclusions Which of these animals descended from diapsids? (Circle your answer.)

a. turtles
b. mammals
c. lizards

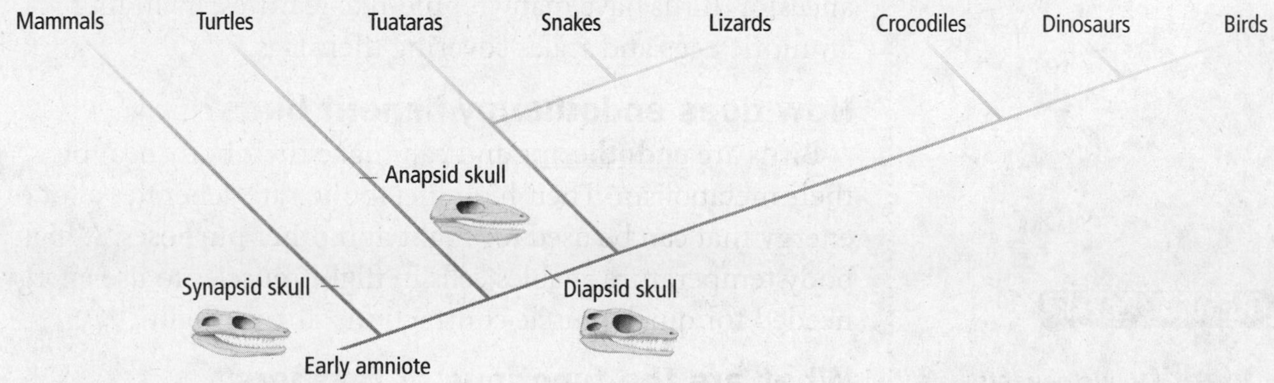

Ecology of Reptiles

Reptiles are both predators and prey. Removal of a reptile species can upset the balance of an ecosystem. The destruction of wetlands and the introduction of nonnative species result in the decline in population of some reptile species.

Think it Over

8. Explain Suppose a snake is removed from an area. What might happen to the population of mice it fed on?

chapter 29

Reptiles and Birds

section 2 Birds

MAIN Idea

Birds have feathers, wings, lightweight bones, and other adaptations for flight.

What You'll Learn

- how adaptations of birds enable them to fly
- the characteristics of different orders of birds

Before You Read

Have you ever found a feather on the ground? On the lines below, describe the feather. Then read the section to learn how feathers do more than enable birds to fly.

____

____

____

Mark the Text

Restate the Main Point Highlight the main point in each paragraph. Then restate each main point in your own words.

Read to Learn

Characteristics of Birds

Birds belong to class Aves. They are the most diverse of all land vertebrates. Birds range in size from hummingbirds to ostriches. They are found in deserts, forests, mountains, prairies, and on all seas. Birds and reptiles have a common ancestor. Birds have many reptile-like features, including amniotic eggs and scales covering their legs.

How does endothermy benefit birds?

Birds are **endotherms** and can make their body heat by their metabolism. Their high metabolic rate generates a lot of energy that can be used for flight and other purposes. A high body temperature enables cells in flight muscles to use energy needed for quick muscle contractions during flight. ☑

Reading Check

1. **Identify** two benefits of the high metabolic rate of birds.

____

____

____

What are the functions of feathers?

Feathers are specialized outgrowths of the skin of birds. Only birds have feathers. Like hair, nails, and the horns of other animals, feathers are made of the protein keratin (KER ah tihn). Feathers have two main functions—flight and insulation. When a bird fluffs its feathers, it creates a dead air space that traps body heat.

Why do birds preen their feathers?

Contour feathers cover the body, wings, and tail of a bird. Barbs branch off from the shaft of these feathers. The barbs are held together by hooks. Barbs that become separated can be rejoined like the teeth of a zipper. Birds can repair separations by preening. A bird preens by running the length of feathers through its beak. A **preen gland** near the base of the tail secretes oil. During preening, birds spread the oil over the feathers to waterproof them.

Down feathers are soft feathers beneath contour feathers. They do not have hooks to hold the barbs together. This looser structure allows down feathers to trap air for insulation.

How are birds' bones adapted for flight?

The bones of birds contain cavities of air, making the skeletons lightweight. Fusion of bones creates skeletons sturdy enough for flight. Large breast muscles provide the power for flight. These muscles connect the wing to the keel of the **sternum** (STUR num), or breastbone.

How does the respiratory system support flight?

Flight muscles use a lot of oxygen. A bird has much more space for air in its respiratory system than a reptile. It also has one-way air circulation.

When a bird inhales, oxygen-rich air moves through the trachea into posterior **air sacs**, as shown in the figure below. Other air already within the respiratory system moves from the lungs, where gas exchange occurs, to the anterior air sacs. When a bird exhales, the oxygen-poor air in the anterior air sacs is expelled from the respiratory system and oxygen-rich air from the posterior air sacs moves into the lungs. As a result of this system, only oxygen-rich air moves through the lungs and it flows in just one direction.

2. Apply Suppose you pick up a feather and find that its parts seem to stick together. What type of feather is it? (Circle your answer.)

a. contour feather

b. down feather

Picture This

3. Highlight in one color the path of air during inhalation. Highlight in another color the path of air during exhalation.

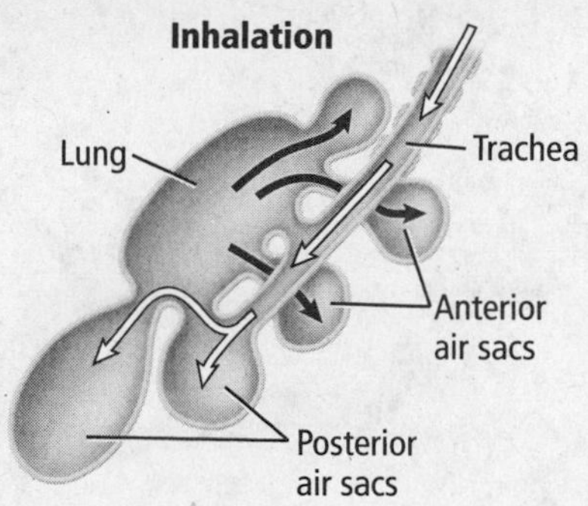

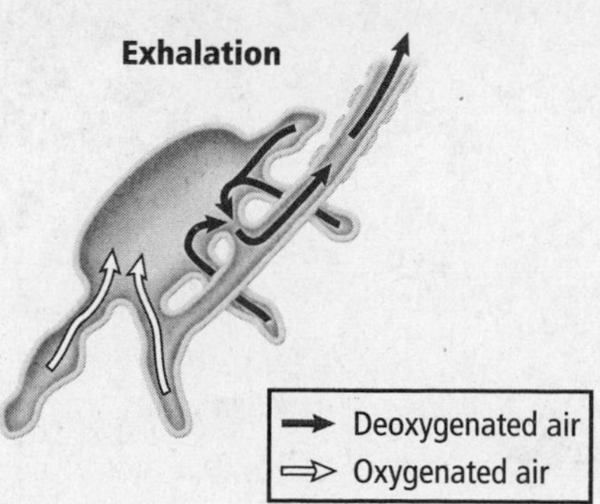

Why are two ventricles important?

Birds have four-chambered hearts. Having two ventricles helps keep oxygen-rich blood separate from oxygen-poor blood. This makes oxygen delivery more efficient. Blood enters the left atrium from the lungs. It is pumped into the left ventricle and out to the body. Blood returns from the body to the right atrium. It then moves into the right ventricle and on to the lungs to pick up more oxygen. ☑

Reading Check

4. Explain why a four-chambered heart is more efficient than a three-chambered heart.

____

____

____

____

How do birds digest their food?

The figure below shows a bird's digestive system. Food travels through the esophagus to a storage chamber called the crop. From the crop, food moves to the stomach. At the base of the stomach is the gizzard (GIH zurd), a thick, muscular sac. Small stones within the gizzard, along with the gizzard's muscular action, crush the food. Digestion and absorption of nutrients occur mostly in the small intestine with the help of digestive juices from the pancreas and liver.

Why do birds lack a urinary bladder?

A bird's kidneys filter waste from blood and convert it to uric acid. Water from uric acid is reabsorbed in the cloaca. A bird lacks a urinary bladder, an organ for storing urine. Stored urine would add weight, making flight difficult. Therefore, not having a urinary bladder is an adaptation for flight.

Picture This

5. Label Circle the structure that is used to crush food.

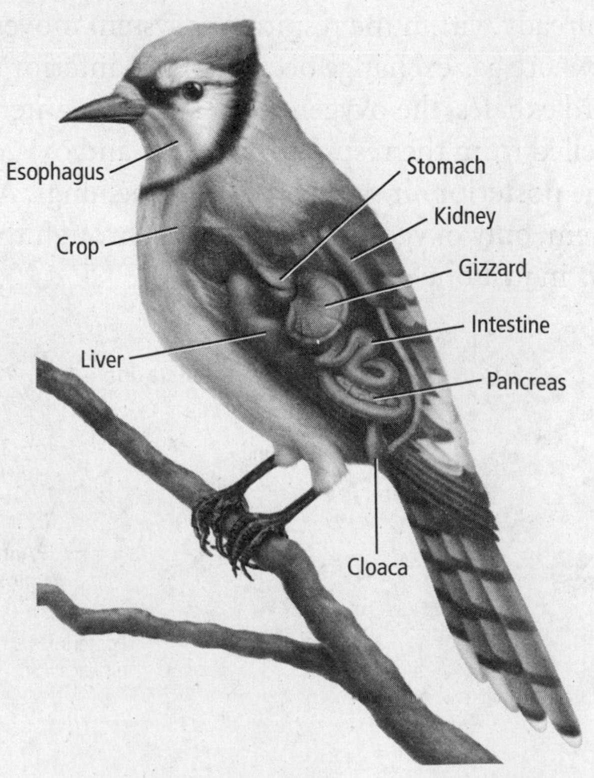

What are other bird adaptations?

A bird has a large brain for its size. The large cerebellum coordinates movement and balance during flight. The optic lobes process visual input. The core of the large cerebrum controls most behaviors. The medulla oblongata controls automatic functions, such as respiration. ☑

The position of a bird's eyes reflect its life habits. A hawk is a predator and needs to focus both eyes on distant prey. Therefore, a hawk's eyes are located in the front of its head. A pigeon needs to be able to see predators coming from any direction as it feeds on grains. This broad view is achieved by eyes positioned on the sides of the pigeon's head.

All birds have internal fertilization. The hard-shelled eggs are released into the nest through the cloaca. One or both parents sit on the eggs to incubate them. To **incubate** means to maintain favorable conditions for hatching.

6. Identify the part of a bird's brain that coordinates movement during flight.

____

Diversity of Modern Birds

The table below summarizes features of the most common orders of birds.

Picture This

7. Highlight the orders of birds that do not fly. Circle the orders of birds likely to be found in or near water.

Order	Members	Characteristics
Passeriformes Perching songbirds	blue jays, crows, nuthatches, finches	feet adapted for perching on branches; many have well-developed vocal organ for singing
Piciformes Cavity-nesters	woodpeckers, toucans, puffbirds	bills specialized for feeding habits; build nests in cavities; feet with two toes that extend forward and two toes that extend backward for clinging to tree trunks
Ciconiiformes Wading birds, vultures	herons, egrets, storks, flamingos, vultures	mostly wading birds with long necks and long legs; live in large colonies in wetlands
Procellariiformes Marine birds	albatrosses, petrels, shearwaters	marine birds; hooked beaks and tube-shaped nostrils on top of beak; many have webbed feet
Sphenisciformes Penguins	penguins	use wings as flippers to swim rather than fly; solid bones without air spaces; live in southern hemisphere
Strigiformes Owls	owls	active at night; large eyes, hooked beak, and sharp talons for capturing prey; many have feathered feet
Struthioniformes Flightless birds	ostriches, kiwis, emus, rheas	reduced wings; flightless; ostrich is largest living bird; all species found in southern hemisphere
Anseriformes Waterfowl	swans, geese, ducks	aquatic; webbed feet; feed on aquatic plants and sometimes small fish and crustaceans; round beaks

Evolution of Birds

Fossils show that birds, crocodiles, and dinosaurs all descended from archosaurs. Fossils of three species of feathered, birdlike dinosaurs were discovered in China. These fossils support the idea that birds evolved from dinosaurs. Scientist Hermann von Meyer discovered the oldest bird fossil—*Archaeopteryx*—in southern Germany. Unlike modern birds, this ancient bird, shown below, had a reptile-like tail, clawed fingers in the wings, and teeth. Yet, like modern birds, feathers covered its body.

Picture This

8. Highlight the features of *Archaeopteryx* that are birdlike.

Ecology of Birds

Birds are both predators and prey in their food chains. They also play an important role in spreading seeds. Some birds eat seeds or fruits and, after digestion, excrete them in different locations. Hummingbirds pollinate flowers as they feed on nectar.

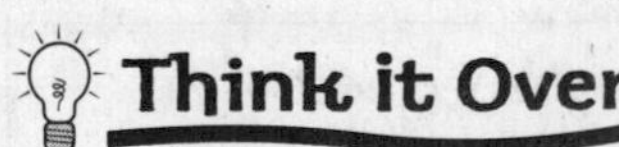

9. Predict Suppose everyone refused to buy birds caught in the wild. How might this affect the illegal trade in wild exotic birds? Why?

___

___

___

How is human activity affecting bird habitats?

Bird habitats are disappearing as wetlands are drained for development and tropical rain forests lose trees through logging. Pesticides and chemical pollutants are also destroying bird habitats.

How does illegal trade affect birds in the wild?

Many pet birds are raised in captivity. Other birds are taken illegally from the wild. Illegal capture and sale of wild exotic birds is increasing because it is profitable. In spite of international laws, illegal capture and sale has led to the disappearance of some rare birds in the wild.

chapter 30 Mammals

section 1 Mammalian Characteristics

Before You Read

Most mammals have bodies covered with protective layers of hair and fur. On the lines below, describe what hair and fur protect mammals from. Are humans the only mammals that are mostly "hairless"?

____

____

____

MAIN Idea

Mammals have hair and mammary glands.

What You'll Learn

- how mammals maintain a constant temperature
- how respiration in mammals differs from other vertebrates

Read to Learn

Mammary Glands and Hair

Members of the class Mammalia (mah MAY lee uh) have two characteristics that no other animals have—mammary glands and hair. <u>**Mammary glands**</u> produce and secrete milk that nourishes developing young. The evolutionary tree for the ancestral chordate shows the two characteristics that separate mammals from other vertebrates.

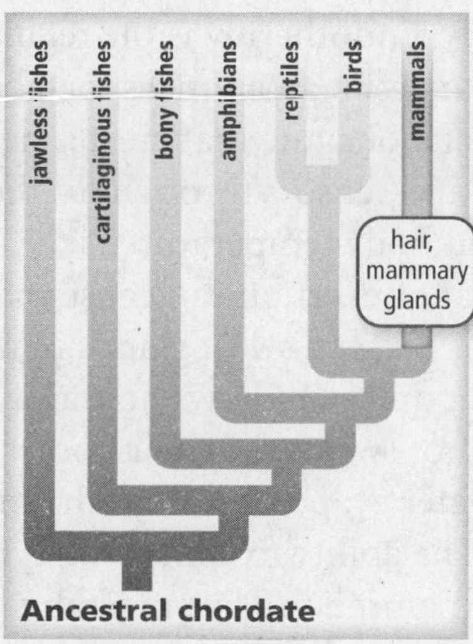

Mark the Text

Identify the Characteristics

Highlight each characteristic of mammals as you read about it. Underline the functions of each characteristic.

Picture This

1. **Highlight** the characteristics that distinguish mammals from other chordates.

Think it Over

2. Draw Conclusions The hair of an arctic fox is white in winter and brown in summer. What function does this change serve?

What are the functions of hair?

A mammal's hair performs the following functions:

1. Insulation—Hair traps body heat, helping to keep the mammal warm.
2. Camouflage—A tiger's striped coat helps it blend into its natural habitat.
3. Sensory devices—Sensitive whiskers, which are modified hairs, help a seal track its prey's movements in dark water.
4. Waterproofing—For an aquatic mammal, hair keeps water from reaching the skin. This helps the mammal maintain a constant body temperature.
5. Signaling—When a white-tailed deer runs, it raises its tail, exposing the white underside so that others can follow.
6. Defense—The sharp quills of a porcupine are modified hairs. They pierce predators that touch the porcupine.

How is hair structured?

Hair contains a tough protein called keratin. A mammal's fur coat is usually made of two kinds of hair—long guard hair and dense, short underhair. The guard hair serves as protection, while the underhair insulates against the cold by trapping air and retaining body heat.

Other Characteristics

Mammals share other characteristics. These include a high metabolic rate which supports endothermy, specialized teeth and digestive systems, a diaphragm, a four-chambered heart, and a highly developed brain.

What is endothermy?

3. Identify the source of a mammal's body heat.

Mammals are endotherms, which means they produce body heat internally. Endothermy is the regulation of body temperature by changes in an animal's metabolism. It is a mammal's high metabolic rate that creates heat. Signals between the brain and sensors throughout the body help maintain a constant body temperature. ☑

For example, heat or exertion can cause mammals to become warm. This triggers sweat glands in the skin to secrete sweat. As sweat evaporates, it draws heat away from the body. Mammals that do not sweat cool their bodies by panting. During panting, water evaporates from the mouth and nose, cooling the body. The ability to control body temperature internally enables mammals to live in a range of ecosystems—from cold polar regions to hot tropics.

How does metabolism influence feeding?

Most of the food an endotherm eats daily is used to generate the body heat it needs. Small mammals, such as shrews, bats, and mice, have high metabolic rates for their size. As a result, they must eat almost constantly in order to fuel their metabolism. Metabolic rates are compared to body mass in the graph below.

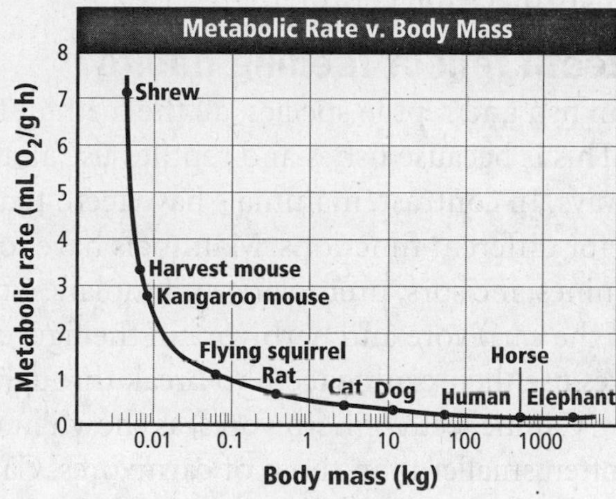

Picture This

4. Determine Which animal has a metabolic rate of approximately 1?

____

How are mammals classified?

Scientists group mammals into four categories based on what they eat. These categories are described as follows:

1. Insectivores, such as moles and shrews, eat insects and other small invertebrates.
2. Herbivores, such as rabbits and deer, eat plants.
3. Carnivores, such as foxes and lions, eat mostly herbivores.
4. Omnivores, such as raccoons and primates, eat both plants and animals.

Mammals have adaptations for finding, catching, chewing, swallowing, and digesting food. These adaptations determine a mammal's structure and function. Plants are more difficult to digest than meat, so an herbivore's digestive tract is longer than a carnivore's.

Think it Over

5. Apply Which of the mammals listed below has the shorter digestive tract? (Circle your answer.)

a. fox
b. cow

How do ruminant herbivores obtain nutrients?

Cellulose found in a plant's cell walls can be a source of nutrition and energy. Mammals do not have enzymes to digest the cellulose. Some herbivores have bacteria in their stomachs that will break down the cellulose. These mammals are called ruminants (REW me nihtz) and have large, four-chambered stomachs. Cattle, sheep, and buffalo are all ruminants.

Think it Over

6. Draw Conclusions What type of relationship do you think exists between a ruminant and the bacteria in its stomach? (Circle your answer.)

a. parasitic
b. mutualistic

Where does digestion occur in ruminants?

Ruminants, including cattle and sheep, have large stomachs with four chambers. As ruminants feed, bacteria in the first two chambers partially digest the grass into a material called cud. Ruminants bring the cud back up into their mouths and chew it for a long time. This further crushes the grass fibers. After they swallow the cud, it eventually reaches the fourth chamber where digestion continues.

How do teeth reflect feeding habits?

Usually, in fish and reptile species, all the teeth in the mouth look alike. This is because fishes and reptiles use all their teeth in similar ways. In contrast, mammals have teeth that are specialized for different functions. Mammals have four types of teeth: canines, incisors, premolars, and molars. Compare the teeth of the carnivore and herbivore in the figure below.

Carnivores use their canine teeth to break through the skin of their prey. Canine teeth of herbivores, as shown in the figure below, are often smaller than those of carnivores. Carnivores use their premolars and molars to cut meat from the bones of their prey. Herbivores use their premolars and molars to crush and grind. The long, curved incisors of insectivores work well as pincers to seize insect prey. The chisel-like incisors of a beaver are modified for gnawing. Biologists can learn a lot about what a mammal eats by looking at its teeth.

Picture This

7. Explain how the shape of a carnivore's canines accomodates the way the carnivore uses these teeth.

____

____

____

What are the functions of kidneys?

The kidneys of mammals filter urea from the blood and excrete it. Urea is a waste product of cellular metabolism. The kidneys also maintain the proper amount of water in the body to maintain homeostasis. They enable animals to live in extreme environments by controlling the body's water.

Molars Premolars
Incisors
Molars Premolars Canines

Carnivore skull

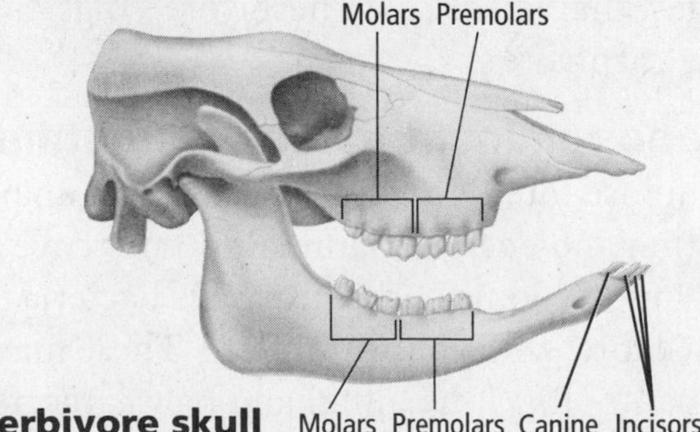

Herbivore skull

How does the diaphragm work?

Although other animals have lungs, mammals are the only animals that have diaphragms. A **diaphragm** is a sheet of muscle located beneath the lungs that separates the chest cavity from the abdominal cavity. As the diaphragm contracts, it flattens. This makes the chest cavity larger, enabling air to enter the lungs. Oxygen then moves by diffusion into blood vessels. High levels of oxygen are needed to maintain the mammal's high metabolism. As the diaphragm relaxes, the chest cavity becomes smaller, and air is exhaled. ☑

Reading Check

8. Summarize When a mammal inhales, what is the diaphragm doing? (Circle your answer.)

a. contracting

b. relaxing

How does blood circulate?

Blood containing oxygen flows through blood vessels into the four-chambered heart, as shown in the figure below. The heart pumps the blood out to the lungs and the body. The blood carrying oxygen is kept separate from the blood without oxygen. This separation makes the delivery of nutrients and oxygen more efficient.

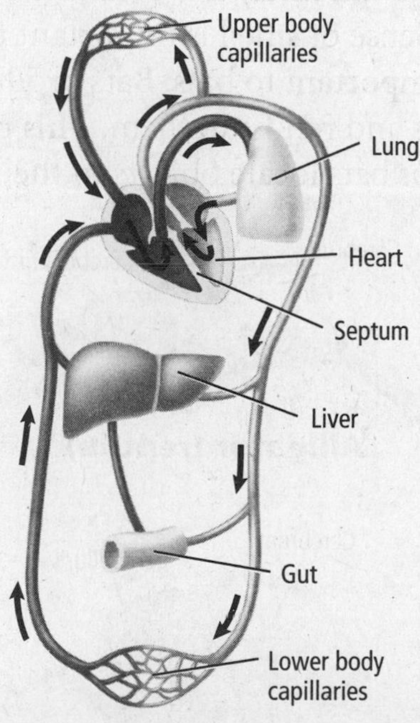

Picture This

9. Identify the structure that separates the chambers in the heart.

____

What happens as body temperatures rise?

The circulatory system of a mammal also helps maintain a constant body temperature. As body temperature increases, blood vessels near the surface of the skin dilate, increasing the amount of blood that flows through them. This increases the loss of body heat. As body temperature decreases, blood vessels near the surface of the skin contract, decreasing the amount of blood that flows through them. This decreases the loss of body heat.

What functions do parts of the brain control?

Mammals have well-developed brains, as shown in the figure below. The **cerebral cortex** (SUH ree brul • KOR teks) is the highly folded outer layer of the cerebrum. The folds increase the surface area for nerve connections and enable the brain to fit into the skull. The cerebral cortex coordinates conscious activities, memory, and the ability to learn. The **cerebellum** (ser uh BE lum) controls balance and coordination. A well-developed cerebellum allows a mammal to move in complex ways. Compare the size and structure of the cerebellums of a reptile, a bird, and a mammal that are shown in the figure below. ☑

Mammals can learn, remember, and use information later. A mother fox will teach her young how to hunt. The ability to learn and teach their young survival skills gives mammals an increased chance of survival.

Different senses are important to different groups of mammals. Vision is extremely important to some mammals, such as humans. Sense of smell is important to a dog. Hearing is most important to bats. Bats produce sounds that bounce off objects and return to them. This process, called echolocation, helps bats locate objects in their path.

Reading Check

10. Identify What functions does the cerebral cortex control?

______

______

______

Picture This

11. Compare the brain structures in the figure. Which animal has the most developed cerebellum?

______

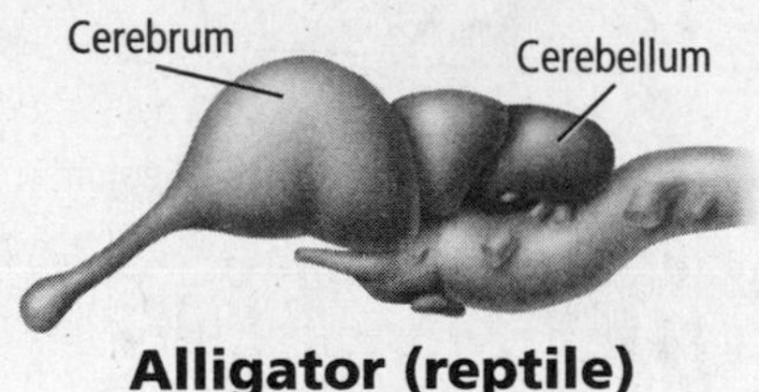

Alligator (reptile)

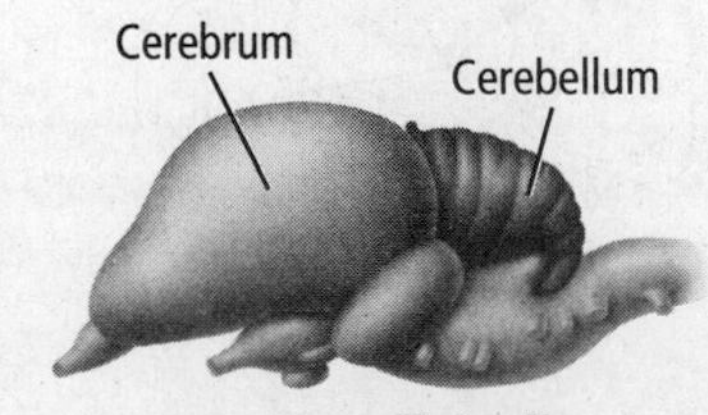

Goose (bird)

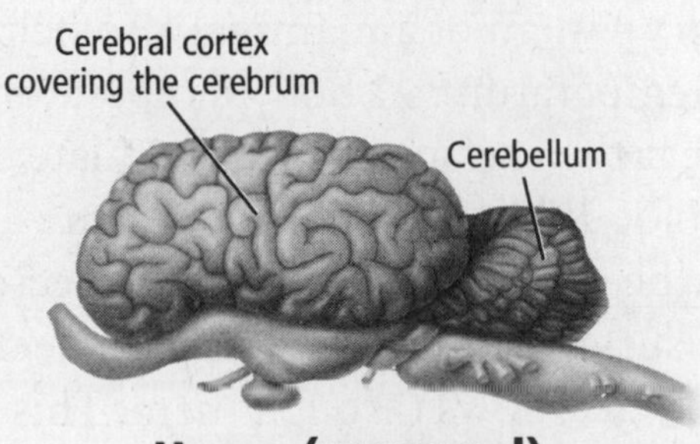

Horse (mammal)

What processes do glands help to control?

A **gland** is a group of cells that secrete fluid to be used elsewhere in the body. Sweat glands help maintain body temperature. Mammary glands produce and secrete milk that nourishes young. Scent glands produce materials that mammals use to mark their territory or attract mates. Oil glands maintain hair and skin quality. Other glands produce hormones that control processes such as growth and the release of eggs from ovaries.

How do mammals reproduce?

The egg is fertilized inside the body of a mammal. In most mammals, the embryo develops within the **uterus**, a saclike muscular organ. The **placenta** provides food and oxygen to the developing young and also removes waste. Different species have different gestation periods. **Gestation** is the amount of time the young develop in the uterus before they are born. In general, the larger the mammal, the longer the gestation period. After birth, the offspring of mammals drink milk from the mother's mammary glands for nourishment.

Think it Over

12. Apply Which animal would you expect to have a longer gestation period, an elephant or a human?

____

What types of movement do mammals display?

Mammals have evolved a variety of limb types that enable them to find food, shelter, and escape from predators. Some mammals, such as coyotes and foxes, run. Cheetahs are the fastest land mammals, reaching speeds of 110 km/h. Other mammals, such as kangaroos, leap. Some mammals, including dolphins, swim. Bats are the only mammals that fly. An animal's skeleton and muscle system reflect the way it moves. The figure below shows the forelimbs of a mole and those of a bat. The mole has powerful, short forelimbs for digging. The bat has thin membranes spread between the elongated arm and hand bones.

Picture This

13. Apply How do the structures of the forelimbs of a mole and a bat reflect their habitats and behaviors?

____

____

____

chapter 30

Mammals

section 2 Diversity of Mammals

MAIN Idea

The three mammal subgroups have distinct characteristics.

What You'll Learn

- how adaptations enable mammals to live in a variety of habitats
- theories of mammal evolution

Before You Read

You might have seen kangaroos in the zoo or on television. On the lines below, write what you know about kangaroos. Then read this section to learn more about mammal subgroups.

____

____

Study Coach

Make an Outline Make an outline of the information you learn in this section. Start with the headings. Include the boldface terms.

Read to Learn

Mammal Classification

Scientists place mammals into three subgroups based on their method of reproduction. The subgroups are monotremes, marsupials, and placental mammals.

How do monotremes reproduce?

Monotremes (MAH noh treemz) are mammals that reproduce by laying eggs. They are mammals because they have hair and mammary glands. However, in some ways, they are like reptiles. They lay eggs, have a lower body temperature than most mammals, and have some reptilelike chromosomes. The only two living monotremes are the duck-billed platypus and the echidna. Both live only in Australia and New Guinea.

FOLDABLES™

Take Notes Make a layered Foldable, as shown below. As you read, take notes and organize what you learn about the classification of mammals, based upon their method of reproduction.

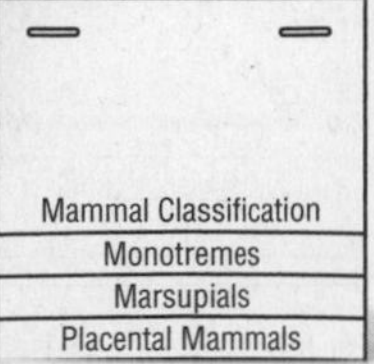

What is the purpose of the marsupial's pouch?

Mammals with pouches are **marsupials** (mar SEW pee ulz). Their young develop in the uterus for a very short time—sometimes just a few days. After birth, the offspring crawl into a pouch made of skin and hair on the outside of the mother's body. There, they continue to develop, feeding on milk from the mother's mammary glands. Marsupials, such as koalas, wallaby, kangaroos, and cuscus, live mostly in Australia. The only marsupial living in North America is the opossum.

Why do most marsupials live in Australia?

Fossil evidence suggests that marsupials originated in North America. They spread to South America, Africa, Australia, Antarctica, and Europe while the continents were still connected as one giant landmass. When movement of Earth's plates caused the continents to move apart, marsupials were isolated to Australia. Marsupials thrived in Australia because they did not have to compete with placental mammals. Elsewhere, placental mammals had competitive adaptive advantages.

How do placentals differ from other mammals?

Most mammals living today, including humans, are placentals. **Placental mammals** are mammals that have a placenta, an organ that provides food and oxygen to and removes waste from developing young. After birth, the young do not need further development within a pouch.

There are 18 orders of placental mammals. Some orders have one or only a few species. Other orders have thousands of species. Placental mammals range from small pygmy shrews to large whales. Some, such as dolphins, can swim. Others, such as moles, live underground.

Order Insectivora The main food source for members of this order is insects. Most members are small and have pointed snouts for capturing insects. Shrews, the smallest mammals, as well as hedgehogs and moles, are insectivores.

Order Chiroptera All members of order Chiroptera (ky RAHP ter uh) are bats. Bats are the only mammals that can fly. Some bats eat insects. Some eat fruit. Others feed on blood. ☑

Order Primates Members of this order, including monkeys, apes, and humans, have the most developed brains of all mammals. Most primates live in trees. The complex movements needed to live in trees might have led to the advanced brain development of primates. Primate hands are adapted for grasping, and most have nails instead of claws.

Order Xenarthra Members of order Xenarthra (zen AR thra), including anteaters, sloths, and armadillos, have simple, peglike teeth or no teeth. Anteaters use their long tongues and sticky saliva to capture ants in their nests. The armadillo is the only member of this order that lives in the United States. The others live in Central America and South America.

Think it Over

1. Predict How might the presence of placental animals affect marsupials in Australia today?

____

____

____

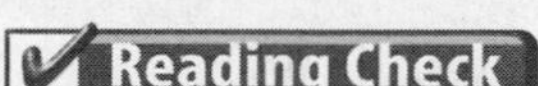

2. Classify The only flying mammals belong to which order?

____

Order Rodentia Members of this order, such as rats, beavers, squirrels, and hamsters, are gnawing mammals. Two pairs of sharp incisor teeth continue to grow throughout a rodent's life. Rodents use their sharp teeth to gnaw through wood, seed pods, and shells to get food. Rodents make up nearly 40 percent of all mammals. Their ability to invade all land habitats and successfully reproduce makes rodents ecologically important in all terrestrial habitats.

Think it Over

3. Contrast How are the teeth of lagomorphs different from those of rodents?

____

____

____

Order Lagomorpha Members of this order, including rabbits, pikas, and hares, have sharp incisors that continue to grow like those of rodents. Lagomorphs also have a pair of peglike incisors that grow behind the first pair. They eat grasses, herbs, fruits, and seeds. Some lagomorphs, such as pikas, harvest grass during warm months and store it to eat in winter when no fresh vegetation is available.

Order Carnivora Members of this order, such as dogs, cats, wolves, bears, seals, skunks, otters, and weasels, are predators. Their teeth are adapted to tear meat.

Order Proboscidea Members of this order, including elephants, have flexible trunks for gathering plants and drinking water. Two upper incisors are modified as tusks for digging up roots and tearing bark from trees. Elephants are the largest land mammals. Extinct relatives of today's elephants include mastodons and mammoths.

Order Sirenia Members of this order, such as manatees and dugongs, are large, slow-moving mammals with big heads and no hind limbs. Their forelimbs are modified into flippers for swimming. These herbivores feed on aquatic plants. Because they move slowly and swim near the surface, they are often injured or killed by boat propellers.

Think it Over

4. Compare What characteristics do members of the orders Perissodactyla and Artiodactyla share?

____

____

____

Order Perissodactyla Members of this order, including horses, zebras, and rhinoceroses, are hoofed mammals that have an odd number of toes (either one or three) on each foot. They eat plants and live on all continents except Antarctica.

Order Artiodactyla Members of this order, such as deer, cattle, sheep, pigs, goats, and hippopotamuses, are also hoofed mammals. Unlike the order Perissodactyla, these mammals have an even number of toes (either two or four) on each foot. They eat plants, and most chew their cuds. Many cattle, sheep, and deer have horns or antlers.

Order Cetacea Members of this order, including whales, dolphins, and porpoises, have front limbs modified into flippers for swimming. They have no hind limbs, and their tails consist of fleshy flukes. Their nostrils are modified into a single or double blowhole on top of their heads.

Members of the order Cetacea have no hair except for a few muzzle hairs. Some whales are predators. Other whales have a baleen, which is a structure inside the mouth used to filter plankton for food.

Evolution of Mammals

The first mammals probably evolved from reptiles. A few lived at the same time as dinosaurs. Mammals did not become common until after the dinosaurs disappeared.

The figure below shows how mammals might have evolved. Refer to this figure as you read the following discussion about the ancestors of mammals.

Picture This

5. Determine Which species evolved before cynodonts?

____

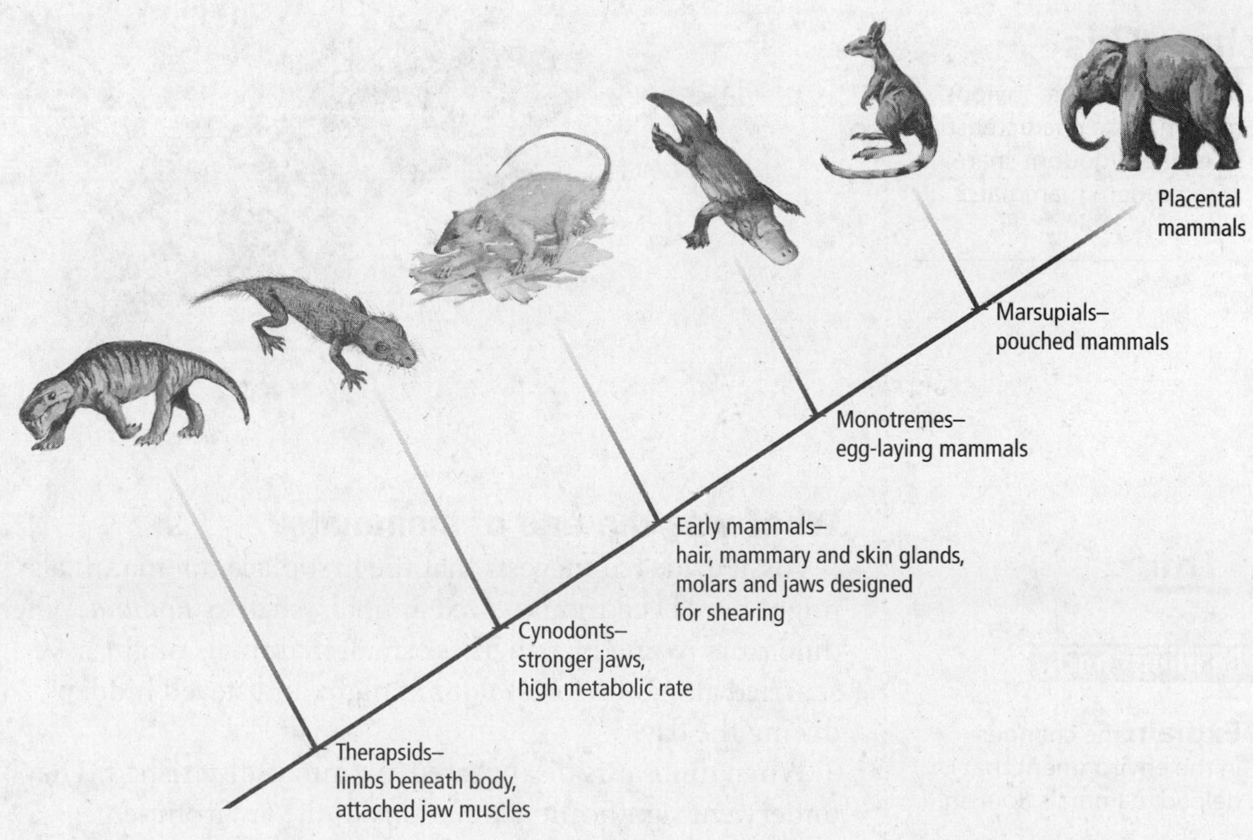

How did mammals evolve from reptiles?

Fossil evidence indicates that the first mammals probably came from a group of mammal-like reptiles called therapsids. A **therapsid** (thur RAP sed) is an extinct vertebrate with features of both mammals and reptiles. Its mammal-like characteristics included a pair of holes in the roof of its skull for the attachment of jaw muscles and limbs positioned beneath the body for more efficient movement.

Evidence shows that therapsids ate more food than their ancestors. This suggests that therapsids might have been able to produce their own body heat. Being endothermic would have given them a competitive advantage. They could be more active during winter than animals without the ability to produce their own body heat.

One group of therapsids continued to evolve more mammal-like characteristics. This group was the cynodonts, shown in the figure below. They developed a high metabolic rate, stronger jaws, and a structure in the mouth that allowed them to breathe while holding food or nursing.

Picture This

6. Identify What obvious mammal-like characteristic does the cynodont share with modern mammals?

What was the age of mammals?

Fossil evidence suggests that the first placental mammals might have been mouse-sized animals, such as *Eomaia*. When dinosaurs roamed Earth, these small mammals might have scurried along the forest floor at night and stayed hidden during the day.

When dinosaurs disappeared, mammals flourished. They underwent significant adaptations to the environment. Flowering plants thrived at this time, providing new sources of nutrition and new habitats for mammals. The Cenozoic Era is sometimes called the "Age of Mammals" because of the dramatic increase in mammal numbers and diversity. ☑

7. Explain the changes in the environment that helped mammals flourish.

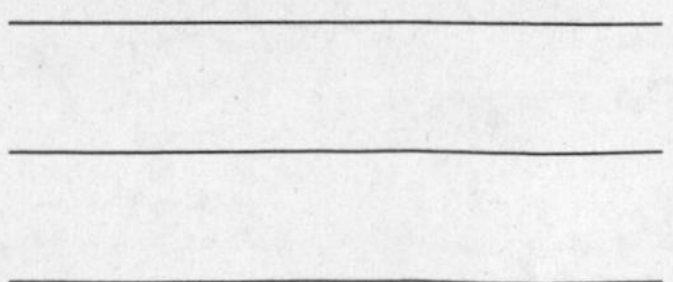

Animal Behavior

section 1 Basic Behaviors

Before You Read

You were born with the ability to do some things, but you needed to learn how to do others. On the lines below, name an activity that you were able to do at birth and an activity that you learned. In this section, you will learn about innate and learned behaviors.

____

____

MAIN Idea

Animal behaviors are innate and learned behaviors that evolve by natural selection.

What You'll Learn

- how animal behaviors relate to evolution by natural selection
- the difference between innate and learned behaviors
- how to identify different animal behaviors

Read to Learn

Behavior

Behavior is the way an animal responds to a stimulus. A stimulus (STIHM yuh lus) is a change in the environment that influences the activity of an organism. Some stimuli are internal. When a lizard is cold, its nervous system detects the low temperature. The lizard responds by lying in the sunlight. Stimuli can also come from outside the body. The lizard could see a predator and respond by running.

What influences behavior?

For many years, scientists have questioned whether behavior is based on genes or experiences. Studies have shown that some behavior is based on genes alone. Today, scientists consider many behaviors to be the results of both genes and experiences. ☑

What triggers an animal to respond to stimuli?

Scientists study the internal biology of an animal to determine what triggers the animal to respond to stimuli. Scientists now know that some male birds sing during breeding season. The levels of the hormone testosterone in the male bird increase during breeding season. This internal stimulus causes the bird to sing.

Study Coach

Create a Quiz After you read this section, create a five-question quiz from what you have learned. Then, exchange quizzes with another student. After taking the quizzes, review your answers together.

Reading Check

1. **List** the two main factors that influence behavior.

____

____

What advantages do certain behaviors provide?

The singing of the male bird during mating season can provide several advantages to the bird. It might keep away other male birds, and it might attract a mate. Scientists can understand the advantages of certain animal behaviors by studying the evolution of behaviors through natural selection.

Animals with traits that give them a competitive advantage over other animals without the traits are more likely to reproduce, passing on their genes. In the past, birds that sang tended to have more offspring than birds that did not sing. Over many generations, birds that sang became the only birds adding to the population's gene pool. The behavior has been selected naturally.

FOLDABLES™

Take Notes Make a two-tab Foldable, as shown below. As you read, take notes and organize what you learn about innate and learned animal behaviors.

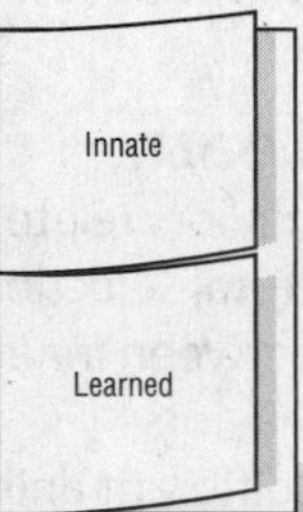

Innate Behavior

Behaviors that are based on genes and not linked to past experiences are called **innate** (ih NAYT) **behaviors**. Behaviors are considered innate, or instinct, if a large number of individuals within a population display the same behavior, even when the environments are different. For example, a newly hatched bird will chirp innately and open its mouth when a parent lands on the nest. The parent's innate response is to feed the chirping bird.

What is a fixed action pattern?

The goose in the figure below is carrying out a specific set of innate behavior actions in sequence in response to a stimulus. The behavior is a **fixed action pattern**. An egg is out of the nest. In response to this stimulus, the goose extends its neck toward the egg and stands. It then rolls the egg back to the nest with the underside of its bill. Even if the egg is removed midway through the process, the goose will continue the behavior without the egg. In a fixed action pattern, the stimulus triggers an innate response. The animal does not control the response. The response is also not affected by environmental conditions or past experiences.

Picture This

2. Predict what would happen if the egg were replaced by a small rubber ball.

____

____

____

A The goose responds to the stimulus of an egg out of the nest.

B The goose rolls the egg back to the nest with the underside of its bill.

C The goose continues to roll the egg until it is in the nest.

Learned Behavior

Learned behaviors result from interactions between innate behaviors and past experiences. They occur within a particular environment.

How does habituation develop?

At first, a baby bird might respond to any object moving overhead by crouching. Some of these objects, such as falling leaves, have no effect on them. Eventually, the birds will stop responding to these stimuli. This is an example of habituation. **Habituation** (huh bit choo AY shun) is a decrease in an animal's response after repeatedly being exposed to a stimulus that has no positive or negative effects. ☑

Habituation is important to an animal's success. It allows the animal to ignore unimportant stimuli and focus on important stimuli, such as the presence of food or a mate.

Reading Check

3. **Explain** how a stimulus must affect the animal for habituation to occur.

____

____

____

What is classical conditioning?

Ivan Pavlov, a Russian scientist, conducted experiments in which he offered meat powder to a dog. The dog responded by producing saliva. Later, Pavlov rang a bell each time he offered the meat powder. After many repetitions, the dog produced saliva when it heard the bell alone, without smelling or tasting the meat powder.

Pavlov concluded that the dog related the ringing bell to the meat powder. This type of learning, illustrated in the figure below, is called classical conditioning. **Classical conditioning** occurs when an association is made between two different types of stimuli. In Pavlov's experiment, the dog connected the ringing bell with the unrelated stimulus of meat powder. The ringing bell could produce saliva—the response.

Picture This

4. **Label** the stimuli and response in the drawing labeled *B*.

A When a dog is presented with food, it salivates.

B A bell is rung each time a dog is presented with food. The dog forms an association with the ringing bell and food.

C Eventually, the dog will salivate to the sound of the bell alone. It has been conditioned to respond to the ringing bell.

Think it Over

5. Apply You study for a test and get a good grade. Identify the stimulus, response, and reward for this behavior.

____

____

____

Think it Over

6. Describe a cognitive behavior you have exhibited today.

____

____

____

____

How does operant conditioning occur?

B.F. Skinner, an American psychologist, experimented with operant (AH pur ent) conditioning. In **operant conditioning**, an animal learns to associate its response to a stimulus with a reward or punishment. Skinner placed a rat in a box. As the rat explored the box, it accidentally hit a lever, releasing a food pellet. At first, the rat ate the pellet but ignored the lever. Eventually, the rat learned that pressing the lever would produce food. The rat was rewarded positively (with food) for its response (pressing the lever) to the stimulus (the lever).

How do negative rewards affect operant conditioning?

Rewards can also be negative. Monarch butterflies are toxic to many predators. When a bird eats a monarch butterfly, it gets sick. The bird soon associates the butterfly with illness. It avoids eating monarch butterflies and other butterflies with similar colors.

Operant conditioning is a powerful and long-lasting kind of learning. It is the main type of learning for humans and other vertebrates.

How does imprinting benefit an animal?

Imprinting is learning that can occur only within a specific time period in an animal's life. Imprinting is permanent. The time during which an animal imprints is called the sensitive period. Often the sensitive period occurs immediately after birth.

For example, salmon imprint on the chemical makeup of the water in which they were hatched. They use this imprint to return to the same location for spawning. A newly hatched bird forms a strong bond with the first object it sees, even if it is an animal of a different species or a cardboard box. In nature, the bird will most likely see its parent first. As a result, the young bird increases its chance for success by being cared for by its parent.

What animals exhibit cognitive behaviors?

Thinking, reasoning, and processing information to understand complex concepts and solve problems are **cognitive behaviors**. Humans solve problems, make decisions, and plan for the future. Scientists have observed other animals, such as ravens and chimpanzees, exhibit cognitive behaviors. For instance, chimpanzees use rocks as tools.

chapter 31 Animal Behavior

section 2 Ecological Behaviors

Before You Read

Have you ever listened to animal sounds, such as the chirps of birds or the croaks of frogs? Explain on the lines below why animals might make such sounds. Read to learn more about cooperative and competitive behaviors.

____

____

MAIN Idea

Animals that engage in complex behaviors might survive and reproduce because they have inherited more favorable behaviors.

What You'll Learn

- types of competitive behaviors
- types of cooperative behaviors

Read to Learn

Types of Behaviors

Ecology is the study of how living things interact with one another and with their environments. Interactions can be between members of the same or different species. Male bighorn sheep fight over a mate. This behavior helps the sheep survive and reproduce. The winner can mate with the female and pass on his genes to future generations. As the species evolves by natural selection, genes that provide adaptive advantages will become more common in the gene pool. Genes that do not help an animal survive or reproduce will become less common in the gene pool.

Mark the Text

Identify Behaviors Highlight each type of behavior as you read about it. Underline the characteristics and advantages of each behavior.

Why do animals exhibit competitive behaviors?

Animals within a population compete for resources such as food, space, and mates. Competitive behaviors enable individuals to establish control of an area or resource. Successful competitors are more likely to get the resources they need to survive and reproduce. ☑

Threatening behavior or fights between individuals of the same species are **agonistic** (ag oh NIHS tihk) **behaviors**. Eventually, one animal stops the behavior and leaves. The winner gains control of a resource, such as food or a mate. Agonistic behaviors do not usually injure or kill either individual.

Reading Check

1. **Identify** the goal of competitive behaviors.

____

____

____

How do ranking systems increase survival?

A hierarchy is a ranking from highest to lowest. Some animals living in groups develop **dominance hierarchies** (DAH muh nunts • HIER ar keez). The top-ranked animal gets access to resources without conflict from other group members. This ranking system helps reduce fights among members. Fights get in the way of important survival tasks, such as finding food and caring for young.

Why are territorial and foraging behaviors important?

A territory is an area that contains resources such as food or potential mates. **Territorial behaviors** are attempts to gain and defend a physical area against other animals of the same species. Territorial behaviors can be verbal such as the singing of birds or chattering of squirrels. They can also be chemical signals, such as marking boundaries with urine. ☑

Finding and eating food are **foraging behaviors**. Foraging skills help animals obtain needed nutrients and avoid predators and poisonous foods. It takes energy to find, pursue, and eat food. Scientists think that natural selection favors individuals whose foraging behaviors use the least energy to gain the most energy. These animals will be most able to reproduce and pass on their genes.

Reading Check

2. Explain why controlling a territory is important to an animal.

____

____

____

Why do animals migrate?

Some animals move long distances to new locations to increase their chances of survival. These animals are performing **migratory behaviors**. Some animals, such as zebras, migrate almost continuously to areas where food is more plentiful. Other animals migrate during certain seasons. Each fall in North America, many bird species fly south to where food is available during winter. They fly north in spring to feed and breed during summer.

How do biological cycles influence behavior?

Many animals, including humans, repeat behaviors in a cycle. A **circadian** (sur KAY dee uhn) **rhythm** is a cycle that occurs daily, such as sleeping and waking. Other biological cycles are seasonal or yearly. Environmental factors, such as changes in temperature and amount of daylight, cause animals to move to another phase of the cycle. Studies have shown that many animals, including humans, have an internal clock, sometimes called a biological clock. The internal clock maintains a daily sleep/wake cycle of about 24 hours.

Think it Over

3. Apply What factors in the environment might tell birds that it is time to start migrating south?

____

____

____

Communication Behaviors

Barks, chirps, howls, growls, and snarls are examples of animal communication. Some animals communicate through odors. They spread chemicals called pheromones (FER uh mohnz) that are specific to their species. Pheromones communicate information to other members of a species, but predators cannot detect them. Pheromones are often used to attract potential mates. ☑

How do animals communicate through sounds?

Sound messages move faster than chemical messages. Howls, hoots, barks, and chirps communicate messages about mating, predators, and territory. Humans communicate through language. **Language** is the use of vocal organs to produce groups of sounds that have shared meanings.

Reading Check

4. Identify a purpose of pheromones.

____

____

Courting and Nurturing Behaviors

The purpose of some behaviors is reproductive success. Two important behaviors related to reproductive success are attracting a mate and caring for offspring.

Why are courting behaviors important?

Courting behaviors are designed to attract a mate. Each species has its own courting behaviors. A bird might display its colorful feathers. An animal might make particular sounds and movements.

In most species, females choose the male. Females often choose to mate with males that look larger and healthier than other males. Males with desired traits have a competitive advantage over other males and usually have a better chance of mating and producing offspring.

Why do animals engage in nurturing behaviors?

Caring for offspring involves **nurturing behaviors** such as providing food and protection. Nurturing costs parents energy. They must take on the extra work of caring for the young until the offspring can survive on their own. For example, the parent bird must make many flights to find food. As a result, animals that nurture their offspring often produce fewer offspring than animals that do not nurture. Cod fishes do not nurture. Instead, they produce millions of eggs, increasing the chance that some will survive. A female chimpanzee gives birth to one offspring. She will nurse and care for the young chimpanzee for up to three years, and watch over it for five to seven years.

Think it Over

5. Synthesize What type of behavior is nurturing behavior? (Circle your answer.)

a. learned behavior
b. innate behavior

Cooperative Behaviors

Cooperative behaviors can exist in groups of animals of the same species. Cooperative behaviors among animals of the same species can benefit all members of the group.

What is altruistic behavior?

An **altruistic** (al trew IHS tihk) **behavior** occurs when one animal of a group performs a behavior that benefits another individual at a cost to itself. It is a self-sacrificing behavior. For example, only the queen mole rat reproduces, and she mates with only a few kings. The other members of the colony exhibit altruistic behaviors. They forage for food and care for the queen, kings, and newborn offspring.

Why do altruistic behaviors evolve?

The theory of kin selection might explain why some altruistic behaviors evolve. According to the theory, the altruistic behavior increases the number of copies of a gene that is common to the population. It does not matter which members of the population pass on the genes. The nonreproductive members of the mole rat colony do not pass on their genes. However, their work to protect and feed the queen ensures that genes similar to theirs will be passed on by the queen.

Advantages and Disadvantages

Picture This

6. Predict Complete the table by filling in the blank space in the last row.

The table below gives advantages and disadvantages of some behaviors to survival and reproductive success. By comparing advantages and disadvantages, you can measure the costs and benefits of each behavior.

Behavior	Advantage	Disadvantage
Migration	Migration increases the chance of survival by positioning an animal in a location that has better climate conditions and more food.	A large amount of energy is required to move long distances. Also, the movement can increase the possibility of capture by predators.
Pheromone communication	Pheromones are species-specific, which reduces a predator's ability to detect them.	Pheromones are limited to specific areas, unlike sound or visual communication.
Nurturing	Offspring have an increased chance of survival, which means genes of parents continue in future generations.	Parents spend a large amount of energy caring for offspring, possibly at the cost of the parents' health or safety.
Foraging		The animal must use energy to find, pursue, and eat the food.

Integumentary, Skeletal, and Muscular Systems

section 1 The Integumentary System

Before You Read

Your entire body is covered by skin. What are the functions of your skin? On the lines below, list your ideas. After you read this section, add other functions to your list.

MAIN Idea

Skin is a multilayered organ that covers and protects the body.

What You'll Learn

- the functions of the integumentary system
- the events that occur when skin is repaired

Read to Learn

The Structure of Skin

The integumentary (ihn TEG yuh MEN tuh ree) system is the organ system that covers and protects the body. Skin is the main organ of the integumentary system.

Skin is made up of four types of tissues. (1) Epithelial tissue covers body surfaces. (2) Connective tissue supports and protects. (3) Muscle tissue enables movement. (4) Nerve tissue is the body's communication network. Refer to the figure on the next page as you read about the two main layers of skin: the epidermis and the dermis.

What are the functions of the epidermis?

The outer layer of skin is the **epidermis**. It consists of a thin layer of epithelial cells. The outer layers of epidermal cells contain keratin. **Keratin** (KER uh tun) is a protein that waterproofs and protects the cells and tissues underneath. These dead outer cells are shed constantly.

The cells of the inner layer of epidermis continually divide by mitosis to replace the cells that are shed. This inner layer produces a pigment called melanin. **Melanin** protects deeper cells from the damaging effects of ultraviolet rays of sunlight. The amount of melanin produced determines skin color.

Study Coach

Make an Outline Make an outline of the information you learn in this section. Start with the headings. Include the boldface terms.

FOLDABLES™

Take Notes Make a layered Foldable, as shown below. As you read, take notes and organize what you learn about the structure and function of skin.

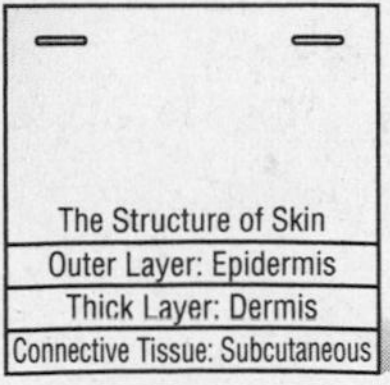

Picture This

1. Highlight the names of the structures that are responsible for acne.

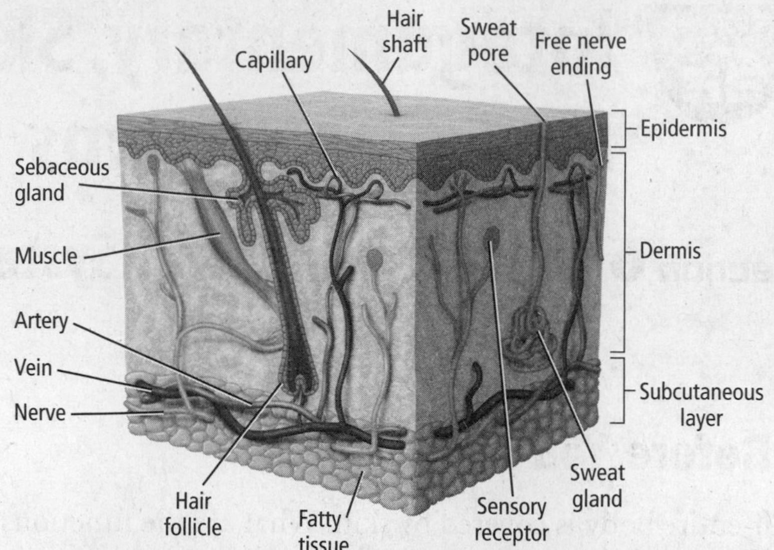

What structures are found in the dermis?

Below the epidermis is the thicker second layer of skin called the **dermis**. The dermis is made up of connective tissue. Connective tissue prevents the skin from tearing and enables the skin to return to its normal state after it is stretched. The dermis contains nerve cells, muscle fibers, sweat glands, oil glands, and hair follicles.

Below the dermis is the subcutaneous layer. This is a layer of connective tissue that stores fat and holds heat.

How do hair and nails develop?

Fingernails, toenails, and hair are part of the integumentary system. Both nails and hair contain keratin and develop from epithelial cells. Hair cells grow out of holes in the dermis called **hair follicles**. Cells at the base of a hair follicle divide and push cells away from the follicle. This causes hair to grow. ☑

Reading Check

2. Define Which of the following best describes a hair follicle? (Circle your answer.)

a. gland
b. shaft of hair
c. hole

Around the hair follicles are **sebaceous glands**. These oil-producing glands lubricate skin and hair. When glands produce too much oil, the follicles can become inflamed and blocked. This can result in a whitehead, a blackhead, or acne—an inflammation of the sebaceous glands.

Fingernails and toenails grow from specialized epithelial cells at their base. The cells at the base divide, and older, dead cells are compacted or pushed out.

Functions of the Integumentary System

Skin serves several important functions. It regulates body temperature, produces vitamin D, protects our bodies, and helps us sense our surroundings.

How does the body regulate its temperature?

When you are hot, your body sweats. The evaporation of sweat cools your body. Evaporation transfers heat energy from your body to your surroundings. When you are cold, your muscles contract causing goose bumps. In animals, these contractions cause hair to stand up and trap air to warm the animal. With little hair to keep us warm, humans depend on fat in the subcutaneous layer for warmth.

Why is vitamin D important?

Skin responds to exposure to the Sun's ultraviolet rays by producing vitamin D. Vitamin D helps the body absorb calcium and is essential for proper bone formation. ☑

How does skin protect and sense?

Unbroken skin keeps microorganisms out of the body. Skin helps maintain body temperature by preventing the loss of too much water. Melanin protects from ultraviolet rays.

Nerves in the skin relay messages about changes in the environment to the brain. The nerves make a person aware of pain, pressure, and changes in temperature.

Damage to the Skin

Skin usually repairs itself. If it did not, the body could be invaded by microbes through breaks in the skin. For minor scrapes, epidermal cells divide and replace the injured cells. Deeper injuries that harm blood vessels result in bleeding. Blood clots form a scab to close the wound. Cells beneath the scab divide and fill the wound, while blood cells help fight infections.

How do the Sun's rays affect the skin?

As people age, their skin becomes less flexible and wrinkles form. Exposure to ultraviolet rays can accelerate this process. Ultraviolet rays can also burn the skin. Burns from any source are classified by severity. First-degree burns only involve epidermal cells. Second-degree burns damage both dermis and epidermis, causing blisters and scars. Third-degree burns damage muscle tissue and nerve cells in both layers, and skin function is lost.

What can put a person at risk for skin cancer?

Exposure to ultraviolet radiation from the Sun or a tanning bed increases the risk of skin cancer. Ultraviolet radiation can damage DNA in skin cells, causing those cells to divide uncontrollably. Clothing and sunscreen of SPF 15+ protect against skin cancer. ☑

✔ Reading Check

3. Explain Why is vitamin D important to humans?

____

____

____

✔ Reading Check

4. Explain why you should avoid using a tanning bed.

____

____

____

Integumentary, Skeletal, and Muscular Systems

section 2 The Skeletal System

MAIN Idea

The skeleton provides a structural framework for the body and protects internal organs.

What You'll Learn

- differences in bones of the axial and appendicular skeletons
- how new bone is formed
- the functions of the skeletal system

Before You Read

Examine your knee. On the lines below, describe ways your knee can move and ways it cannot move. Read the section to learn how joints allow different kinds of motion.

____

____

Study Coach

Make Flash Cards Make a flash card for each underlined term in this section. Write the term on one side of the card. Write the definition on the other side. Use the flash cards to review what you have learned.

Read to Learn

Structure of the Skeletal System

You have 206 bones in your body. There are two divisions of the human skeleton. The **axial skeleton** includes the skull, vertebral column, ribs, and sternum. The **appendicular skeleton** includes bones of the arms, hands, legs, feet, shoulders, and hips.

How are bones structured?

Bone is connective tissue. Bones are classified as long, short, flat, or irregular. Arm and leg bones are long bones. Wrist bones are short bones. The skull is made of flat bones. The bones of the face and vertebrae are irregular bones.

All bones have a dense outer layer of **compact bone** for strength and protection. As shown in the figure on the next page, tubelike osteons run the length of compact bone. Osteons contain nerves and blood vessels. The blood vessels bring oxygen and nutrients to living bone cells called **osteocytes**. ☑

Less-dense **spongy bone** is found at the center of short and flat bones and at the ends of long bones. Spongy bone is surrounded by compact bone and does not have osteons. Instead, it has cavities that contain bone marrow.

Reading Check

1. **Describe** the role of the blood vessels in osteons.

____

____

____

Types of Bone

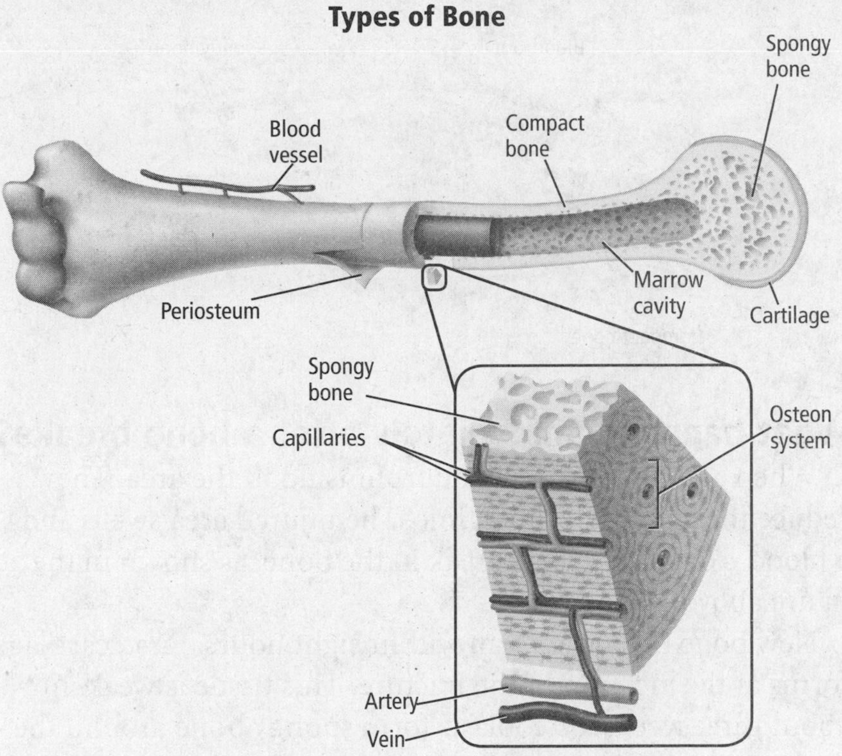

What are the two types of bone marrow?

The two types of bone marrow are red and yellow. **Red bone marrow** produces red and white blood cells and platelets. Red bone marrow is found in the humerus, femur, sternum, ribs, vertebrae, and pelvis. Children's bones have more red bone marrow than adult bones. **Yellow bone marrow** in other bones contains stored fat.

How is bone formed?

The skeletons of embryos are made of cartilage. As the fetus develops, the cells in cartilage become bone-forming cells called **osteoblasts**. Bone forms from osteoblasts through **ossification**. Except for places such as the nose, ears, and vertebrae disks, the human skeleton is bone. Osteoblasts are also responsible for bone growth and repair.

How do bones repair themselves?

Bones are constantly being remodeled, which means old bone cells are replaced with new cells. Cells called **osteoclasts** break down bone cells, which are replaced with new cells. Nutrition and exercise are important for bone growth. ☑

When a bone breaks but does not come through the skin, it is a simple fracture. When the bone breaks and does come through the skin, it is a compound fracture. A stress fracture is a thin crack in the bone.

Picture This

2. Determine What is the importance of the osteon system?

______

______

______

Reading Check

3. Define the role of osteoclasts.

______

______

______

Picture This

4. Comprehend When a bone breaks how long does it take for new bone to form?

Think it Over

5. Draw Conclusions Why is it important to keep broken bones in place until compact bone forms?

Bone Fracture

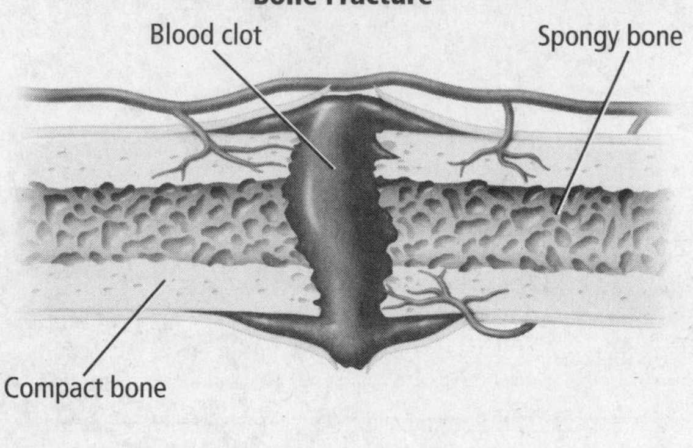

What happens immediately after a bone breaks?

When a bone is broken, endorphins flood the area. They reduce the pain for a short time. The injured area swells and a blood clot forms at the break in the bone, as shown in the figure above.

New bone begins to form within eight hours. First, cartilage forms at the location of the fracture. This tissue is weak. In about three weeks, osteoblasts form spongy bone around the fracture. Later, osteoclasts remove the spongy bone while osteoblasts create stronger compact bone to replace it. Splints and casts help keep broken bones in place until new bone forms.

What is osteoporosis?

Calcium in the diet is important to bone growth and repair. Too little calcium can lead to a condition called osteoporosis, which results in weak bones that break easily.

Joints

Two or more bones meet at a joint. The table on the next page describes the five types of joints: ball and socket, pivot, hinge, gliding, and sutures. Each allows a certain type of movement. The joints between some skull bones are the exception. They do not move.

Ligaments hold the bones of joints together. **Ligaments** are tough bands of connective tissue that attach one bone to another.

What is osteoarthritis?

Cartilage covers the ends of bones in movable joints. The cartilage cushions the bones in the joint and allows smooth movement. Osteoarthritis (ahs tee oh ar THRI tus) is a painful condition that affects joints and results when cartilage in the joints deteriorates. Injury to a joint can result in osteoarthritis later in life.

Joint Movements					
Joint name	Ball-and-Socket	Pivot	Hinge	Gliding	Sutures
Example	hip joint, shoulder joint	elbow joint	knee joint	wrist joint, ankle joint, and vertebra	skull joint
Description	ball-like surface fits into cuplike depression in another bone; allows widest range of motion, such as swinging arms	primary movement is rotation, such as twisting lower arm	outward curve of one bone fits into inward curve of another bone; allows back and forth movement	allows side-to-side and back-and-forth movement	joints in the skull that are not movable

How does rheumatoid arthritis affect joints?

Rheumatoid (roo MAH toyd) arthritis affects the joints but does not result from deteriorating cartilage. Affected joints are swollen and painful. The joints lose both strength and function.

What is bursitis?

Fluid-filled sacs called bursae surround shoulder and knee joints. Bursae reduce friction and act as cushions between bones and tendons. Bursitis is an inflammation of the bursae, causing pain, swelling, and reduction in movement.

What causes a sprain?

A sprain is damage to the ligaments of a joint. Sprains result from overstretching a joint and cause pain and swelling.

Functions of the Skeletal System

Support is a major function of the skeleton. Bones of the legs, pelvis, and vertebral column support the body. The mandible supports teeth. Almost all bones support muscles. The pull of muscles on bones causes movement.

The skeleton has other functions as well. The skull protects the brain. Vertebrae protect the spinal cord. The rib cage protects the heart, lungs, and other organs.

The outer layers of bone protect the bone marrow. Red bone marrow produces red and white blood cells and platelets for blood clotting. Yellow bone marrow stores fat for energy.

Bones store minerals. When blood calcium levels are too low, bones release calcium. When the levels are too high, bones store calcium. This helps maintain homeostasis.

Picture This

6. **Highlight** the type of movement allowed by each joint described in the table.

Think it Over

7. **Summarize** Identify three major functions of the skeletal system.

____

____

____

____

Integumentary, Skeletal, and Muscular Systems

section 3 The Muscular System

MAIN Idea

The three major types of muscle tissue differ in structure and function.

What You'll Learn

- the events involved in muscle contraction
- the difference between slow-twitch and fast-twitch muscle fibers

Before You Read

On the lines below, list several physical activities that you like to do. Based on your list, do you prefer activities that require endurance or quick bursts of speed? Record your answer. Read the section to learn how fast-twitch and slow-twitch muscle fibers affect movement.

____

____

Mark the Text

Identify Main Ideas As you read the section, highlight the main ideas in each paragraph.

Read to Learn

Three Types of Muscle

A muscle is made up of groups of fibers or muscle cells that are bound together. The figure below illustrates the three types of muscle: skeletal muscle, smooth muscle, and cardiac muscle. Muscles are grouped according to their structure and function.

Where is smooth muscle found?

Smooth muscle lines the inside of hollow organs, such as the stomach, intestines, and bladder. Smooth muscle is **involuntary muscle** because it cannot be controlled consciously. For example, smooth muscle lining the esophagus, stomach, and intestines moves food through the digestive tract. Smooth muscle has no striation (stripes). ☑

Reading Check

1. Explain Why is smooth muscle also called involuntary muscle?

____

____

____

What muscle makes the heart beat?

Cardiac muscle is involuntary muscle that is only found in the heart. Cardiac muscle cells are arranged in a web that enables the heart to beat in a rhythm. Cardiac muscle appears striated (striped) with light and dark bands.

Types of Muscle

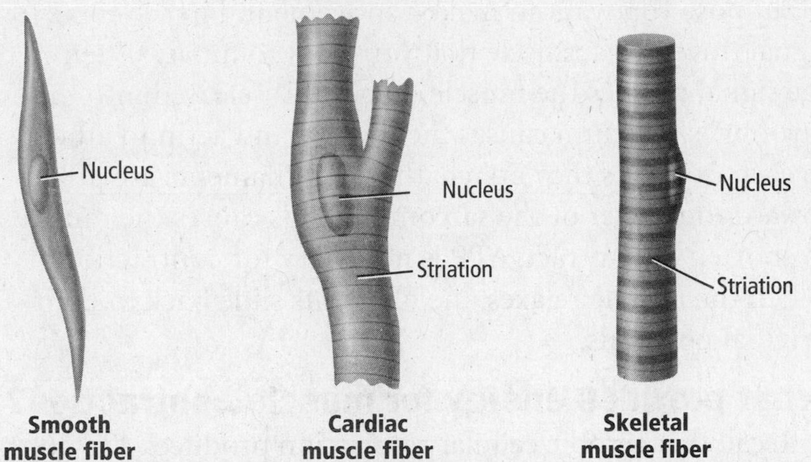

Which muscles are voluntary?

Skeletal muscles are muscles attached to bones by tendons. **Tendons** are tough bands of connective tissue. When contracted, skeletal muscles cause movement. Skeletal muscles are **voluntary muscles** because they can be controlled consciously. Skeletal muscles appear striated.

Skeletal Muscle Contraction

Skeletal muscles are made up of fibers that are fused muscle cells. Muscle fibers are made of smaller units called **myofibrils**. Myofibrils are made of protein filaments called **myosin** and **actin**. Myofibrils are arranged in sections called sarcomeres. A **sarcomere** is the basic unit of a muscle and is the part of the muscle that contracts.

Most skeletal muscles are arranged in opposing pairs. The sarcomeres shorten and relax the muscle, causing movement. The figure below illustrates how opposing muscles enable you to raise and lower your arm.

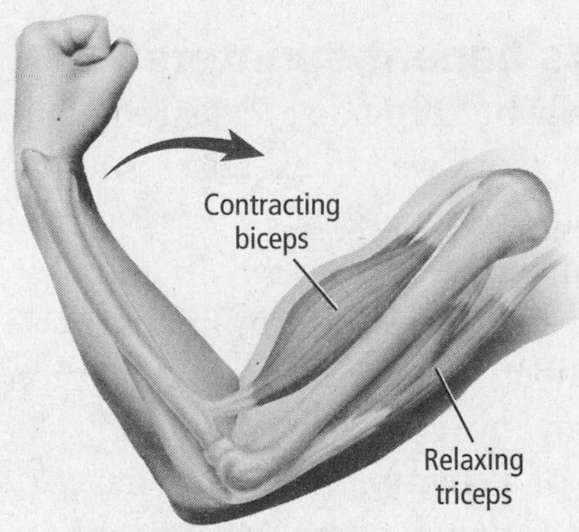

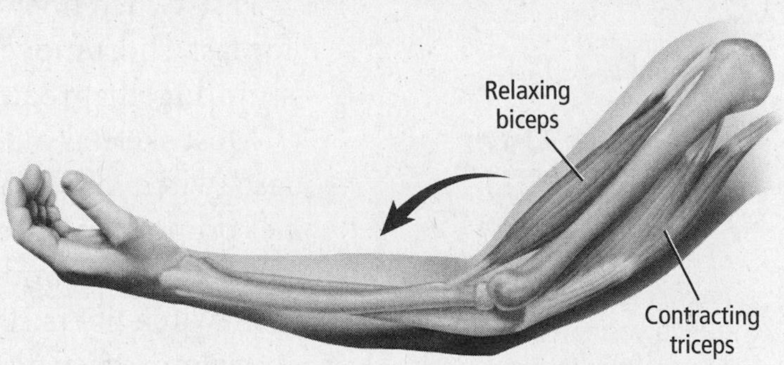

Picture This

2. Label Next to each type of muscle, write where it is found in the body.

Picture This

3. Describe what happens to the lower arm when the tricep muscle contracts.

___

___

What is the sliding filament theory?

Suppose you want to make a movement. First, a nerve signal travels to a muscle that you want to move. When the signal reaches the muscle, calcium is released into the myofibrils. Calcium causes the myosin and actin to attach to each other. As they attach, the actin filaments are pulled toward the center of the sarcomere. This causes the muscle to shorten or contract. ATP is necessary for contraction. When the muscle relaxes, the filaments slide back to their original positions.

What provides energy for muscle contraction?

Recall that aerobic cellular respiration produces ATP for energy. After intense exercise, muscles might not get enough oxygen to maintain cellular respiration. This limits the amount of ATP available to provide energy to the muscles. Muscles must get the energy they need from lactic-acid fermentation, an anaerobic process. ☑

Reading Check

4. Identify the anaerobic process that produces energy during intense exercise.

____

____

____

During exercise, lactic acid builds up in muscle cells, causing fatigue. Resting restores the needed oxygen, and lactic acid is broken down.

Skeletal Muscle Strength

Every person's muscles have both slow-twitch and fast-twitch fibers. However, some people have more of one type of fiber than the other.

What are characteristics of slow-twitch fibers?

Slow-twitch fibers contract more slowly than fast-twitch fibers. However, slow-twitch fibers have more endurance. Slow-twitch fibers work well in long-distance running because they resist fatigue. These fibers contain a substance that can hold more oxygen for respiration.

How do fast-twitch fibers benefit sprinters?

Fast-twitch fibers fatigue easily but provide great strength for fast, short movements. They work well in exercises, such as sprinting, that require short bursts of energy.

Most skeletal muscles contain a mix of slow-twitch and fast-twitch muscle fibers. A person's genes determine his or her proportion of each type of fiber. The muscles of a champion cross-country runner have many more slow-twitch than fast-twitch fibers. The muscles of a champion sprinter have many more fast-twitch than slow-twitch fibers. Most people have a similar number of each.

Think it Over

5. Apply Which type of muscle fiber is more important for weightlifting? (Circle your answer.)

a. fast-twitch fiber

b. slow-twitch fiber

Nervous System

section 1 Structure of the Nervous System

Before You Read

On the lines below, describe how you reacted the last time you touched a hot object such as a pan heating on the stove. Read about the structures that help you react quickly to your environment.

____

____

MAIN Idea

Neurons conduct electrical impulses that allow cells, tissues, and organs to detect and respond to stimuli.

What You'll Learn

- the major parts of a neuron and their functions
- how a nerve impulse is similar to an electrical signal

Read to Learn

Study Coach

Make Flash Cards Write an underlined term on one side of a flash card. Write the definition for the term on the other side of the card. Use the flash cards to quiz yourself on the terms and their definitions.

Neurons

A <u>neuron</u> (NOOR ahn) is a specialized nerve cell that helps you gather information about your environment, interpret the information, and react to it. Neurons make up the nervous system. The nervous system is a huge communication network that runs throughout your body.

As you can see in the figure below, a neuron has three regions: the dendrites, a cell body, and an axon. <u>Dendrites</u> receive signals called impulses. They funnel the signals to the cell body. The nucleus and other organelles of the neuron are found in the <u>cell body</u>. The <u>axon</u> carries the impulse from the cell body to other neurons and muscles.

Picture This

1. Label Add a title that describes the figure.

____

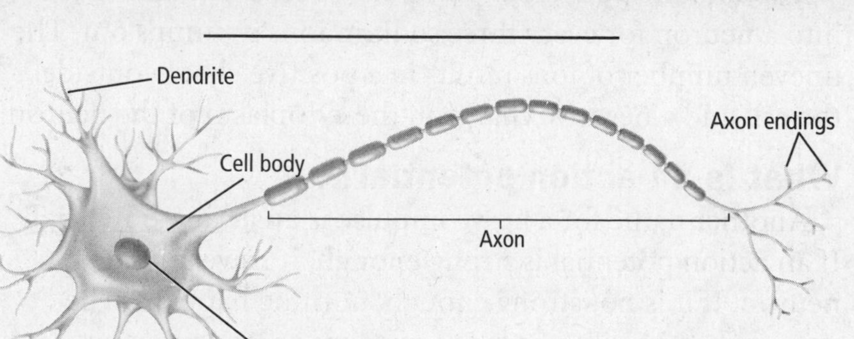

What are the three types of neurons?

There are sensory neurons, interneurons, and motor neurons. All neurons have the same three regions. However, each type of neuron performs a specific function. Sensory neurons send signals from receptors in your skin and sense organs to your brain and spinal cord. Interneurons are located in the brain and spinal cord. They receive the signals sent by the sensory neurons. Interneurons also send signals to the motor neurons. The motor neurons are located in your glands and mucles and cause movement.

When you stub your toe, sensory neurons in your foot send impulses to the interneurons. The interneurons signal the motor neurons to move your foot. The nerve pathway that consists of a sensory neuron, an interneuron, and a motor neuron is called a **reflex arc**. A reflex arc is the basic structure and function of the nervous system.

Think it Over

2. Predict What might happen if your interneurons could not function?

____

____

____

____

____

A Nerve Impulse

A nerve impulse is an electrical charge traveling the length of a neuron. Any stimulus, such as a touch or a loud noise, can cause an impulse.

What is a sodium-potassium pump?

When a neuron is not conducting an impulse, it is at rest. When a neuron is at rest, there are more sodium ions outside the cell than inside the cell. In addition, there are more potassium ions inside the cell than outside the cell.

Ions diffuse from an area of high concentration to an area of low concentration. Proteins in the neuron's plasma membrane work against diffusion of sodium ions and potassium ions. These proteins are called the sodium-potassium pump. They work against the normal flow of ions by actively transporting sodium ions out of the cell and potassium ions into the cell.

The sodium-potassium pump moves two potassium ions into a neuron for every three sodium ions it pumps out. The uneven number of ions results in a positive charge outside the cell and a negative charge in the cytoplasm of the neuron.

What is an action potential?

Another name for a nerve impulse is an **action potential**. If an action potential is strong enough, it travels along a neuron. If it is not strong enough, nothing happens. The minimum stimulus needed to produce an action potential is a **threshold**. ✔

✔ Reading Check

3. Identify What produces an action potential?

____

____

What happens when threshold is reached?

When a stimulus reaches threshold, channels in the plasma membrane open. A channel is a path along which an electrical signal passes. As the channels open, sodium ions rapidly move into the neuron's cytoplasm. The inside of the cell now has a positive charge. ☑

The positive charge causes other channels in the membrane to open. Potassium ions leave the cell through these channels, and the cytoplasm returns to a negative charge. This change in charge, shown below, moves like a wave down the length of the axon. In the figure, sodium ions are labeled Na^+ and potassium ions are K^+. The + and − signs indicate positive and negative charges inside and outside the cell.

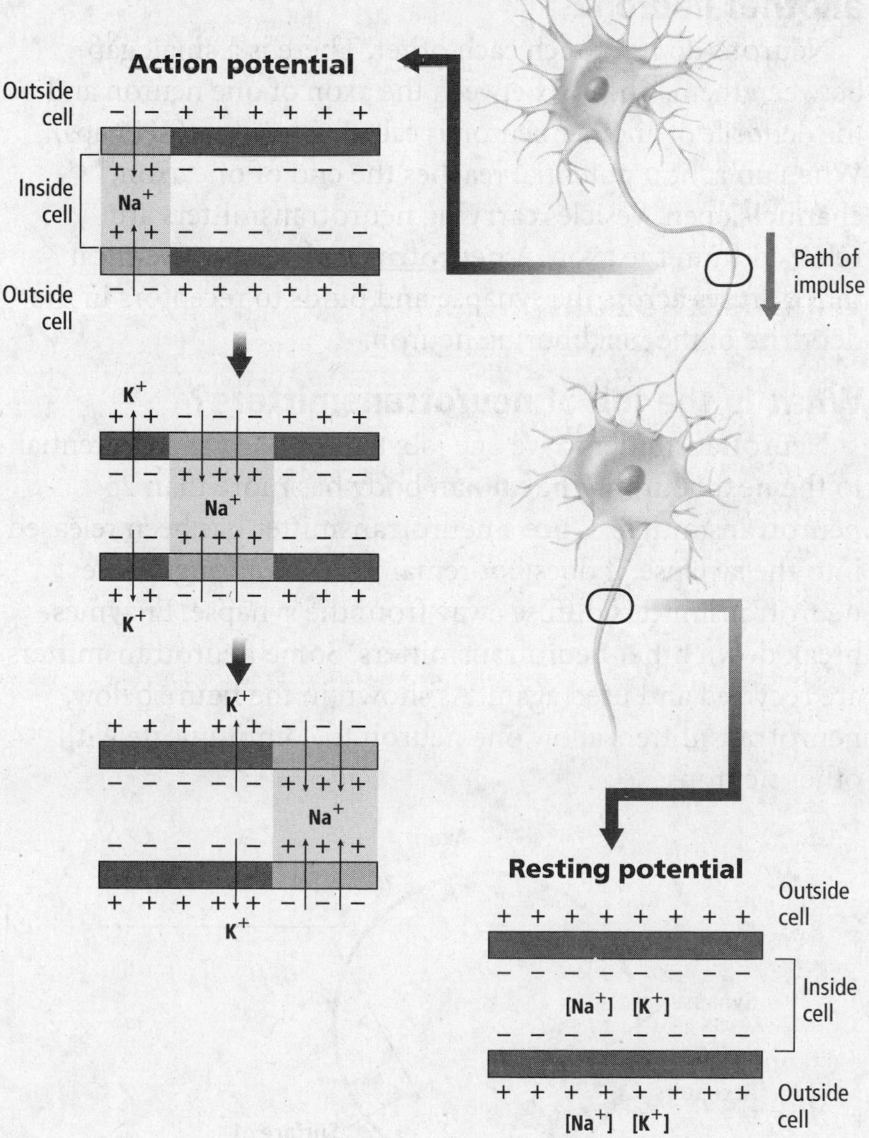

Reading Check

4. Define What is a channel?

Picture This

5. Explain Use the figure to explain to a partner what happens when a stimulus reaches threshold.

What is the speed of an action potential?

The speed of an action potential varies. Many axons are covered in a myelin sheath, which is a lipid (fat) layer that protects the axon. Sodium and potassium cannot diffuse through the myelin sheath. The myelin sheath, however, has many gaps called **nodes**. The ions reach the cell's plasma membrane at the nodes. The action potential jumps from node to node, increasing speed as it moves along the axon. ☑

Some neurons in the human body have a myelin sheath, and other neurons are not protected by myelin. Neurons with myelin carry impulses that signal sharp pain, such as the pain felt when you stub your toe. Neurons that do not have myelin are associated with dull, throbbing pain.

Reading Check

6. Describe What is the purpose of myelin?

____

____

____

How do impulses move from one neuron to another neuron?

Neurons do not touch each other. There is a small gap between them. The gap between the axon of one neuron and the dendrite of another neuron is called a **synapse** (SIH naps). When an action potential reaches the end of one axon, channels open. Vesicles carrying neurotransmitters are released from the axon. A **neurotransmitter** is a chemical that diffuses across the synapse and binds to receptors in the dendrite of the neighboring neuron.

What is the job of neurotransmitters?

Neurotransmitters have one job: to send the action potential to the next neuron. The human body has more than 25 neurotransmitters. Once a neurotransmitter has been released into the synapse, it does not remain there for long. Some neurotransmitters diffuse away from the synapse. Enzymes break down other neurotransmitters. Some neurotransmitters are recycled and used again. As shown in the figure below, neurotransmitters allow one neuron to communicate with other neurons.

Picture This

7. Identify Add a label to identify how one neuron communicates with other neurons.

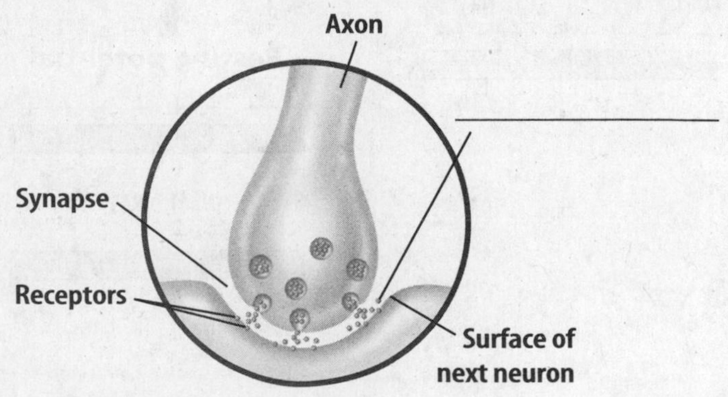

chapter 33

Nervous System

section 2 Organization of the Nervous System

Before You Read

Electrical impulses travel throughout your body, sending information from place to place. On the lines below, list three examples of information that can be sent from place to place in your body. Read the section to learn about how the brain controls the flow of information in your body.

____

____

MAIN Idea

The nervous system is divided into the central nervous system and the peripheral nervous system.

What You'll Learn

- the major divisions of the nervous system
- the somatic nervous system and the autonomic nervous system

Read to Learn

The Central Nervous System

The two major divisions of the nervous system are the central nervous system (CNS) and the peripheral nervous system (PNS). The brain and the spinal cord make up the **central nervous system**. The CNS coordinates all of the body's activities and mostly consists of interneurons. Functions of the CNS include sending messages, processing information, storing information, and analyzing responses.

What makes up the peripheral nervous system?

The **peripheral nervous system** consists of sensory neurons and motor neurons. The neurons of the PNS send information to and receive information from the CNS.

Sensory neurons send information about the environment to the interneurons in the spinal cord. The interneurons relay that information to the brain. The brain responds by sending a message to the interneurons. The interneurons send the message to the motor neurons. Your body responds appropriately to the messages received by the motor neurons. The brain also stores some of the information received from sensory neurons for later use. ✓

Study Coach

Create a Quiz After you have read the section, create a quiz based on what you learned. After writing the quiz questions, be sure to answer them.

Reading Check

1. **Summarize** Complete the flowchart to show how information about the environment travels in the body.

Interneurons

What is the largest part of the brain?

More than 100 billion neurons are found in the brain. The brain is involved in most of the body's activities and serves as the body's control center as it works to maintain homeostasis.

The largest part of the brain is the **cerebrum** (suh REE brum). The two halves of the cerebrum are called hemispheres. The hemispheres are connected by a bundle of nerves. The cerebrum controls most of the body's voluntary activities, memory, language, speech, and the senses. Most higher thought processes occur on the surface of the cerebrum. The many folds and grooves of the cerebrum increase the surface area. This large surface area enables more complex thinking.

Think it Over

2. Compare How is the cerebrum similar to the ER you read about in the chapter about cell organelles?

What does the cerebellum control?

The cerebellum (ser uh BE lum) is at the back of the brain, as shown below. It controls balance, posture, and coordination. When you are playing a musical instrument or riding a bike, your cerebellum is hard at work.

The brain stem connects the brain to the spinal cord. The two regions of the brain stem are the medulla oblongata (muh DEW luh • ahb long GAH tuh) and the pons. The **medulla oblongata** controls breathing, heart rate, and blood pressure. The **pons** also helps control the rate of breathing.

A small structure called the hypothalamus (hi poh THA luh mus) is located between the brain stem and the cerebrum. The **hypothalamus** helps the body maintain homeostasis by regulating body temperature, thirst, appetite, and water balance. The hypothalamus is also important in controlling blood pressure, sleep, aggression, fear, and sexual behavior.

Picture This

3. Circle the names of the parts of the brain that control breathing.

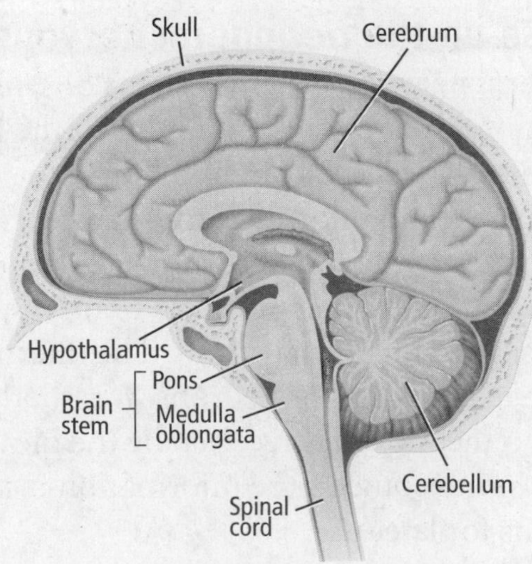

Where is the spinal cord located?

The spinal cord extends from the brain to the lower back. It is a column of nerves protected by vertebrae. Many pairs of spinal nerves reach out from the spinal cord to all parts of the body. This nerve network connects the body to the CNS. ☑

4. Identify What protects the spinal cord?

____

The Peripheral Nervous System

A nerve is a bundle of axons. Many nerves contain both sensory neurons and motor neurons. The 12 cranial nerves, the 31 spinal nerves, and their branches are part of the CNS. All of the neurons that are not part of the CNS make up the PNS. The neurons of the PNS are either part of the somatic nervous system or part of the autonomic nervous system.

What does the somatic nervous system do?

Nerves in the **somatic nervous system** send information from sensory receptors in the skin to the CNS, and motor nerves relay information from the CNS to skeletal muscles. This pathway of information is voluntary.

Sometimes a stimulus results in an automatic, unconscious response within the somatic system. When you touch something hot, you automatically jerk your hand away. Such an action is a reflex, an automatic response to a stimulus. A reflex impulse travels only to the spinal cord or brain stem, and an impulse is sent directly back to a muscle.

What is the autonomic nervous system?

Have you ever heard scary sounds in the middle of the night? Maybe your heart began to pound and your palms became sweaty. This type of reaction is involuntary—you do not think about it, it just happens. The autonomic nervous system is responsible for this reaction. The **autonomic nervous system** carries impulses from the CNS to the heart and other internal organs and glands. The body responds involuntarily. This response is often called a fight-or-flight response.

There are two branches of the autonomic nervous system that act together. The **sympathetic nervous system** controls internal body reactions in times of stress. The **parasympathetic nervous system** controls many of the body's internal functions when the body is at rest. After a stressful experience, the parasympathetic nervous system helps restore the body to its resting state. Both systems send impulses to the same organs. The resulting activity of the organ depends on the strength of the opposing signals.

5. Apply If the smoke alarm goes off in your house, which branch of the autonomic nervous system responds?

____

____

chapter 33

Nervous System

section 3 The Senses

MAIN Idea

Sensory receptors allow you to detect your surroundings.

What You'll Learn

- different sensory structures
- how each sense organ is able to send nerve impulses
- the relationship between smell and taste

Before You Read

Think of the last time you had a bad cold. On the lines below, describe how your appetite was affected by your cold. Read the section to learn how your senses of taste and smell work together.

____

____

____

Study Coach

Make a Chart Create a three-column chart. Label the columns: *Sense, Sensory Receptors,* and *Location of Receptor*. As you read, record information in each column to help you organize what you are reading.

Read to Learn

Taste and Smell

Specialized neurons in your body allow you to taste, smell, hear, see, and touch. You also use specialized neurons to sense motion and temperature. Specialized neurons that allow you to detect your surroundings are known as sensory receptors.

Taste buds are sensory receptors on the tongue that identify the tastes of sweet, sour, salty, and bitter. Taste buds work by sensing chemical combinations in food. They send this information to the brain. ☑

Reading Check

1. **Define** What is the name for the sensory receptors that allow you to taste?

____

Sensory receptors designed for taste and smell are located in your mouth and nasal cavity. These receptors work together. If you hold your nose while you eat, the food will seem to have less flavor.

Sight

Light first enters your eyes through the cornea. The cornea helps focus the light through an opening called the pupil. The size of the pupil changes to let in more or less light. Muscles in the iris, the colored part of your eye, control the size of the pupil opening.

Where are the eye's sensory receptors located?

Behind the iris is the **lens**, which inverts the image and projects it onto the retina. Between the lens and the retina is a gelatinlike liquid called the vitreous humor. The eye's sensory receptors—rods and cones—are located in the **retina**. **Rods** are sensory receptors that respond to low levels of light. **Cones** provide information about color to the brain. The structures of the eye are shown in the figure below. Action potentials travel from rods and cones to neurons in the optic nerve and finally to the brain. The brain interprets the signals and forms a visual image.

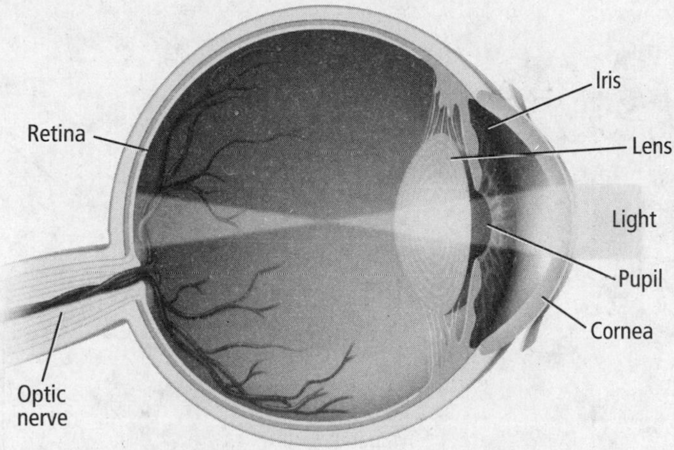

Hearing and Balance

Hearing and balance are the two major functions of the ear. Sensory receptors in the ear detect the volume of sounds and the pitch of sounds. The pitch is the highness or lowness of sounds in your environment. The inner ear controls your sense of balance.

What structures in the ear detect sound waves?

Particles in the air vibrate as a result of sound waves. Sound waves enter the auditory canal of the ear. The waves cause the tympanic membrane (eardrum) at the end of the canal to vibrate. These vibrations travel through three bones in the middle ear, the malleus (hammer), the incus (anvil), and the stapes (stirrup). As the stapes vibrates, it causes a membrane separating the middle ear from the inner ear to move back and forth.

The **cochlea** (KOH klee uh) in the inner ear is filled with fluid and lined with tiny hair cells. When the vibrations reach the inner ear, the fluid in the cochlea moves like waves against the hair cells. The hair cells produce nerve impulses in the auditory nerve and send the impulses to the brain. ☑

Picture This

2. Explain Add the label *rods and cones* beside the structure where rods and cones are located.

Reading Check

3. Identify The sensory receptors for sound are located in what structure?

____

How do structures in the ear monitor balance?

There are three semicircular canals in the inner ear, as shown in the figure below. **Semicircular canals** send information about body position and balance to the brain. The canals are positioned at right angles to each other. Like the cochlea, the semicircular canals are filled with fluid and lined with hair cells. As you move your head, fluid moves within the canals. The fluid movement causes the hair cells to bend, sending nerve impulses to the brain. The brain uses this information to determine your body position. The brain can also tell whether or not your body is in motion.

Picture This

4. Name The structures that monitor balance are located in which part of the ear? (Circle your answer.)

a. outer ear
b. middle ear
c. inner ear

Touch

The epidermis and dermis layers of the skin contain many sensory receptors. These sensory receptors respond to temperature, pressure, and pain.

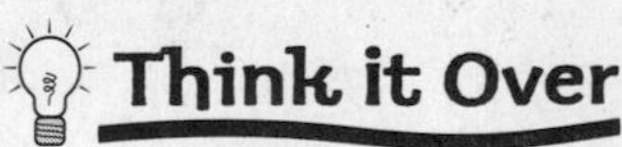

Think it Over

5. Draw Conclusions Why do you think your fingertips have more sensory receptors for light touch than your elbows?

____

____

____

____

____

Where are sensory receptors located?

Sensory receptors are not distributed evenly throughout your body. The tips of your fingers have many sensory receptors to detect light touch. The soles of your feet have sensory receptors that respond to heavy pressure. However, the soles of your feet have few sensory receptors to respond to a light touch. Consider how this difference helps you respond to your environment.

What is the purpose of pain receptors?

Pain receptors are found in all body tissues except the brain. Pain receptors are free nerve endings. They send pain signals to the brain, and the brain responds to help ease the pain.

Nervous System

section 4 Effects of Drugs

MAIN Idea

Some drugs alter the function of the nervous system.

What You'll Learn

- ways drugs can harm the body or cause death
- how a person can become addicted to a drug

Before You Read

On the lines below, list some reasons that you might take legal drugs. Read the section to learn about the helpful and harmful effects of drugs on the body.

____

____

____

Read to Learn

Mark the Text

Identify Main Ideas As you read this section, highlight the main point in each paragraph. State each main point in your own words.

How Drugs Work

A **drug** is a substance that alters the function of the body. Some drugs come from natural sources. Other drugs are made from artificial products. Legal and illegal drugs affect the body in many ways. Some drugs, such as pain killers, affect the nervous system. Other drugs have no effect on the nervous system. The drugs that cause changes in the nervous system work in one of four ways.

1. A drug can cause an increase in the amount of a neurotransmitter that is released into a synapse.
2. A drug can block a receptor site on a dendrite. This prevents the neurotransmitter from binding.
3. A drug can stop a neurotransmitter from leaving a synapse.
4. A drug can act like a neurotransmitter.

What does dopamine control?

Dopamine is a neurotransmitter found in the brain that helps control body movements. Many drugs that affect the nervous system influence the amount of dopamine released by a neuron. The normal action of dopamine is shown in the figure on the next page.

Think it Over

1. **State** the job of neurotransmitters.

____

____

____

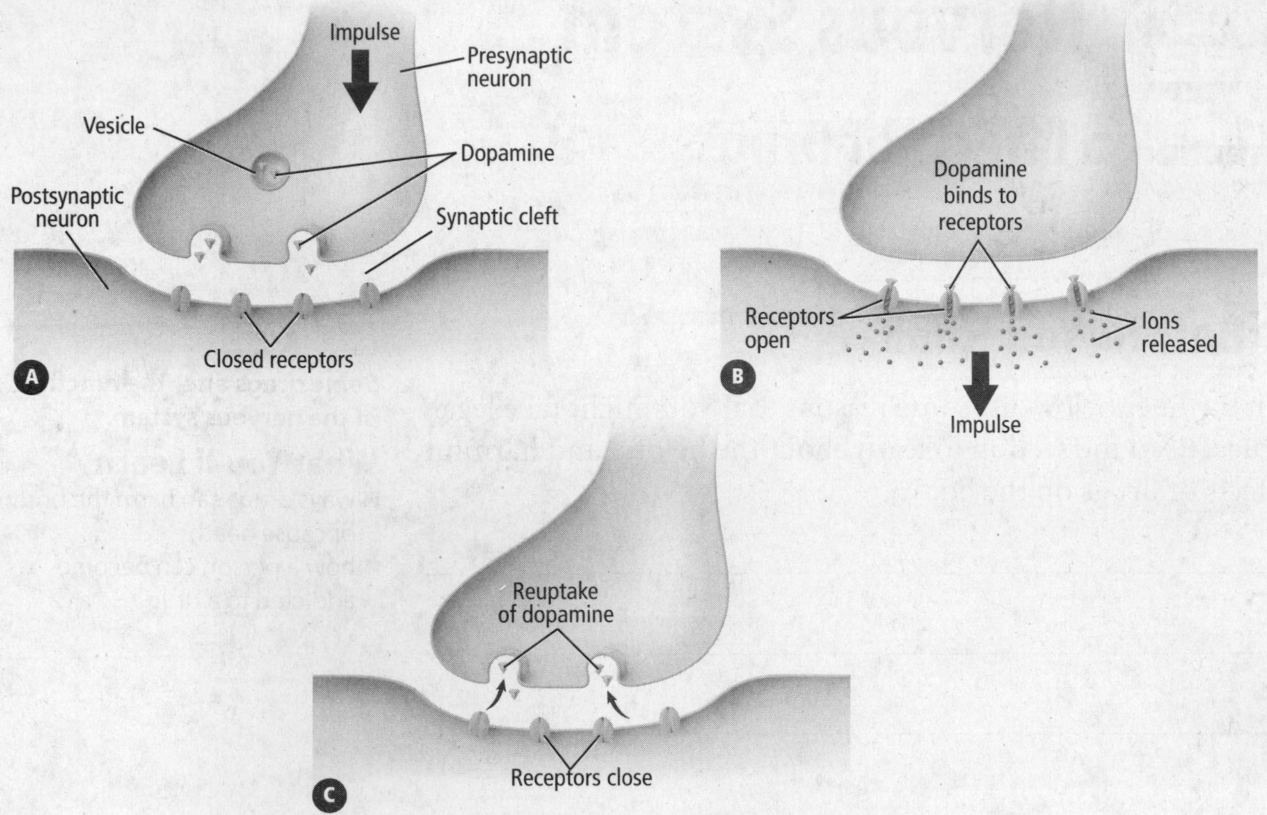

Picture This

2. Label Draw an arrow on the portion of the figure labeled A to show the direction that dopamine travels.

Think it Over

3. Generalize What effect do many people get from drinking coffee when they get up in the morning?

____

____

____

Classes of Commonly Abused Drugs

Both legal and illegal drugs can be abused. Drug abuse is using a drug for any reason other than a valid medical purpose.

What are stimulants?

Stimulants are drugs that increase alertness and physical activity. Common stimulants include nicotine and caffeine.

Nicotine in cigarette smoke increases the amount of dopamine released into a synapse. Nicotine also narrows blood vessels, raises blood pressure, and causes the heart to work harder. Cigarette smoking contributes to about 90 percent of all lung cancer cases.

Caffeine is found in coffee, tea, some soft drinks, and even some foods such as chocolate. It is a widely used and often abused stimulant. Caffeine binds to adenosine receptors on nerve cells in the brain. Adenosine slows down nerve cell activity and causes drowsiness. When caffeine binds to the adenosine receptors, it causes a feeling of heightened alertness. Caffeine also raises adrenaline levels briefly and gives a quick burst of energy that soon wears off.

What are depressants?

Drugs that tend to slow down the central nervous system are **depressants**. These drugs can lower blood pressure, affect breathing, and slow the heart rate.

How does alcohol use affect humans?

Alcohol is a depressant that is abused widely. It affects at least four different neurotransmitters. Alcohol use harms a person's judgment, coordination, and reaction time. Continued abuse of alcohol has long-lasting effects on the body. These effects include a reduction in brain mass, liver damage, ulcers, and high blood pressure. Alcohol use by the mother during pregnancy can harm the fetus. Fetal alcohol syndrome damages a baby's brain and nervous system.

What effects do inhalants have on the nervous system?

Inhalants are chemical fumes that affect the nervous system. Most inhalants slow down the nervous system. Inhalants might produce a feeling of intoxication and can cause nausea and vomiting. Inhalants can cause death. Long-term effects of inhalants include memory loss, hearing loss, vision problems, and permanent nerve and brain damage.

What illegal drugs affect the nervous system?

Amphetamines and cocaine keep dopamine from being reabsorbed. This leaves dopamine to build up in the synapse. Some amphetamines can also increase the amount of dopamine released from a neuron. As the levels of dopamine in the brain increase, a person can feel pleasure and a sense of well-being.

What are the effects of amphetamines and cocaine?

Amphetamines and cocaine have both short-term and long-term effects on the body. Amphetamines can increase the heart rate, cause an irregular heartbeat, and increase blood pressure. Permanent damage to the small blood vessels in the brain can occur. Amphetamine use can also affect behavior. Abusers can experience periods of violent behavior, anxiety, confusion, paranoia, and insomnia. An amphetamine overdose can cause death. It can take a year or longer for the drug's effects to cease after a user stops taking the drug. Cocaine abuse can cause heart attacks, irregular heart rhythms, chest pain, respiratory failure, strokes, seizures, headaches, abdominal pain, and nausea. ☑

FOLDABLES™

Take Notes Make a concept map Foldable, as shown below. As you read, take notes and organize what you learn about commonly abused legal and illegal drugs.

Commonly Abused Drugs

Legal	Illegal

Reading Check

4. Name both a short-term and long-term physical effect of amphetamine use.

____

____

____

____

____

What problems can marijuana cause?

Marijuana is the most-used illegal drug in the United States. Smoking marijuana releases the chemical tetrahydrocannabinol (THC) into the bloodstream. This chemical travels to the brain and binds to receptors in the neurons. The immediate effect is a strong feeling of pleasure.

In the short-term, marijuana use can cause problems with memory and learning, poor coordination, increased heart rate, anxiety, paranoia, and panic attacks. Long term marijuana use can lead to lung cancer. ☑

Reading Check

5. List three problems associated with marijuana use.

____

____

____

____

Tolerance and Addiction

Tolerance of a drug occurs when a person needs more and more of the same drug to get the same effect. Tolerance can lead to addiction.

Addiction is the physical or psychological dependence on a drug. Physical dependence occurs when a drug affects the normal functions of the body's systems. Psychological dependence means that a person has a strong emotional desire for a drug. Marijuana and similar drugs cause psychological addiction. The desire to keep taking the drug is strong, making it difficult to quit.

What neurotransmitter is involved in addiction?

A physical dependence occurs when the drug affects normal body functions. Researchers suggest that the neurotransmitter dopamine is involved in most types of addiction. An addicted person gets pleasure from the increased levels of dopamine. A tolerance to the drug builds up and the person takes more of the drug to achieve the same sense of pleasure. When the person tries to quit using the drug, dopamine levels decrease and make it difficult to resist taking the drug.

How is addiction treated?

People who are psychologically or physically dependent on a drug experience serious withdrawal symptoms without it. Many people who are addicted to a drug have trouble quitting on their own. They might quit for short periods of time. However, they find it hard to resist using the drug again.

The best way to avoid addiction is not to use drugs. People who do use drugs should seek treatment for drug dependency. Counseling might be needed to break an addiction. Physicians, nurses, counselors, clergy, and social workers are trained to help people deal with addictions.

Think it Over

6. Summarize What are two reasons to avoid drug use?

____

____

____

____

Circulatory, Respiratory, and Excretory Systems

section 1 Circulatory System

Before You Read

Press the tips of two fingers to the inside of your wrist, at a point just below your thumb. Can you feel the regular pulsing of your blood? Count the number of beats you feel in fifteen seconds. Record that number on the line below. Multiply the number by four. Then read the section to learn what the number means and how your heart creates its regular rhythm.

____

MAIN Idea

Blood delivers substances, such as oxygen, to cells and removes wastes, such as carbon dioxide, from cells.

What You'll Learn

- the main functions of the circulatory system
- how blood flows through the heart and body
- the major components of blood

Read to Learn

Functions of the Circulatory System

The circulatory system is the body's transport system. It delivers oxygen and nutrients to the cells and removes waste products. The parts of the circulatory system are blood, the heart, blood vessels, and the lymphatic system. These parts work together to maintain homeostasis in the body. The heart pumps blood through tubes inside your body called blood vessels. You will learn about the lymphatic system, which also is part of the immune system, in a different chapter.

In addition to oxygen and nutrients, the circulatory system transports disease-fighting materials produced by the immune system. The blood contains cell fragments and proteins for blood clotting. It also distributes heat throughout the body to help to control body temperature.

Blood Vessels

Blood vessels circulate blood throughout the body. They help to keep blood flowing to and from the heart. The three major types of blood vessels are arteries, capillaries, and veins. ☑

Study Coach

Make Flash Cards Make a flash card for each key term in this section, with the term on one side and the definition on the other side. Use the flash cards to study.

Reading Check

1. **Name** the three major types of blood vessels.

____

____

____

FOLDABLES™

Take Notes Make a folded table Foldable, as shown below. As you read, take notes and organize what you learn about the circulatory, respiratory, and excretory systems in this chapter.

Chapter 34	Function	Organs	Disorders
Circulatory			
Respiratory			
Excretory			

Why do arteries have a thick inner layer?

Arteries (AR tuh reez) are large blood vessels that carry oxygen-rich, or oxygenated, blood away from the heart. Arteries are made of three layers: an outer layer of connective tissue, a middle layer of smooth muscle, and an inner layer of endothelial tissue. The endothelial layer of an artery is thicker than that of other blood vessels because blood is under higher pressure when it is pumped from the heart.

What is the function of capillaries?

Capillaries (KAP uh ler eez) are microscopic blood vessels where the exchange of important substances and wastes occurs. These vessels are so small that red blood cells move single-file through them. Capillary walls are only one cell thick. As a result, the blood and body cells can easily exchange materials through the capillary walls.

Where do veins carry blood?

After blood moves through the capillaries, it enters the veins—the largest blood vessels. **Veins** (VAYNZ) carry oxygen-poor, or deoxygenated, blood back to the heart. The endothelial walls of veins are thinner than those of arteries because by the time blood reaches the veins, the heart's original pushing force has lessened. The contractions of skeletal muscles keep the blood moving. Larger veins have flaps of tissue called **valves** that prevent blood from flowing backward. Breathing movements squeeze against veins in the chest, forcing blood back to the heart.

The Heart

The **heart** is a hollow, muscular organ that pumps blood throughout the body. It is located in the center of the chest. The heart performs two pumping functions at the same time—it pumps oxygenated blood throughout the body, and it pumps deoxygenated blood to the lungs.

What are the parts of the heart?

The heart is made of cardiac muscle. This unique muscle can create and conduct electrical impulses for muscular contractions. The heart is divided into four chambers, as shown in the figure on the next page. The two chambers in the top half of the heart are the right atrium (plural, atria) and left atrium. The atria receive returning blood. The right and left ventricles, below the atria, pump blood away from the heart. Valves keep blood flowing in one direction. ☑

Reading Check

2. Identify the heart chambers that push the blood through the body. (Circle your answer.)

a. atria

b. ventricles

How does the heart beat?

First, the atria fill with blood. Next, the atria contract, filling the ventricles with blood. Once the ventricles are full, they contract to pump the blood out of the heart and into the lungs and body.

The heart works in a regular rhythm. A group of cells in the right atrium, called the **pacemaker** or sinoatrial (SA) node, send out signals that tell the heart muscle to contract. The SA node receives signals about the body's need for oxygen. It then responds by adjusting the heart rate. The signal from the SA node causes both atria to contract. This signal then travels to the atrioventricular (AV) node, causing both ventricles to contract. This two-step contraction is one complete heartbeat.

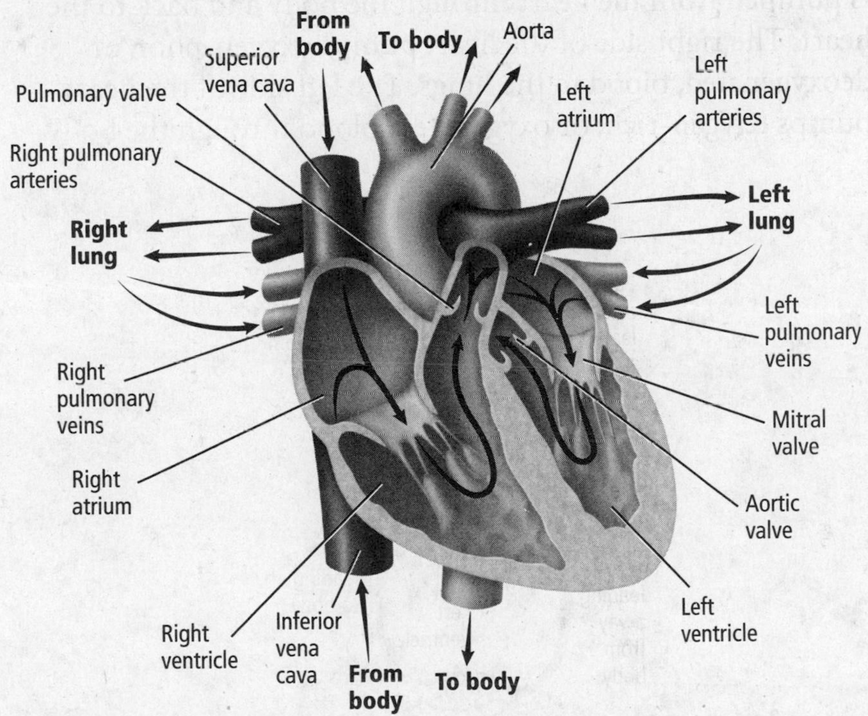

What causes a pulse?

During the *Before You Read* activity, the beat you felt in your wrist was your pulse. As your left ventricle contracts, it pushes blood through your arteries, causing the arteries to expand. Between contractions, the arteries relax. The pulse is the alternating expansion and relaxation of the artery wall. The number of times your artery pulses is the number of times your heart beats. The heart beats approximately 70 times per minute.

Think it Over

3. Apply Suppose you are running hard as you play soccer. How do you think the SA node will respond to this situation?

Picture This

4. Determine When blood is returning from the body to the heart, which chamber of the heart does the blood enter first?

Applying Math

5. Calculate Suppose Cory's blood pressure is 125 at its highest point. To return his blood pressure to normal, Cory must reduce it by what percentage? (Show your work.)

______

______

______

What does a blood pressure reading mean?

Blood pressure is a measure of how much pressure the blood is applying against the vessel walls. Blood pressure readings provide information about the health of arteries. The contraction of the heart, or systole (SIS tuh lee), causes blood pressure to rise to its highest point. Relaxation of the heart, or diastole (di AS tuh lee), causes blood pressure to drop to its lowest point. A normal blood pressure reading for a healthy adult is about 120 (systolic pressure)/80 (diastolic pressure).

How does blood flow through the body?

In the figure below, notice that blood flows in a figure eight pattern. In the first loop, blood travels from the heart to the lungs and back to the heart. In the second loop, blood is pumped from the heart through the body and back to the heart. The right side of the heart pumps oxygen-poor, or deoxygenated, blood to the lungs. The left side of the heart pumps oxygen-rich, or oxygenated, blood through the body.

Picture This

6. Highlight the blood's path from the heart to the body and back.

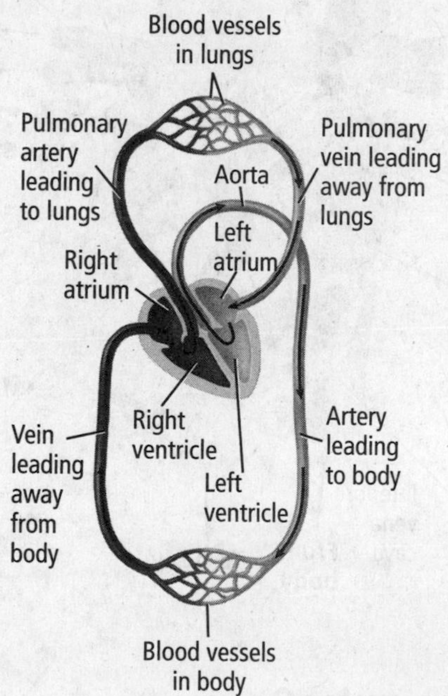

To the lungs and back When blood from the body flows into the right atrium, it contains a little oxygen and a lot of carbon dioxide. From the right atrium, the oxygen-poor blood flows into the right ventricle and into the lungs. The air in the lungs has a lot of oxygen. Oxygen diffuses through the capillaries of the lungs into the blood. At the same time, carbon dioxide diffuses from the blood into the capillaries of the lungs and then into the air. Oxygen-rich blood then flows to the left atrium of the heart to be pumped through the body.

To the body and back The second loop of the figure eight begins as the left atrium fills with oxygen-rich blood from the lungs. The blood moves from the left atrium to the left ventricle. The left ventricle pumps the blood into the largest artery in the body called the aorta. From there, the blood flows into the capillaries throughout the body. The capillaries are in close contact with body cells. Oxygen is released from the blood into the body cells. Carbon dioxide moves from the cells into the blood. The oxygen-poor blood then flows back to the right atrium through the veins.

7. Draw Conclusions Which best describes the role of carbon dioxide in the body? (Circle your answer.)

a. nutrient

b. waste product

Blood Components

Blood contains living cells. It is made up of plasma, red and white blood cells, and cell fragments called platelets.

What is the role of plasma?

Plasma is the clear, yellowish fluid part of blood. Plasma is mostly water. It carries the products of digested food, such as glucose and fats. It also transports vitamins, minerals, and chemical signals. Waste products are carried away by plasma.

What do red blood cells transport?

Red blood cells carry oxygen to all body cells. They develop in the bone marrow. Red blood cells do not have a nucleus, and are made mostly of a protein called hemoglobin. Hemoglobin binds with oxygen and carries it to the body's cells. Some carbon dioxide is carried by the hemoglobin, but most carbon dioxide is carried by plasma. ☑

Reading Check

8. Explain the importance of hemoglobin.

____

____

____

How do white blood cells fight disease?

White blood cells are the body's disease fighters. Some recognize disease-causing organisms and alert the body. Other white blood cells produce chemicals to fight the invaders. Still others surround and kill the invaders. There are many more red than white blood cells.

Why does the body need platelets?

Platelets (PLAYT luts) are cell fragments that play an important part in forming blood clots. When a blood vessel is cut, platelets collect and stick to the vessel at the site of the wound. Platelets release chemicals that produce a protein called fibrin, also known as a clotting factor. Fibrin weaves fibers across the cut that trap platelets and red blood cells. As more platelets and blood cells get trapped, a blood clot or scab forms, slowing and then stopping the flow of blood.

Blood Types

There are four types of blood. They are A, B, AB, and O.

What determines blood type?

Marker molecules attached to red blood cells determine blood type. Type A blood has A markers. Type B blood has B markers. Type AB has both A and B markers. Type O has neither A nor B markers.

Why is blood type important?

If you need a blood transfusion, you can only receive certain blood types, as shown in the table below. This is because plasma contains antibodies that recognize "foreign" markers and cause those red blood cells to clump together. For example, if your blood is type B, the antibodies in your plasma will cause red blood cells with A markers to clump, blocking blood flow.

Picture This

9. Identify the blood type that can be transfused into anyone.

____

Blood Type	Marker Molecules	Can Donate Blood To:	Can Receive Blood From:
A	marker molecule: A antibody: anti-B	A or AB	A or O
B	marker molecule: B antibody: anti-A	B or AB	B or O
AB	marker molecules: AB antibody: none	AB	A, B, AB, or O
O	marker molecules: none antibodies: anti-A, anti-B	A, B, AB, or O	O

How does Rh factor affect blood transfusion?

The Rh factor is another marker on the surface of red blood cells. Clumping will result if someone without the Rh factor (Rh-negative) receives a transfusion of blood with the Rh factor (Rh-positive).

Circulatory System Disorders

Blood clots and fats can block blood flow through arteries. The condition of blocked arteries is called **atherosclerosis** (a thuh roh skluh ROH sus). Signs include high blood pressure and high cholesterol levels. Atherosclerosis can lead to heart attack or stroke, two leading causes of death. Heart attacks occur when blood does not reach the heart muscle. Strokes occur when clots form in blood vessels supplying oxygen to the brain. ☑

✔ Reading Check

10. Name two causes of death that can result from atherosclerosis.

____

____

Circulatory, Respiratory, and Excretory Systems

section 2 Respiratory System

Before You Read

Breathing happens automatically. You do not think about every breath you take. Look at the clock and count how many breaths you take in a minute. Write that number on the lines below. Then write one sentence describing a time when you did think about your breathing. In this section you will learn what happens in your body as you breathe.

____

____

____

MAIN Idea

The respiratory system exchanges oxygen and carbon dioxide between the atmosphere and the blood and between the blood and the body's cells.

What You'll Learn

- the difference between internal and external respiration
- the path of air through the respiratory system
- the changes that occur in the body during breathing

Read to Learn

The Importance of Respiration

Your body's cells need oxygen. Recall that cells use oxygen and glucose to produce energy-rich ATP molecules needed for cellular metabolism. This process is called cellular respiration. Cellular respiration releases energy. It also releases carbon dioxide and water.

Mark the Text

Main Ideas As you read the section, highlight the main ideas in each paragraph.

How is breathing different from respiration?

The respiratory system supports cellular respiration by supplying oxygen to body cells and removing carbon dioxide waste from cells. Two processes make up the respiratory system: breathing and respiration.

First, air enters the body. **Breathing** is the mechanical movement of air into and out of the lungs. Second, gases are exchanged. **External respiration** is the exchange of gases between the atmosphere and the blood. **Internal respiration** is the exchange of gases between the blood and the body's cells.

Think it Over

1. **Explain** why one form of respiration is called "external."

____

____

____

The Path of Air

As you read about the path air travels through your body, follow along in the figure below. First, air enters your mouth or nose. Hairs in your nose filter out dust in the air. Hairlike cilia that line your nasal passages trap particles from the air and sweep them toward the throat. This keeps particles from entering the lungs. Mucous membranes beneath the cilia warm and moisten the air, while trapping foreign particles. ☑

Reading Check

2. Identify three filters through which air passes on its way through your nose and nasal passages.

____

____

____

What structures does air pass through as it travels to the lungs?

Filtered air then passes through the upper throat, or pharynx (FER ingks). A flap called the epiglottis covers the opening to the larynx (LER ingks). The epiglottis allows air to pass while keeping food out of the respiratory tubes. Air moves through the larynx to a tube in the chest called the **trachea** (TRAY kee uh), or windpipe. The trachea branches into two large tubes, called **bronchi** (BRAHN ki). The bronchi lead to the **lungs**, where gas exchange takes place. Each bronchus branches into smaller bronchioles (BRAHN kee ohlz). Branching continues until each branch ends in an air sac called an **alveolus** (al VEE uh lus) (plural, alveoli). Alveoli have walls that are one cell thick and are surrounded by capillaries.

How does gas exchange occur?

Oxygen in the air diffuses across the thin walls of the alveoli into capillaries and then into red blood cells. The blood carries the oxygen to the cells. At the same time, carbon dioxide moves from the blood into the capillaries. It diffuses into the alveoli to be returned to the atmosphere.

Picture This

3. Identify Circle the name of the structure in which oxygen diffuses.

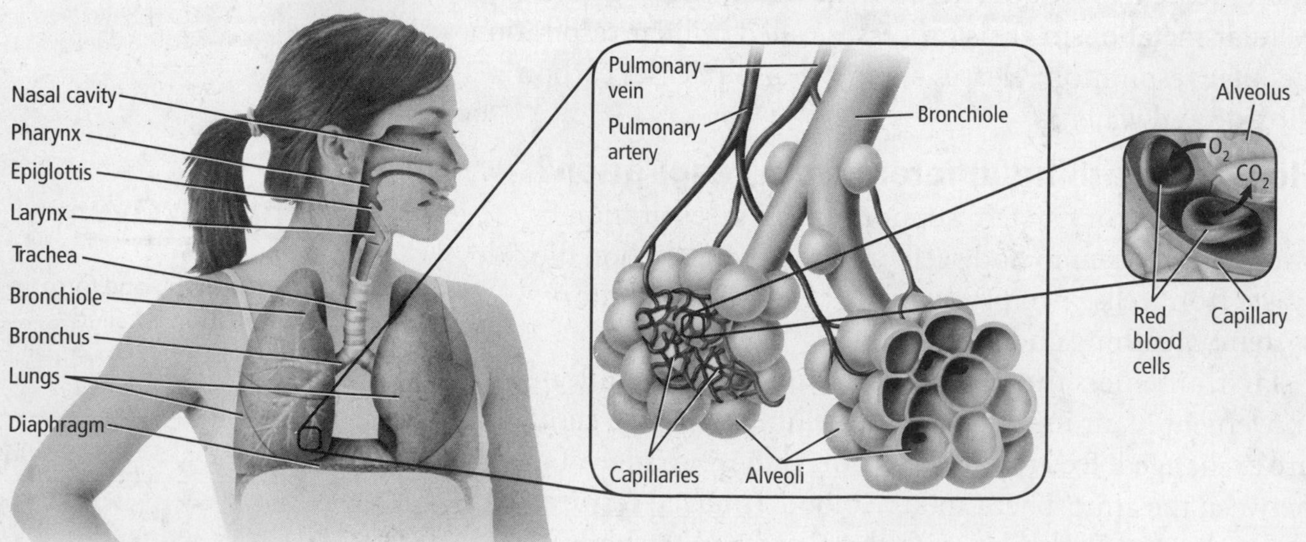

Breathing

Your brain directs the rate of your breathing. If you have a lot of carbon dioxide in your blood, you need more oxygen, so you breathe faster.

As shown in the figure below, the rib and diaphragm muscles contract during inhalation. This increases the size of the chest cavity, allowing air to move into the lungs. During exhalation, the rib and diaphragm muscles relax. This reduces the size of the chest cavity, forcing air to flow out.

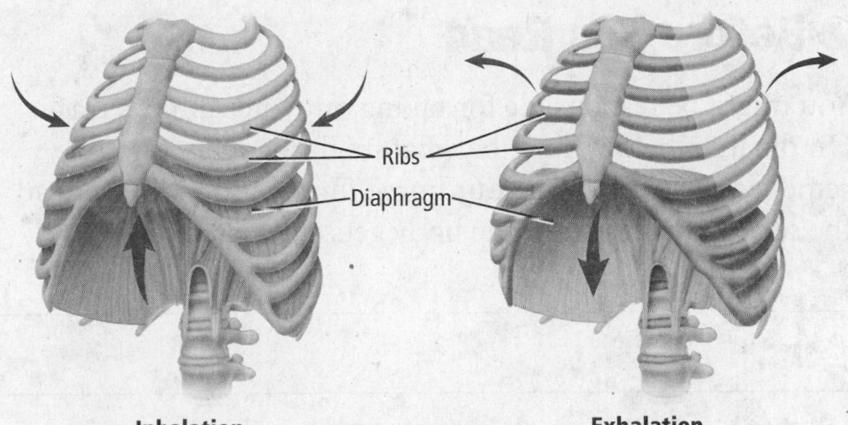

Picture This

4. Compare how the diaphragm moves during inhalation and exhalation.

____

____

____

Respiratory Disorders

The table below lists common disorders that affect the respiratory system. Smoking irritates respiratory tissues and inhibits cellular metabolism. Allergic reactions to particles in the air can also lead to respiratory problems.

Lung Disorder	Description
Asthma	Respiratory pathways become irritated and bronchioles constrict.
Bronchitis	Infected respiratory pathways result in coughing and production of mucus.
Emphysema	Alveoli break down, resulting in reduced surface area needed for gas exchange.
Pneumonia	Infection in the lungs causes alveoli to collect mucus.
Pulmonary tuberculosis	A bacterium infects the lungs, harming the capillaries surrounding the alveoli and inhibiting gas exchange.
Lung cancer	Uncontrolled cell growth in the lungs can lead to persistent cough, shortness of breath, bronchitis or pneumonia, and death.

Picture This

5. Identify the lung disorder that damages the alveoli.

____

Circulatory, Respiratory, and Excretory Systems

section 3 Excretory System

MAIN Idea

The kidneys help maintain homeostasis in the human body.

What You'll Learn

- the steps of the excretion of wastes from the Bowman's capsule to the urethra
- the difference between filtration and reabsorption

Before You Read

You might be responsible for taking out your family's trash. On the lines below, describe what might happen if no one removed the trash from your home for several months. Read the section to learn how the body gets rid of wastes.

____

____

____

Mark the Text

Read for Understanding

As you read this section, highlight any sentences that you do not understand. Reread the highlighted sentences to make certain that you understand their content. Ask your teacher to help you with anything that you still do not understand.

Read to Learn

Parts of the Excretory System

The lungs, skin, and kidneys make up the excretory system. The lungs excrete carbon dioxide. The skin excretes water and salts in sweat. The main excretory organs are the kidneys.

What is the purpose of the excretory system?

The body produces wastes, such as toxins and carbon dioxide, during metabolism. The excretory system removes these wastes. The excretory system also controls the amount of fluids and salts in the body and maintains the pH of the blood. All of these activities help maintain homeostasis.

Reading Check

1. **Describe** the function of the kidneys.

____

____

____

The Kidneys

The <u>**kidneys**</u> are two bean-shaped organs that filter out wastes, water, and salts from the blood. The kidneys are divided into two regions: the outer renal cortex and the inner renal medulla. The body's filters are found in the renal pelvis in the center of the kidney. ☑

Functioning of Nephron

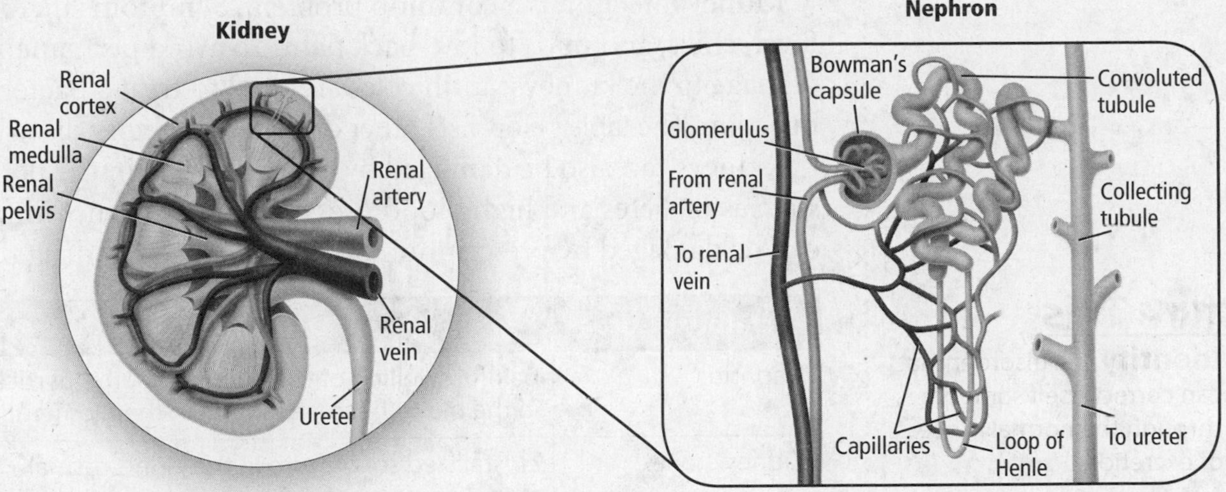

How do the nephrons filter the blood?

Nephrons, shown above, are the kidney's filters. Each kidney contains approximately one million nephrons. Blood enters each nephron through a long tube. A ball of capillaries called the glomerulus (gluh MER uh lus) (plural, glomeruli) surrounds the tube. The glomerulus lies within the Bowman's capsule.

The renal artery transports nutrients and wastes to the kidney. This artery branches into smaller blood vessels, eventually reaching the capillaries in the glomerulus. The walls of the capillaries are very thin. The force of the blood pushes water and substances dissolved in water, such as the nitrogenous waste product **urea** (yoo REE uh), through the capillary walls into the Bowman's capsule. Larger molecules, such as red blood cells and proteins, remain in the bloodstream.

How is urine formed?

Materials collected in the Bowman's capsule flow through the renal tubule. Water and useful materials, such as glucose and minerals, return to the capillaries in a process called reabsorption. Urine, which is waste and unneeded fluids, leaves the kidney through ducts called ureters (YOO ruh turz). The urine is stored in the urinary bladder until it exits the body through the urethra (yoo REE thruh). The process of filtering wastes and reabsorbing useful materials requires large amounts of energy. The kidneys account for only 1 percent of a person's body weight, but they use 20 to 25 percent of the body's oxygen to generate the energy needed to function properly.

Picture This

2. Highlight each structure in the figure as you read about it.

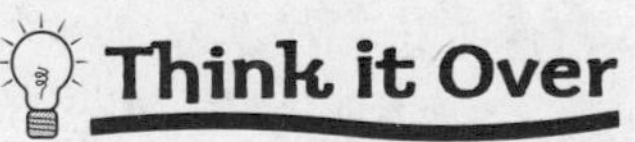

Think it Over

3. Predict what might happen if the excretory process did not include reabsorption.

___

___

___

Kidney Disorders

Kidney infection is a common problem. Symptoms include fever, chills, and mid- to low-back pain. To avoid permanent damage to the kidneys, antibiotics are used to treat a bacterial infection. The table below lists other common excretory disorders.

Kidneys can also be damaged by other diseases in the body, such as diabetes and high blood pressure. In addition, kidneys can be damaged by prescription drugs and by illegal drug use.

Picture This

4. Identify the disorder that can correct itself sometimes through the normal process of excretion.

Excretory Disorder	Description
Nephritis	painful swelling of the glomeruli; large particles in the blood become lodged in the glomeruli
Kidney stones	crystallized solids form in the kidneys; small stones pass out of the body in urine; larger stones can block urine flow
Urinary tract blockage	abnormal formation at birth can block urine flow
Polycystic kidney disease	genetic disorder in which many fluid-filled cysts grow in the kidneys; can reduce kidney function or lead to kidney failure
Kidney cancer	uncontrolled cell growth that often begins in the lining of kidney tubules; can spread to other organs; can lead to death

Kidney Treatments

If kidney problems are not treated or kidney damage occurs, wastes accumulate in the body, leading to coma, seizure, and death. Modern medicine offers two possible treatments.

How is dialysis performed?

Dialysis is a procedure that filters out wastes from the patient's blood. In one type of dialysis, the patient's blood passes through a machine that filters the blood and returns it to the patient's body. This procedure requires three sessions a week.

A second type of dialysis uses the membrane lining the abdomen as an artificial kidney. A special fluid is injected through a tube attached to the body. The patient's waste fluid is drained. This procedure is performed daily.

5. Explain why kidney transplants do not occur more often.

What is a kidney transplant?

In a kidney transplant, a healthy kidney from a donor is placed in the patient's body during surgery. Transplants are becoming more successful. However, the supply of donated kidneys is far below the number of kidneys needed. ☑

Digestive and Endocrine Systems

section 1 The Digestive System

Before You Read

Have you ever had food "go down the wrong pipe"? On the lines below, describe how your body responded. Explain what purpose you think your body's response serves. Then read the section to learn about reflexes in the digestive process.

____

____

____

MAIN Idea

The digestive system breaks down food so nutrients can be absorbed by the body.

What You'll Learn

- the structures of the digestive system and their functions
- the process of chemical digestion

Read to Learn

Functions of the Digestive System

The digestive system performs three main functions. It takes in food. Then it breaks down food so nutrients can be absorbed. Finally, it gets rid of what cannot be digested.

What are mechanical and chemical digestion?

Mechanical digestion is the action of breaking down food into smaller pieces by chewing and by the mixing action of smooth muscles in the stomach and small intestine. **Chemical digestion** is the action digestive enzymes of breaking down large molecules into smaller molecules that cells can absorb. **Amylase** (AM uh lays), an enzyme in saliva, starts chemical digestion by breaking down starches into sugars. ☑

How is food forced through the esophagus?

The tongue pushes chewed food to the back of the mouth which stimulates the swallowing reflex. Chewed food enters the **esophagus** (ih SAH fuh gus), a muscular tube that connects the pharynx, or throat, to the stomach. Smooth muscles that line the esophagus contract in a rhythm to move food through the digestive system in a process called **peristalsis** (per uh STAHL sus).

Mark the Text

Identify Digestive Structures Highlight each structure involved in digestion as you read about it. Underline the functions of each structure.

Reading Check

1. **Compare** What type of digestion uses enzymes to break down food?

____

____

What is the function of the epiglottis?

The epiglottis is a small plate of cartilage that covers the opening to the trachea. If the opening is not closed, food can enter the trachea, which will trigger a coughing reflex. The body coughs to keep the food from entering the lungs.

How does digestion continue in the stomach?

Refer to the figure below as you follow the path of food through the digestive system. Food moves through the esophagus, passes through a circular muscle called a sphincter and into the stomach. The stomach walls are made of three layers of smooth muscle. During mechanical digestion, these smooth muscles contract. These wavelike contractions, called peristalsis, break food into smaller pieces and mix food with acid that is secreted by stomach glands, called gastric glands. The condition commonly known as heartburn is the result of some acid leaking through the sphincter back into the esophagus.

Think it Over

2. Draw Conclusions What part of the digestive tract could be damaged by constant heartburn?

____

The acidic environment in the stomach aids the action of **pepsin**, an enzyme involved in the chemical digestion of proteins. Mucus secreted by the lining of the stomach helps protect the stomach from the acid and pepsin.

Contractions of the muscular walls of the stomach push food farther along the digestive tract. Food passes through the pyloric sphincter at the lower end of the stomach and enters the small intestine.

Picture This

3. Circle the structures that secrete digestive juices but do not hold food on its way through the body.

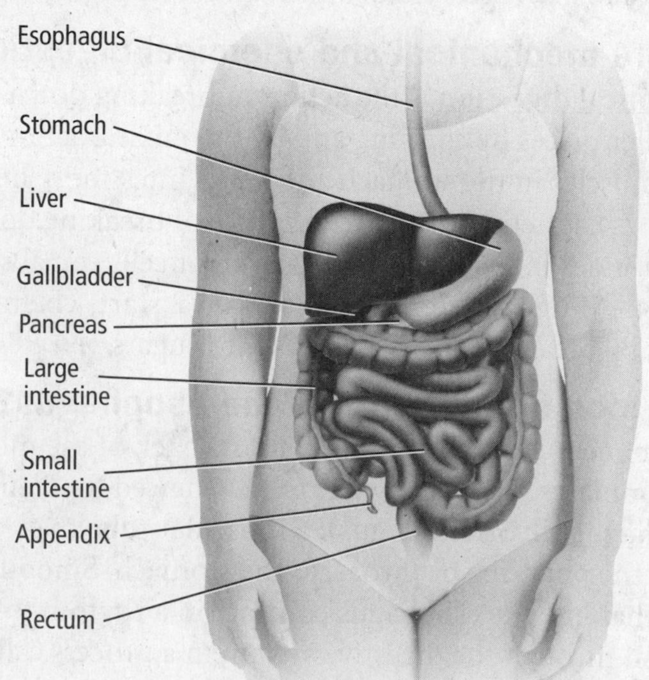

What is the role of the small intestine?

The **small intestine** is a muscular tube that connects the stomach and the large intestine. Smooth muscles in the wall of the small intestine continue mechanical digestion and move food farther along the digestive tract.

The small intestine completes chemical digestion with the help of the pancreas, liver, and gallbladder, as illustrated in the figure below. The pancreas makes enzymes that digest carbohydrates, proteins, and fats. It also makes hormones, which you will learn about later. The **liver** is the organ that makes bile. Bile helps break down fats. Extra bile is stored in the gallbladder to be released into the small intestine when needed.

The small intestine is lined with fingerlike structures called **villi** (VIH li) (singular, villus). Chemical digestion is completed and most of the nutrients from food are absorbed through the villi. Villi increase the surface area of the small intestine. Food that cannot be digested moves into the large intestine as a thick liquid substance called chyme (KIME).

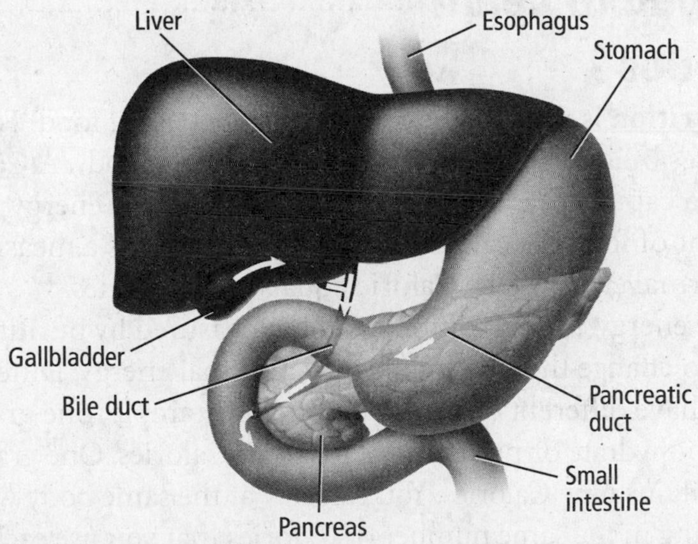

Picture This

4. Label the structure that produces bile and the structure that stores bile.

What is the main function of the large intestine?

The **large intestine** is the end part of the digestive tract. It includes the colon, rectum, and appendix. The appendix is a saclike structure that has no known function.

Some bacteria normally live in the colon. They make vitamin K and certain B vitamins that the body can use.

The main function of the colon is to absorb water from the chyme. The remaining material is a more solid material called feces. Smooth muscle contractions, called peristalsis, move feces toward the rectum. Feces exit the body through the anus. ☑

✓ Reading Check

5. Explain the function of the colon.

____

____

Digestive and Endocrine Systems

section 2 Nutrition

MAIN Idea

The body needs certain nutrients to function properly.

What You'll Learn

- how proteins, carbohydrates, and fats are broken down
- roles of vitamins and minerals in homeostasis
- how to eat healthfully

Before You Read

On the lines below, describe foods that are healthful. In this section you will learn how to use the food pyramid and food labels to help you choose nutritious foods.

____

____

Study Coach

Create a Quiz After you read this section, create a five-question quiz from what you have learned. Then, exchange quizzes with another student.

Read to Learn

Calories

Nutrition is the process of taking in and using food. Food supplies building materials and energy for the body. A **Calorie** (with a capital C) is the unit used to measure the energy content of foods. A calorie (with a lowercase c) is a measure of thermal energy. One Calorie equals 1000 calories.

The energy content of food can be measured by burning the food to change the stored energy to thermal energy. Different foods have different energy content. For example, one gram of carbohydrate or protein contains four Calories. One gram of fat contains nine Calories. You can stay at the same body weight by taking in the same number of Calories that you use each day. To lose weight, you need to use more Calories than you eat. The table below compares Calorie usage in different activities. Physical activity is a key part of good health.

Picture This

1. **Highlight** the activity in the table that interests you most. Then calculate the number of Calories you would use in a week if you did it every day.

____

____

____

Activity	Calories Used per Hour	Activity	Calories Used per Hour
baseball	282	hiking	564
basketball	564	jogging	740–920
bicycling	240–410	skating	300
football	540	soccer	540

Carbohydrates

Sugars such as glucose, fructose, and sucrose are simple carbohydrates. They are found in fruits and candy. Complex carbohydrates such as starches are long chains of sugars. Foods such as potatoes, cereals, and bread are starches.

During digestion, complex carbohydrates are broken down into simple sugars. Simple sugars are absorbed through villi in the small intestine into blood capillaries and then circulated to provide energy for cells. Extra glucose is stored in the liver.

Cellulose, or dietary fiber, is also a complex carbohydrate found in plants. Fiber is important in the diet, even though humans cannot digest it. Fiber helps keep food moving through and out of the digestive tract. Bran, whole-grain breads, and beans are good sources of fiber.

2. Apply Which food would be the best addition to your diet if you wanted to increase your fiber intake? (Circle your answer.)

a. potatoes
b. oatmeal
c. white bread

Fats

Fats are the most concentrated energy source. In proper amounts, fats are needed in a healthful diet. Fats are broken down in the small intestines into fatty acids and glycerol. Fatty acids are absorbed and circulated in the blood through the body for energy. Fats help maintain homeostasis by providing energy and by storing and transporting vitamins.

Not all fats are healthful. Recall that fats are either saturated or unsaturated. Meats, cheeses, and other dairy products contain saturated fats. Plants are the main source of unsaturated fats.

A diet high in saturated fats might result in high blood levels of cholesterol. This can lead to high blood pressure and other heart problems.

Proteins

During digestion, proteins in foods are broken down into amino acids. Amino acids are absorbed into the bloodstream and carried to the cells. By a process called protein synthesis, the cells assemble the amino acids into proteins needed for body structures and functions.

Twenty amino acids are needed by the body for protein synthesis. The body can make 12 of the amino acids. The other eight, known as essential amino acids, need to come from foods. Animal products such as meat, fish, eggs, and dairy products provide all eight essential amino acids. Vegetables, fruits, and grains do not contain all eight. However, they can be combined to provide all eight essential amino acids.

3. Describe an important consideration for vegetarians as they plan their meals.

____

____

____

____

Picture This

4. Identify the three food groups that should make up the largest portion of your diet.

____

____

Food Pyramid

In 2005, the Department of Agriculture published MyPyramid, shown in the figure below. This diagram provides general guidelines for a healthful diet. Notice that the bands representing food groups are not the same width. The message is that people need more nutrients from grains and vegetables than from meats and oils.

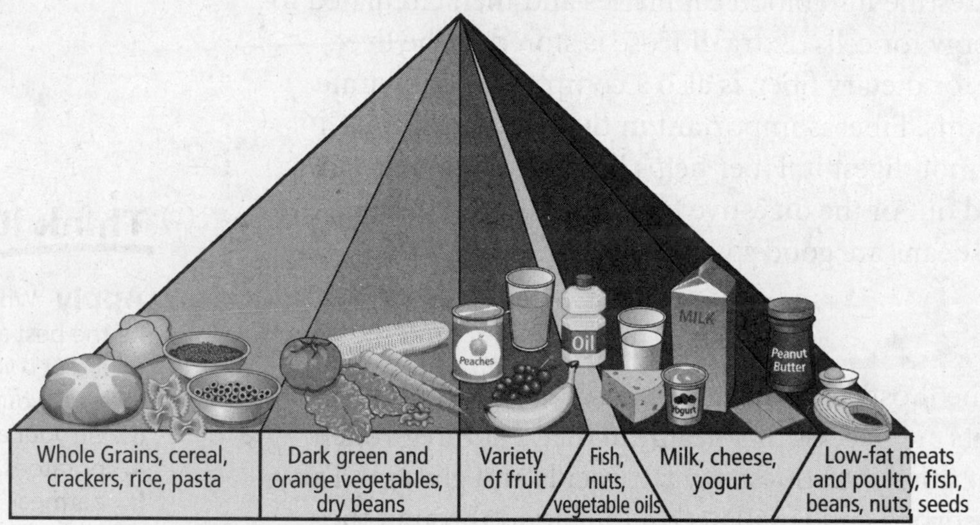

Vitamins and Minerals

<u>**Vitamins**</u> are organic compounds that are needed in small amounts for metabolism. Many vitamins help enzymes function. Vitamin D is made by cells in skin. Bacteria living in the large intestines make vitamin K and certain B vitamins. However, most vitamins need to come from a balanced diet.

Fat-soluble vitamins can be stored in the liver and fatty tissues. Water-soluble vitamins cannot be stored. They need to be eaten regularly.

<u>**Minerals**</u> are inorganic compounds used as building materials and are involved in metabolism. For example, iron is used to make hemoglobin. Calcium is in bones and is involved in muscle and nerve function.

Applying Math

5. Calculate Suppose a cereal label shows 7 g of sugar. The serving size is 1.5 cups. You eat 3 cups of the cereal. How much sugar did you consume? Show your work.

____

____

What information do nutrition labels provide?

The FDA requires food labels to list the name of the food; product's weight or volume; name and address of the manufacturer, distributor, or packager; ingredients; and nutrition content. The percent daily values are based on an individual serving, not the entire package. Use the daily requirement percentages as a guide.

Digestive and Endocrine Systems

section 3 The Endocrine System

Before You Read

On the lines below, describe how your body felt during a stressful situation. Then read the section to learn how the endocrine system participates in the fight or flight response.

MAIN Idea

Systems of the human body are regulated by hormonal feedback mechanisms.

What You'll Learn

- the function of glands that make up the endocrine system
- how the endocrine system helps to maintain homeostasis

Read to Learn

Action of Hormones

The endocrine system is composed of glands and works as a communication system. **Endocrine glands** produce hormones, which are released into the bloodstream and distributed to body cells. A **hormone** acts on certain target cells and tissues to produce a specific response.

Hormones are classified as steroid hormones and nonsteroid, or amino acid, hormones based on their structure and how they do their job.

How do steroid hormones produce a response?

All steroid hormones cause target cells to start protein synthesis. Steroid hormones diffuse through the plasma membrane of target cells and bind to receptors inside the cell. The hormones and receptors move together into the cell nucleus, where they bind to DNA, activating certain genes.

How do amino acid hormones work?

Nonsteroid hormones are made of amino acids. They cannot diffuse into a cell. Instead, they bind with receptors on the plasma membrane of target cells, activating an enzyme inside the membrane that produces the desired response.

Study Coach

Make an Outline Make an outline of the information you learn in this section. Start with the headings. Include the underlined terms.

FOLDABLES™

Organize Information

Make a Foldable chart, as shown below. As you read, include information about glands or organs, the hormones they produce, and their functions.

Gland/ Organ	Hormones	Function

Negative Feedback

An internal feedback process called negative feedback maintains homeostasis in the body. When a body system is too different from a set point, negative feedback returns the system to the set point.

For example, parathyroid hormone maintains the proper amount of calcium in the blood. If blood calcium drops below a certain level, the parathyroid glands respond by releasing more hormone. The hormone causes calcium to be released from the bones. This raises the amount of calcium in the blood. If the amount of calcium in the blood rises too high, parathyroid glands stop making the hormone, causing the opposite effect. ☑

Reading Check

1. Apply What happens to the blood calcium level when the parathyroid glands produce more hormones? (Circle your answer.)

a. increases
b. decreases
c. stays the same

Endocrine Glands and Their Hormones

The endocrine system includes all the glands that secrete hormones. The locations of one of these glands is shown in the figure on the next page.

Why is the pituitary gland called the master gland?

The **pituitary gland** secretes hormones that control many body functions as well as other endocrine glands. The pituitary gland is sometimes called the "master gland." It is located at the base of the brain. Primarily during childhood and adolescence human growth hormone (hGH) secreted by the pituitary gland stimulates cell division in muscle and bone tissue.

How do parathyroid and thyroid glands work together?

The thyroid gland makes thyroxine and calcitonin. **Thyroxine** does not act on specific organs. Instead, it causes body cells to have a higher metabolic rate. **Calcitonin** (kal suh TOH nun) is partly responsible for controlling blood clotting, nerve function, and muscle contraction. Calcitonin also lowers blood calcium levels by signaling bones to absorb more calcium and also by signaling the kidneys to excrete more calcium.

When blood calcium levels are too low, the parathyroid glands make more parathyroid hormone. **Parathyroid hormone** increases blood calcium levels by stimulating the bones to release calcium. They also cause the kidneys to reabsorb more calcium and the intestines to absorb more calcium from food. The hormones of the thyroid and parathyroid work together to maintain homeostasis. ☑

Reading Check

2. Identify the gland that produces more hormones when blood calcium levels are too high.

____

What hormones does the pancreas secrete?

The pancreas secretes insulin and glucagon, which work together to maintain homeostasis. When blood glucose levels are high, the pancreas secretes insulin. **Insulin** signals body cells to convert more glucose to glycogen, which is stored in the liver. When blood glucose levels are low, the pancreas secretes glucagon. **Glucagon** (GLEW kuh gahn) signals liver cells to convert glycogen to glucose and release the glucose into the blood. The process is shown in the figure below.

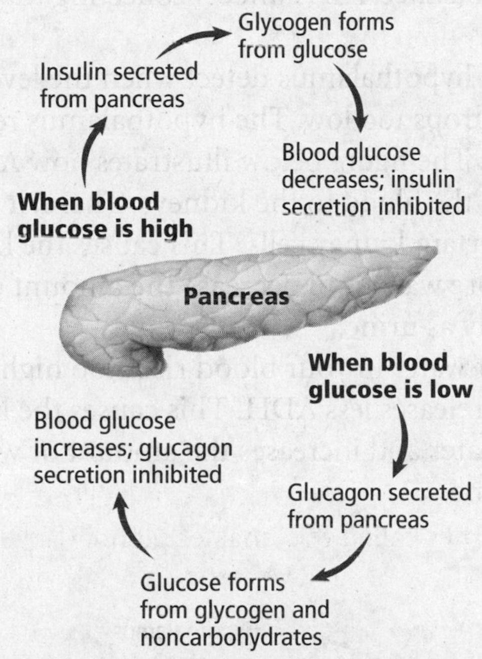

Picture This

3. Circle the names of the materials that are secreted by the pancreas.

What do adrenal hormones affect?

The adrenal glands are located just above the kidneys. The adrenal cortex, or outer part, makes steroid hormones. The hormone **aldosterone** (al DAWS tuh rohn) affects the kidneys and aids sodium reabsorption. **Cortisol** raises blood glucose levels and reduces inflammation. ☑

Reading Check

4. Name the hormone that adjusts the level of sodium in the body.

____

How do adrenal glands respond to stress?

During a stressful situation, the adrenal glands create a sudden burst of energy. The inner part of the adrenal glands secretes epinephrine (eh puh NEH frun), also called adrenaline, and norepinephrine. These hormones work together to increase heart rate, blood pressure, breathing rate, and blood sugar levels. All of these responses increase the activity of body cells as part of the fight or flight response.

Think it Over

5. Generalize How do hormones function as a communication system?

____

____

____

____

Link to the Nervous System

Both the nervous and endocrine systems regulate the activities of the body and help maintain homeostasis. The hypothalamus, located in the brain, serves as a link between the nervous system and the endocrine system. The hypothalamus produces two hormones—antidiuretic hormone (ADH) and oxytocin (ahk sih TOH sun). Oxytocin is produced during childbirth. ☑

Reading Check

6. Identify What two hormones are produced by the hypothalamus?

____

____

____

The **antidiuretic** (AN ti DY yuh REH tic) **hormone** (ADH) controls water balance. ADH affects collecting tubules in the kidneys.

Cells in your hypothalamus detect when the level of water in your blood drops too low. The hypothalamus responds by releasing ADH. The figure below illustrates how ADH works. ADH travels in the blood to the kidneys. There, it binds with receptors on certain kidney cells. This causes the kidneys to reabsorb more water and decrease the amount of water leaving the body as urine.

If the level of water in your blood rises too high, the hypothalamus releases less ADH. This causes the kidneys to reabsorb less water and increases the amount of water leaving the body as urine.

Picture This

7. Identify ADH helps maintain homeostasis by controlling the balance of which important substance in the body? (Circle your answer.)

a. calcium
b. glucose
c. water

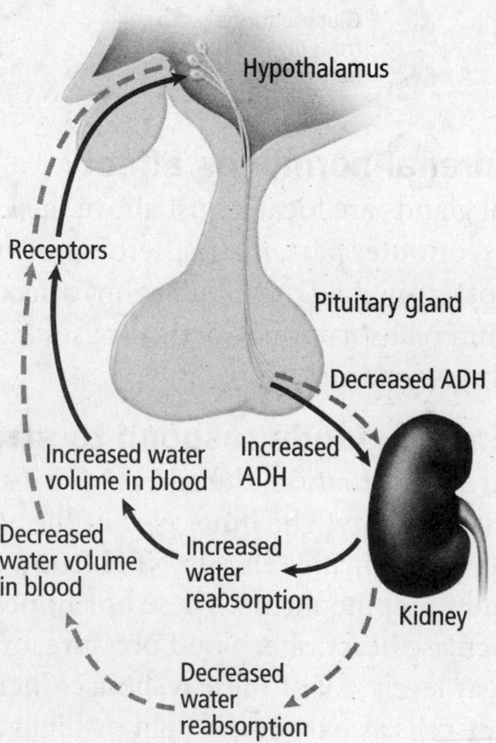

Human Reproduction and Development

section 1 Reproductive Systems

Before You Read

On the lines below, describe two visible ways in which the male body differs from the female body. In this section, you will learn about the sexual differences between males and females.

____

____

MAIN Idea

The structures of the male and female reproductive systems differ.

What You'll Learn
- how hormones regulate reproductive systems
- the events of the female menstrual cycle

Read to Learn

Human Male Reproductive System

Reproduction is needed for a species to continue to exist. The human reproductive process consists of the union of an egg cell and a sperm cell, the development of a fetus, and the birth of an infant.

Sperm cells are the male reproductive cells. They are produced in the testes (tes TEEZ) (singular testis), the male reproductive glands. The testes are located outside the body cavity in a pouch called the scrotum (SKROH tum), as shown below.

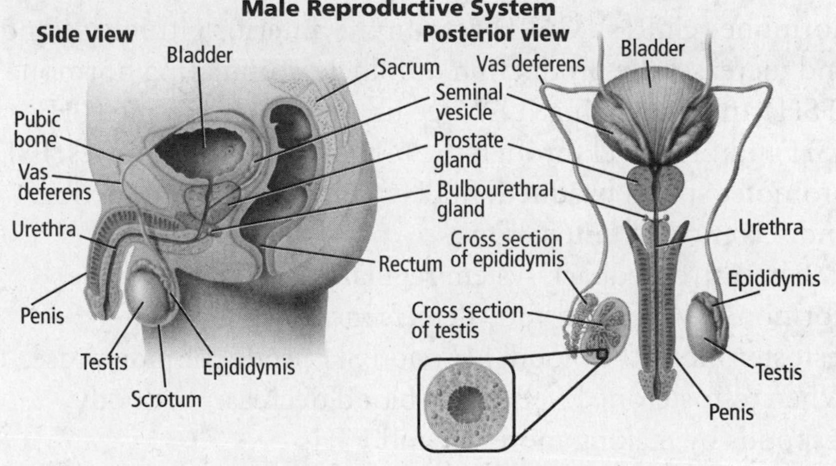

Study Coach

Make Flash Cards For each paragraph, think of a question your teacher might ask on a test. Write the question on one side of a flash card. Write the answer on the other side. Quiz yourself until you know the answers.

Picture This

1. **Underline** the name of the structure that produces sperm.

Where do sperm cells develop?

The scrotum provides a good environment for normal sperm development because the temperature in the scrotum is slightly lower than the normal body temperature. Sperm develop in the **seminiferous tubules** (se muh NIHF rus • TEW byulz) of the testes. Sperm travel to the top of each testis where they mature and are stored in the **epididymis** (eh puh DIH duh mus).

When the sperm are released from the body, they travel away from the testis through a duct called the **vas deferens** (VAS • DEF uh runz). The vas deferens, one for each testis, joins and enters the urethra. The **urethra** (yoo REE thruh) is the tube that carries both semen and urine outside the body through the penis.

Semen (SEE mun) is the fluid that contains the sperm and their nourishment. Energy for the sperm comes from sugars produced by seminal vesicles. The prostate gland and the bulbourethral glands produce an alkaline solution that becomes part of the semen. This solution reduces the effect of acidic conditions sperm might encounter in the female reproductive tract.

FOLDABLES™

Take Notes Make a table Foldable, as shown below. As you read, take notes and organize what you learn about the male and female reproductive systems.

Human Reproduction	Reproductive Cells	Hormones
Male		
Female		

What is the role of testosterone?

Testosterone (tes TAHS tuh rohn) is a steroid hormone needed to produce sperm. Testosterone also influences the development of male secondary sex characteristics that appear at puberty. **Puberty** is the period of growth when sexual maturity is reached. Secondary sex characteristics in males include hair on the face and chest, broad shoulders, increased muscle mass, and a deeper voice.

What hormones influence testosterone production?

The hypothalamus produces a gonadotropin-releasing hormone (GnRH). GnRH acts on the anterior pituitary gland and increases the production of follicle-stimulating hormone (FSH) and luteinizing (LEW tee uh ni zing) hormone (LH). FSH and LH travel through the bloodstream to the testes. FSH promotes sperm production. LH stimulates the production and secretion of testosterone.

A negative feedback system regulates the levels of male hormones. When the body detects increased levels of testosterone in the blood, LH and FSH production decrease. When testosterone levels in the blood decrease, the body responds by making more LH and FSH.

Think it Over

2. **Name** another hormone that is controlled by a negative feedback system.

____

____

____

Human Female Reproductive System

The female reproductive system has three main functions. The first is to produce the female sex cells called egg cells. The second is to receive sperm. The third function is to provide an environment in which a fertilized egg can develop.

Where are egg cells produced?

Egg cells are produced in the ovaries, as shown in the figure below. Inside each ovary are immature eggs known as **oocytes** (OH uh sites). Eggs, called ovum, form when oocyte development is stimulated.

When an egg is released from the ovary, it travels through a tube called an **oviduct** (OH vuh duct). The oviduct connects to the uterus, which is where a baby develops before birth. The lower end of the uterus is called the cervix. A narrow opening in the cervix leads to the vagina. The vagina leads to the outside of the female's body.

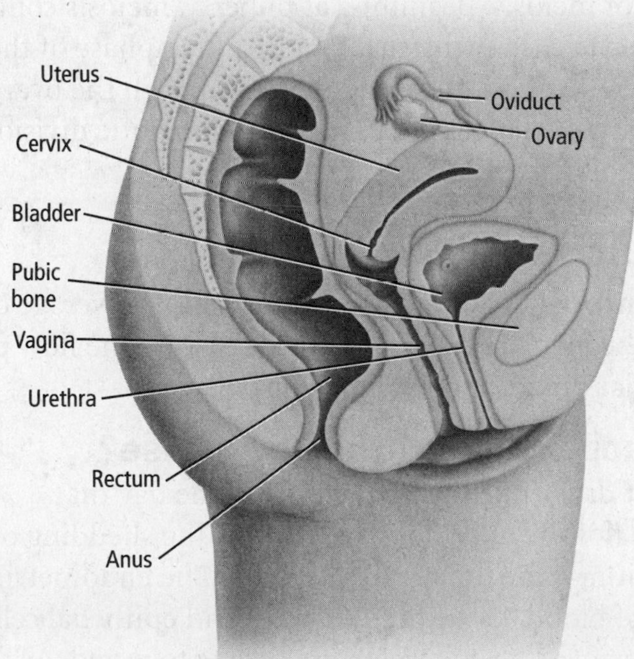

Where are female hormones produced?

Estrogen and progesterone (proh JES tuh rohn) are steroid hormones made by cells in the ovaries. At puberty, estrogen levels increase and cause a female's breasts to begin to develop, hips to widen, and the amount of fat tissue to increase. A female experiences her first menstrual cycle during puberty. The **menstrual** (MEN stroo ul) **cycle** is the monthly series of events that help prepare the female body for pregnancy.

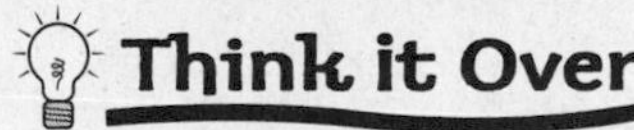

3. Compare Name one similarity between the male and female reproductive systems.

______

______

Picture This

4. Highlight the name of the structure that releases eggs into the oviduct.

Sex Cell Production

In the human male, sperm production begins at puberty and continues for the rest of the male's life. Sex cell production in males follows the general pattern of meiosis.

Sex cell production is different in females. Beginning at puberty, the female releases one mature egg each month. The egg is formed through a unique meiotic process. The structures at the end of the first meiotic division are unequal in size. The smaller structure is called a **polar body**. The chromosomes separate following the general pattern of meiosis, but the cytoplasm divides unequally. Most of the cytoplasm goes to the cell that will eventually become the egg. The polar body disintegrates.

The second meiotic division follows the same pattern as the first meiotic division. In the end stage, the larger portion of the cytoplasm results in the egg, and the polar body disintegrates.

Think it Over

5. Contrast Name one difference between sex cell production in males and females.

What is ovulation?

Before a female is born, cells in the ovaries divide up to the first stage of meiosis. Beginning at puberty, meiosis continues for one oocyte each menstrual cycle. At metaphase of the second meiotic division, an egg breaks through the ovary wall in a process called ovulation. The second meiotic division is completed only if fertilization takes place.

The Menstrual Cycle

The menstrual cycle varies in length, but the average cycle is 28 days. The menstrual cycle has three phases: the flow phase, the follicular phase, and the luteal phase.

What happens during the flow phase?

The first day of the menstrual cycle is the day that menstrual flow begins. Menstrual flow is the shedding of the endometrium—the lining of the uterus. The endometrium is made up of blood, tissue fluid, mucus, and epithelial cells. Around day five, the endometrium begins to mend. As the menstrual cycle continues, the endometrium thickens. ☑

Reading Check

6. Explain What is menstrual flow?

What happens during the follicular phase?

Hormone levels change throughout the menstrual cycle through a negative feedback system. At the beginning of the cycle, hormone levels are relatively low. The anterior pituitary increases production of LH and FSH. This stimulates a few follicles to begin to mature in the ovary. Inside each follicle is an oocyte.

When does ovulation occur?

Cells in the follicles produce estrogen and progesterone. Only one follicle continues to grow and secrete estrogen. The amount of estrogen in the blood keeps the production of LH and FSH low. On approximately day 12 of the menstrual cycle, the high level of estrogen stimulates the anterior pituitary to release a large quantity of LH. At this point, ovulation occurs.

What is the luteal phase?

After ovulation, the follicle transforms into a structure called the corpus luteum (KOR pus • LEW tee um). The corpus luteum produces high amounts of progesterone and some estrogen. These hormones keep the levels of LH and FSH low through negative feedback.

LH and FSH stimulate new follicle development. When the levels of LH and FSH are low, new follicles cannot mature. Toward the end of the luteal phase, the corpus luteum breaks down and stops producing estrogen and progesterone. The rapid decrease in these hormone levels triggers menstrual flow. At this point, the flow phase of a new menstrual cycle begins, as shown below.

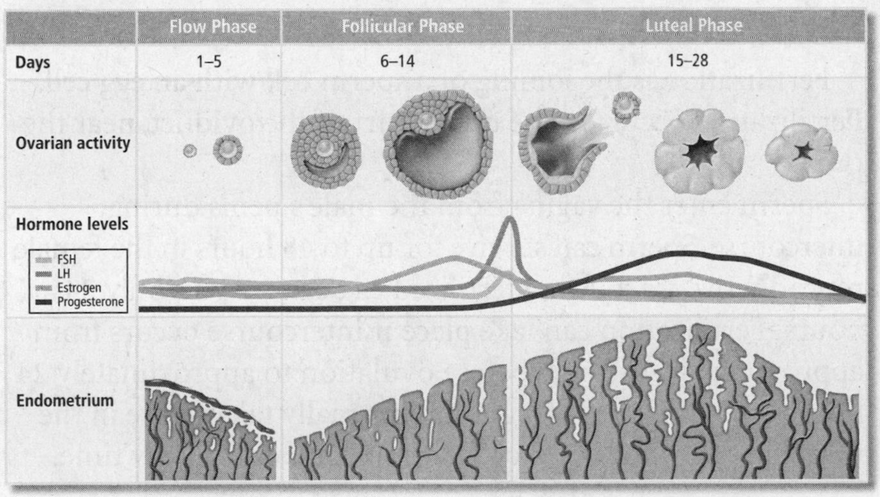

What happens when an egg is fertilized?

When an egg is fertilized, a new menstrual cycle does not begin. Progesterone levels remain high and increase the blood supply to the endometrium. The corpus luteum does not break down, and hormone levels do not drop. The endometrium begins to secrete a fluid rich with nutrients to feed the embryo.

Think it Over

7. Sequence the events at approximately day 12 of the menstrual cycle.

Picture This

8. Identify At approximately day 14, which hormone level is the highest?

Human Reproduction and Development

section 2 Human Development Before Birth

MAIN Idea

Human development involves a series of changes.

What You'll Learn

- the events that take place during the first week after fertilization
- how female hormone levels change during pregnancy

Before You Read

On the lines below, describe one physical change that occurs in a female who is pregnant. In this section, you will read about the development of humans before birth.

Mark the Text

Locate Information As you read this section, highlight the portions of the text that describe changes to an embryo and fetus during pregnancy.

Read to Learn

Fertilization

Fertilization is the joining of a sperm cell with an egg cell. Fertilization occurs in the upper part of the oviduct, near the ovary.

Sperm enter the vagina from the male's penis during intercourse. Sperm can survive for up to 48 hours in the female reproductive tract. An unfertilized egg can survive only 24 hours. Fertilization can take place if intercourse occurs from approximately 48 hours before ovulation to approximately 24 hours after ovulation. Ovulation normally takes place in the middle of the menstrual cycle, but it can occur at any time.

Reading Check

1. Explain Why must many sperm reach the egg for fertilization to take place?

How many sperm fertilize the egg?

About 300 million sperm are released into the vagina during intercourse. Many never leave the vagina, and many more die on the way to the oviduct.

Only one sperm fertilizes the egg, but hundreds are needed to weaken the egg's plasma membrane. The tips of these sperm release enzymes that weaken the membrane. Eventually, the membrane becomes weak enough that one sperm can penetrate it. Immediately, the egg forms a barrier to prevent other sperm from entering the newly fertilized egg.

Early Development

Human sperm and eggs are haploid cells, and each has 23 chromosomes. The fertilized egg is called a zygote (ZI goht) and has 46 chromosomes. The zygote is forced through the oviduct by involuntary smooth muscle contractions and by the cilia that line the oviduct.

The zygote undergoes its first mitosis and cell division approximately 30 hours after fertilization. Mitosis continues. By the third day, the embryo leaves the oviduct and enters the uterus. The embryo at this point is called a **morula**—a solid ball of cells.

What is a blastocyst?

By the fifth day, the morula has developed into a **blastocyst**, or hollow ball of cells. Inside the blastocyst is a group of cells that will become the embryo. If this inner group of cells splits, identical twins might form.

The blastocyst attaches to the endometrium around the sixth day after fertilization, becoming more firmly implanted over the next few days. Implantation is shown in the figure below.

What four extraembryonic membranes develop?

Early human development includes the formation of four extraembryonic membranes—the amnion, the chorion, the yolk sac, and the allantois. The chorion (KOR ee ahn) and allantois (uh LAN tuh wus) help form the placenta. Red blood cells form in the yolk sac.

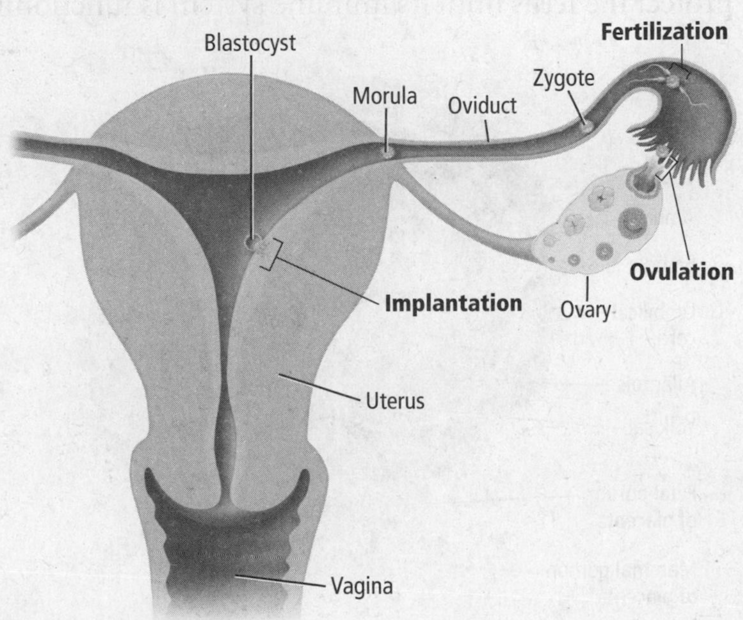

2. Predict How many chromosomes do most human cells have?

____

Picture This

3. Trace the route that the fertilized egg takes to reach the uterus.

What is the purpose of amniotic fluid?

The amnion is a thin layer that forms a sac around the developing embryo. **Amniotic fluid** (am nee AH tihk • FLU id) inside the amnion protects and cushions the embryo. This fluid also helps regulate the embryo's body temperature.

What is the placenta?

About two weeks after fertilization, tiny fingerlike projections of the chorion begin to grow into the wall of the uterus. These projections are called villi (VIH li). The villi combine with part of the uterine lining to form the placenta. ☑

Reading Check

4. Identify the two structures that form the placenta.

____

____

____

The placenta (pluh SEN tuh) is the organ that delivers nutrients to the embryo and carries away wastes from the embryo. It is formed completely by the tenth week of pregnancy. The placenta has two surfaces. The fetal side forms from the chorion, and the maternal side forms from uterine tissue. The umbilical cord connects the fetus and the mother. It contains many blood vessels.

The placenta regulates what passes from the mother to the fetus and from the fetus to the mother. Oxygen and nutrients travel from the mother to the fetus. Harmful substances such as alcohol and drugs can also pass through the placenta to the developing fetus.

Metabolic waste products and carbon dioxide travel through the placenta from the fetus to the mother, as shown in the figure below. The fetus and the mother have separate circulatory systems, and blood cells do not pass through the placenta. The mother's antibodies are passed to the fetus. The antibodies help protect the fetus until its immune system is functioning.

Picture This

5. Underline the name of the structure that connects the fetus to the mother.

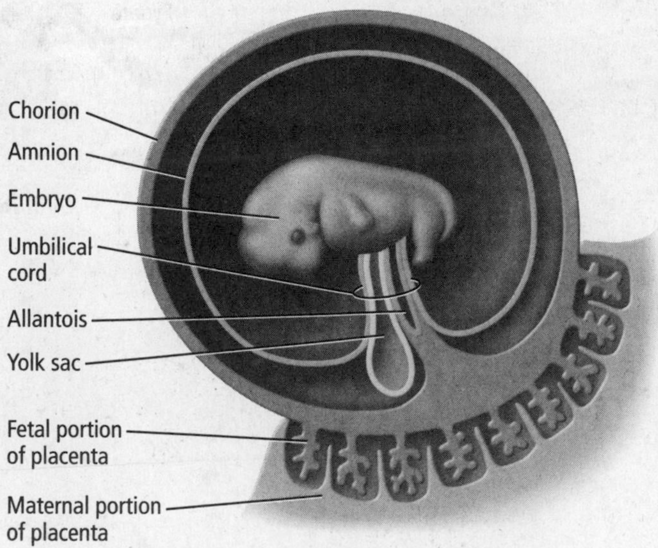

What role do hormones play in pregnancy?

Once the blastocyst is implanted in the endometrium during the first week of development, the embryo begins to secrete a hormone called human chorionic gonadotropin (hCG) (kor ee AH nihk • go na duh TROH pen). This hormone keeps the corpus luteum from breaking down. As long as the corpus luteum remains active, progesterone and estrogen levels remain high. As a result, a new menstrual cycle will not begin. Two to three months into the pregnancy, the placenta secretes the proper amounts of estrogen and progesterone to maintain the conditions needed for pregnancy. ☑

Reading Check

6. Define the purpose of hCG.

Three Trimesters of Development

Human development takes approximately 266 days from fertilization to birth. This time span is divided into three trimesters. Each trimester lasts about three months. During this time, the zygote grows from a single cell to a baby with trillions of cells.

What changes occur during the first trimester?

All tissues, organs, and organ systems begin to develop during the first trimester. At the end of eight weeks, the embryo is called a fetus. At the end of the first trimester, the fetus can move its arms, fingers, and toes. It can make facial expressions and has fingerprints.

During the first two weeks of pregnancy, a woman might not realize she is pregnant because she has not missed a menstrual period. The fetus is especially vulnerable during the first trimester. Alcohol, drugs, tobacco, and pollutants can harm the fetus. A lack of important nutrients might cause damage that cannot be reversed.

Think it Over

7. Evaluate Why would a baby born during the second trimester likely remain in the hospital for many weeks?

What changes occur during the second trimester?

The second trimester marks a period of rapid body growth. The developing fetus can suck its thumb and open its eyes. Hair usually forms during this period. The mother might feel light kicks as the fetus moves its arms and legs.

At the end of the second trimester, the fetus might be able to survive outside the mother's uterus with medical help. If born this early, the baby cannot maintain a constant body temperature and might be placed in an incubator. Because its lungs are not fully developed, the baby might need assistance to breath. The baby is also at risk for serious illness because its immune system is not fully formed.

What changes occur during the third trimester?

The fetus continues to grow rapidly during the third trimester. Fat that will help insulate the newborn baby accumulates under the skin. The developing fetus needs protein to fuel brain growth. New nerve cells in the brain form at a rate of 250,000 cells per minute. The developing fetus might respond to sounds in the environment during the third trimester.

Diagnosis in the Fetus

Many medical conditions can be diagnosed before a baby is born. Identifying problems before birth helps medical personnel treat the problems early. This helps a newborn baby have the highest possible quality of life. ☑

Reading Check

8. List two goals of diagnosing medical conditions before a baby is born.

____

____

____

____

____

What does ultrasound show?

Some medical conditions can be identified using ultrasound. Ultrasound is a procedure in which sound waves are bounced off the fetus. The sound waves are changed to light images that can be viewed on a monitor. Ultrasound helps doctors know if the fetus is growing properly. The position of the fetus in the uterus and the gender of the fetus can be determined by ultrasound.

What else is used to diagnose problems with the fetus?

Amniocentesis (am nee oh sen TEE sus) and chorionic villi sampling are prenatal tests. Both tests carry a small risk of miscarriage.

Amniocentesis is usually performed during the second trimester if a problem is suspected. Fluid from the amniotic sac is removed by a needle inserted in the abdomen of the pregnant female. Enzyme tests and DNA analysis can be performed on the fluid to detect certain medical conditions. For example, fetal cells can be examined by a karyotype to detect unusual chromosome numbers. Recall that a karyotype is a chart of chromosome pairs used to identify unusual chromosome numbers or the sex of the fetus.

Chorionic villi sampling might be conducted during the first trimester. A small tube is inserted through the vagina and cervix of the mother. Cells are removed and analyzed by karyotyping. The chromosomes in the cells of the chorion are identical to those in the cells of the fetus. Genetic problems might be identified with either form of testing. ☑

Reading Check

9. State What type of problems are identified by amniocentesis and chorionic villi sampling?

____

____

Human Reproduction and Development

section 3 Birth, Growth, and Aging

Before You Read

Compare a picture of you as a young child to how you look today. Write the differences you notice on the lines below.

MAIN Idea

Developmental changes continue throughout a person's life.

What You'll Learn

- the events that occur during the three stages of birth
- the hormones needed for growth

Read to Learn

Mark the Text

Highlight Main Ideas Read the paragraphs under each question heading. Underline the part of the text that answers the question.

Birth

The beginning of the birthing process is called **labor**. Labor begins when the posterior pituitary gland releases the hormone oxytocin (ahk sih TOH sun), which stimulates involuntary uterine contractions.

The three stages of birth are dilation, expulsion, and the placental stage. **Dilation** (die LAY shun) is the opening of the cervix to approximately 10 cm.

When the uterine contractions become strong, the mother contracts her abdominal muscles to help push out the baby. The baby comes through the vagina in the **expulsion stage**.

When the baby is out of the mother's body, the umbilical cord is clamped and cut. During the **placental stage**, the placenta and the extraembryonic membranes detach from the uterus and leave the mother's body through the vagina.

A baby is called a newborn during the first four weeks of life. The average human newborn has a mass of 3300 g and is 51 cm long.

What is a cesarean section?

Complications might prevent the baby from being born vaginally. During a cesarean section, the mother's abdomen walls and uterus are cut, and the baby is removed.

Think it Over

1. **Compare** Name one way in which a cesarean section differs from a vaginal birth.

Growth and Aging

Newborns continue to grow and develop. They become infants and then children. **Adolescence** (a dul ES unts) is a major phase of human development that begins at puberty and ends at adulthood. ☑

Reading Check

2. Generalize How long does adolescence last?

____

____

____

What hormones affect growth?

Hormones influence growth. The human growth hormone stimulates growth in most parts of the body. This hormone works by increasing the rate of protein synthesis and the rate at which fats break down. Thyroxine is produced by the thyroid. It increases metabolism and is essential for growth.

Steroid hormones, such as estrogen and testosterone, are also important for growth. These hormones pass through the plasma membrane of cells. They trigger genes in a cell's nucleus to form proteins, causing an increase in cell size.

When does infancy occur?

The first two years of life are known as **infancy**. Physical and mental development occur rapidly during this period. Most infants learn to walk and talk. Infants grow taller and increase in weight. Growth rates slow after the second year.

Childhood is the period between infancy and adolescence. Children progressively learn to reason and solve problems during childhood.

What marks the beginning of adolescence?

Puberty marks the beginning of adolescence. Puberty begins between ages 8 and 13 in girls and between ages 10 and 15 in boys. You have already learned about the sexual development that occurs during adolescence. An adolescent also experiences growth spurts and other physical changes. **Adulthood** begins when physical growth is complete.

What happens as adults age?

As the body ages, there are distinct, noticeable changes. Graying hair marks a decline in pigment production. Adults become shorter as they age because the discs between their vertebrae flatten. Metabolism slows, muscle mass decreases, and the skin loses its elasticity. In women, the ability to have children ends with menopause (MEN uh pawz). Sperm production decreases in men. However, despite the changes, many people remain physically and mentally active as they grow older. ☑

Reading Check

3. List three physical changes that occur as humans age.

____

____

____

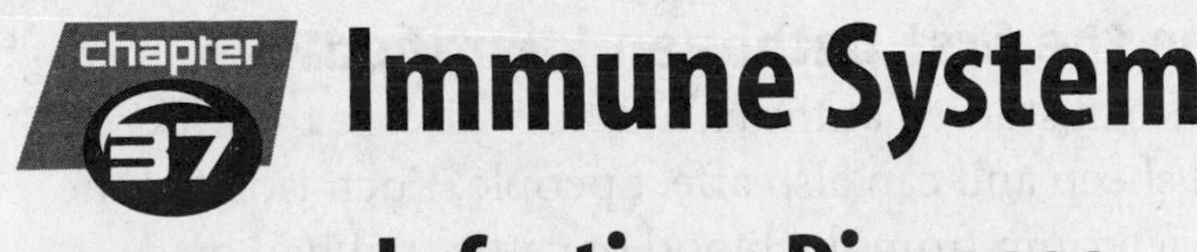

chapter 37 Immune System

section 1 Infectious Diseases

Before You Read

Think of a disease that you have had. On the lines below, write the name of the disease and answer the following questions. Is the disease contagious? How is it spread? What causes the disease? Is there a cure? These are the types of questions that scientists ask about diseases.

____

____

MAIN Idea

Pathogens are spread by people, other animals, and objects.

What You'll Learn

- how to construct a flow chart of Koch's postulates
- how diseases are transmitted and the role of reservoirs in the spread of disease
- the symptoms of bacterial infectious disease

Read to Learn

Pathogens Cause Infectious Disease

An **infectious disease** is a disease caused when a pathogen is passed from one organism to another, disrupting homeostasis. Colds and athlete's foot are examples of infectious diseases. **Pathogens** (PA thuh junz) are the cause of infectious diseases. Some, but not all, types of bacteria, viruses, protozoans, fungi, and parasites are members of this group.

Not all members of these groups are harmful. Some types of bacteria and protozoans normally live inside your body and on your skin. These organisms help protect your body from pathogens.

Study Coach

Create a Quiz After you read this section, create a five-question quiz from what you have learned. Exchange quizzes with another student. After taking the quizzes, review your answers together.

Germ Theory and Koch's Experiments

With the invention of the microscope, scientists discovered microorganisms. Louis Pasteur showed that microorganisms from the air could grow in nutrient solutions. With knowledge gained from discoveries like these, scientists began to develop the germ theory. The germ theory states that some microorganisms are pathogens. Scientists were not able to prove this theory until Robert Koch, a German physician, performed experiments on anthrax (AN thraks). ☑

Reading Check

1. Explain What does the germ theory state?

____

____

____

How was the first pathogen identified?

Robert Koch studied anthrax, a deadly disease that affects cattle and sheep and can also affect people. Koch isolated the anthrax bacterium from the blood of cattle that had died from the disease. After growing the bacterium in a laboratory, Koch injected it into healthy cattle. These cattle developed the disease anthrax. Koch then isolated the anthrax bacterium from the blood of the newly infected cattle and grew it in a laboratory. Koch compared the characteristics of the two sets of cultures. They were identical. This showed that the same type of bacterium caused the illness in both sets of cattle.

What are the steps in Koch's postulates?

Koch's experimental steps became known as Koch's postulates. **Koch's postulates** are rules for showing that an organism causes a disease. The figure below lists and illustrates these rules. Scientists still follow Koch's postulates to identify a specific pathogen as the cause of a specific disease.

Artificial media are gels containing nutrients that the bacteria need to live and reproduce. Scientists use artificial media to grow bacteria in the laboratory. Some pathogens, including viruses, will not grow on artificial media. Scientists need cultured cells to grow viruses.

Picture This

2. Sequence On the line below, write the letters of the following items in the correct sequence.

- **a.** pathogen injected into healthy animal causes the disease
- **b.** pathogen from second animal shows same characteristics as the first
- **c.** pathogen grown in laboratory
- **d.** pathogen isolated from animal with the disease

____

____

Postulate 1

The suspected pathogen must be isolated from the diseased host in every case of the disease.

Postulate 2

The suspected pathogen must be grown in pure culture on artificial media in the laboratory.

Postulate 3

The suspected pathogen from the pure culture must cause the same disease when placed in a healthy new host.

Postulate 4

The suspected pathogen must be isolated from the new host, grown again in pure culture, and shown to have the same characteristics as the original pathogen.

Spread of Disease

Of the many microorganisms that surround us, only a few cause disease. The table below lists some common human infectious diseases. To spread, a pathogen needs a reservoir and a way to spread. A disease **reservoir** is a source of the pathogen in the environment. Reservoirs might be animals, people, or objects such as soil or a dirty countertop.

How do humans act as reservoirs of pathogens?

Humans are the main reservoir for pathogens that infect humans. People can pass pathogens directly or indirectly to other people. Individuals can be capable of passing a pathogen, even if they have no symptoms of the disease. These people are called carriers. Humans can spread colds, influenza (flu), and sexually transmitted diseases, such as human immunodeficiency (ih MYEWN nuh dih fih shun see) virus (HIV), without knowing they are infected.

Can other animals pass pathogens to humans?

Other animals can also carry pathogens, such as influenza and rabies, that can infect humans. Pigs and birds can spread influenza pathogens. Rabies can pass to humans from dogs and wild animals such as bats, foxes, and skunks.

3. Evaluate Why can keeping a countertop clean help prevent the spread of disease?

____

____

____

Picture This

4. Draw Conclusions Use of insect repellent can help protect you from which disease? (Circle your answer.)

a. tetanus
b. athlete's foot
c. West Nile virus

Human Infectious Diseases

Disease	Cause	Affected Organ System	How Disease is Spread
Tetanus	bacteria	nervous system	soil in deep puncture wound
Strep throat	bacteria	respiratory system	droplets/direct contact
Meningitis	bacteria or virus	nervous system	droplets/direct contact
Lyme disease	bacteria	skeletal and nervous system	vector (tick)
Chicken pox	virus	skin	droplets/direct contact
Rabies	virus	nervous system	animal bite
Colds	virus	respiratory system	droplets/direct contact
Influenza	virus	respiratory system	droplets/direct contact
Hepatitis B	virus	liver	direct contact with exchange of body fluids
West Nile	virus	nervous system	vector (mosquito)
Giardia	protozoan	digestive tract	contaminated water
Malaria	protozoan	blood and liver	vector (mosquito)

What are some other reservoirs?

Some bacteria normally found in soil, such as tetanus, can cause disease in people. Infection can result if the contaminated soil gets into a deep wound.

Contaminated water and food are reservoirs of pathogens. Sewage treatment plants prevent human feces from contaminating the water supply. Food can become contaminated through contact with humans or insects.

What are the main methods of transmission?

Pathogens are mainly transmitted to humans in four ways: (1) direct contact, (2) indirectly through the air, (3) indirectly through touching contaminated objects, and (4) by vectors, organisms that carry pathogens.

Direct Contact Direct contact with other humans can spread diseases such as colds, herpes (HUR peez), infectious mononucleosis (mah noh new klee OH sus) (also known as "mono" or the "kissing disease"), and sexually transmitted diseases. ☑

Reading Check

5. Explain how "mono" is spread.

____

____

____

Indirect Contact Some pathogens can be passed through the air. When an infected person sneezes or coughs, pathogens can pass to another person or to an object in tiny mucous droplets. Many pathogens can survive on objects touched by humans. Careful hand washing and cleaning of cooking surfaces and utensils can help prevent the spread of pathogens.

Vectors The most common vectors are arthropods, which include biting insects such as mosquitoes and ticks. Flies can pick up pathogens by landing on infected material, such as feces. Flies then spread the pathogens by landing on materials handled or eaten by humans.

Symptoms of Disease

Symptoms such as aches, coughing, and sneezing result when a pathogen invades cells in your body. Recall that, as the virus multiplies, it damages tissues and kills some cells.

Bacteria that invade the body sometimes produce harmful chemicals or toxins. The toxins travel in the bloodstream and can cause damage in different parts of the body. ☑

Reading Check

6. Define toxins.

____

____

____

All viruses and some types of bacteria and protozoans invade and live inside cells. They damage and sometimes kill the cells, causing symptoms in the host.

Disease Patterns

Health agencies constantly observe disease patterns to help control the spread of disease. The Centers for Disease Control and Prevention (CDC) receives information from doctors and clinics and publishes a weekly report about the number of cases of specific diseases. The World Health Organization (WHO) watches disease patterns worldwide.

Some diseases, such as the common cold, are known as endemic diseases. **Endemic diseases** are continually found in small amounts within the population. A large outbreak of the same disease in an area is an **epidemic**. If an epidemic is widespread through a large region, such as a country, a continent, or worldwide, it is called a **pandemic**.

Think it Over

7. Contrast How is a pandemic different from an epidemic?

____

____

____

Treating and Fighting Diseases

Doctors might prescribe a drug, such as an antibiotic, to help the body fight a disease. A substance that can kill or slow the growth of other microorganisms is an **antibiotic** (an ti bi AH tihk). Many fungal secretions, such as penicillin, are used as antibiotics. Other antibiotics are manufactured.

Over the last 60 years, the widespread use of antibiotics has caused many bacteria to become resistant to particular antibiotics. Resistance develops through natural selection. Some bacteria in a population might have a trait that enables them to survive a particular antibiotic. These surviving bacteria reproduce and pass on this trait. Because bacteria can reproduce quickly, the number of resistant bacteria in a population can increase quickly as well. The figure below shows the rise in the United States of penicillin-resistant gonorrhea (gah nuh REE uh).

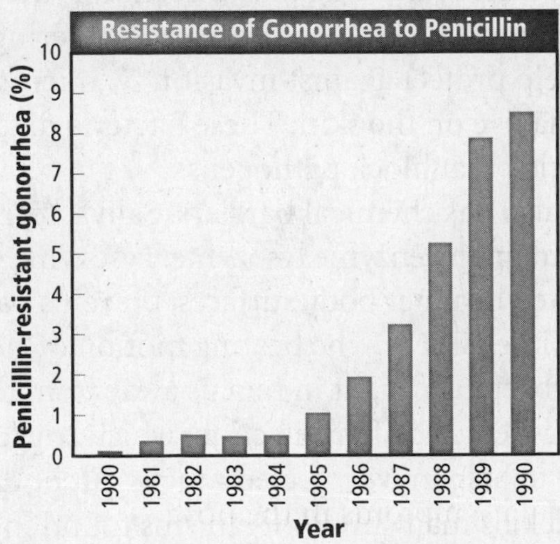

Picture This

8. Compare Penicillin-resistant gonorrhea increased most rapidly between which two years? (Circle your answer.)

a. 1981 and 1982
b. 1988 and 1989
c. 1989 and 1990

Chapter 37 Immune System

section 2 The Immune System

MAIN Idea

Components of the immune system are nonspecific immunity and specific immunity.

What You'll Learn

- the structure and function of the lymphatic system
- the difference between passive and active immunity

Before You Read

Think about the last time you had a cut or scrape. On the lines below, describe how your body responded to the injury. What did the site look like? How did it feel? Read the section to learn how your body fights against infection.

____

____

Mark the Text

Restate the Main Point As you read the section, highlight the main point in each paragraph. Then restate each main point in your own words.

Read to Learn

Nonspecific Immunity

At birth, the body's immune system has defenses to fight pathogens. These defenses are nonspecific— they do not target a specific pathogen. They protect against any pathogen.

Nonspecific immunity helps prevent disease. It also helps to slow the progress of a disease while specific immunity begins to develop its defenses. Nonspecific immunity works quickly, but specific immunity is more effective.

What barriers protect the body from pathogens?

The body's main barrier against infection is the unbroken skin and skin oils. The layers of dead cells covering the skin's living cells help protect against invasion by microorganisms. Many bacteria live on the skin. These bacteria digest skin oils, producing acids that block pathogens.

The body also has chemical barriers. Saliva, tears, and nasal secretions contain the enzyme lysozyme. Lysozyme kills bacteria. Mucus, secreted by inner body surfaces, prevents bacteria from sticking to epithelial cells. The beating motion of cilia inside the airway sends bacteria caught in mucus away from the lungs. Extra mucus is secreted after infection, which triggers coughing and sneezing to help move infected mucus out of the body. Stomach acid kills many microorganisms found in food. ☑

Reading Check

1. **Explain** how coughing benefits the body.

____

____

____

What if pathogens get past the barriers?

When pathogens get through the barriers, the body's nonspecific immunity continues the defense.

Cellular Defense If foreign microorganisms enter the body, the cells of the immune system defend the body. One defense is phagocytosis (fa guh si TOH sus) by which white blood cells, especially neutrophils and macrophages, engulf and absorb foreign microorganisms. The phagocytes then release chemicals that destroy the microorganisms. ☑

Blood plasma contains approximately 20 complement proteins. **Complement proteins** aid phagocytosis by helping phagocytes bind more efficiently to pathogens, activating the phagocytes and enhancing the destruction of the pathogen's membrane. Phagocytes are activated by materials in the bacteria's cell walls.

Interferon When a virus enters the body, interferon helps prevent the virus from spreading. **Interferon** is a protein secreted by virus-infected cells. Interferon binds to neighboring cells and causes these cells to produce antiviral proteins. Antiviral proteins can prevent virus cells from multiplying.

Inflammatory Response When pathogens damage tissue, both the invading pathogen and the cells of the body release chemicals. These chemicals attract phagocytes, increase blood flow to the infected area, and make blood vessels more permeable to allow white blood cells to escape into the infected area. The result is an accumulation of white blood cells to fight the infection. Pain, heat, and redness in the infected area are the result of the inflammatory response.

Specific Immunity

When pathogens get past the nonspecific defenses, the body has a second line of defense—the lymphatic (lim FA tihk) system. The lymphatic system is the body's specific immunity. It attacks the pathogen directly. Specific immunity is more effective but takes more time to develop. ☑

What are the functions of the lymphatic system?

The lymphatic system filters lymph and blood and destroys foreign microorganisms. Lymph leaks out of capillaries to bathe body cells. After the fluid circulates among the cells, lymphatic vessels collect and return it to the veins near the heart.

Reading Check

2. Describe the process of phagocytosis.

____

____

____

____

Reading Check

3. Identify the system that provides specific immunity.

____

____

What organs are part of the lymphatic system?

The figure below shows the organs of the lymphatic system. The organs of the lymphatic system contain lymph, lymphocytes, a few other cell types, and connective tissue. **Lymphocytes** are a type of white blood cell that is produced in red bone marrow. The blood carries them to lymphatic organs. Lymphocytes play a role in specific immunity. ☑

Reading Check

4. Identify the role of lymphocytes.

The lymph nodes are located in different places along the path of the lymphatic vessels. Lymph nodes filter the lymph and remove foreign materials. The tonsils form a protective ring between the nose and mouth. The spleen stores blood and destroys damaged red blood cells. Lymphatic tissue within the spleen also responds to foreign substances in the blood. The thymus gland, located above the heart, helps to activate a special kind of lymphocyte called a T cell. T cells are produced in the bone marrow, but they mature in the thymus gland.

Picture This

5. Label Write these three labels next to the appropriate structures in the diagram.
Label: produces T cells
Label: activates T cells
Label: destroys damaged red blood cells

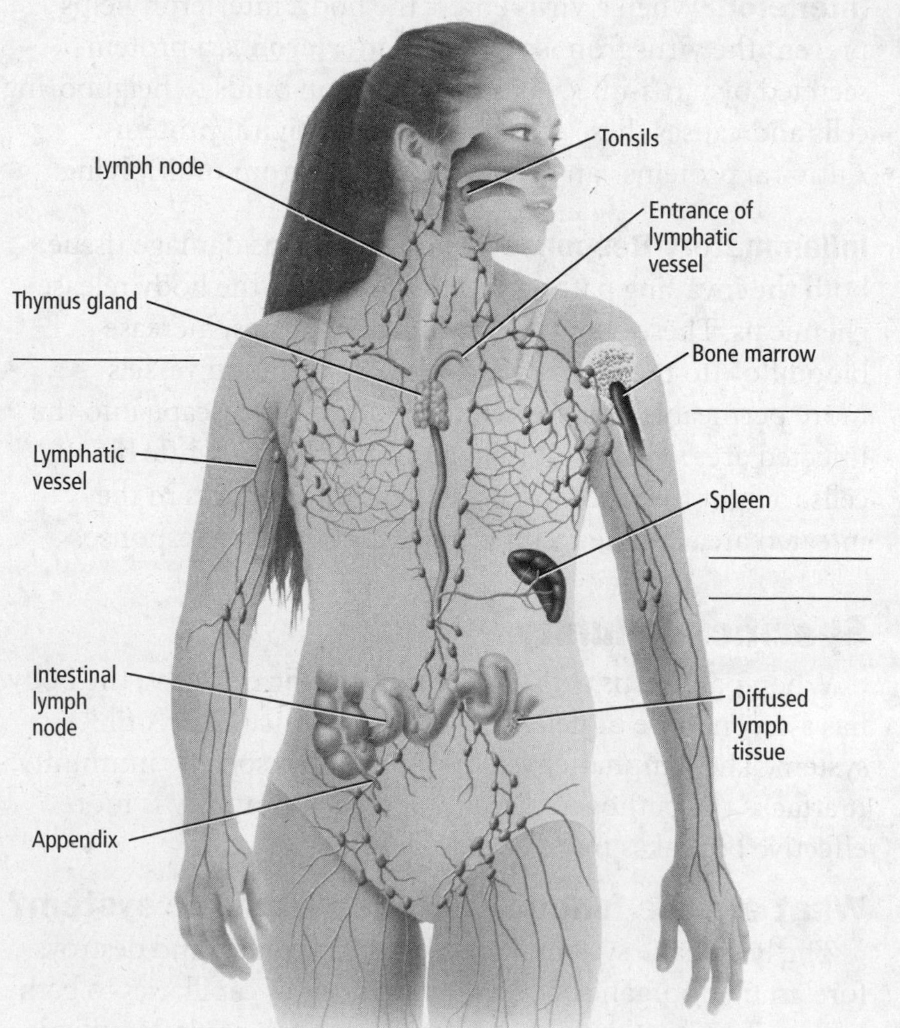

B Cell Response

When an antigen, or foreign substance, enters the body, it triggers the production of antibodies. **Antibodies** are proteins produced by B lymphocytes that specifically react with the antigen. B lymphocytes, or **B cells**, are antibody factories located in all lymphatic tissues. The figure below shows how B cells are activated to produce antibodies.

When a macrophage engulfs, absorbs, and digests a pathogen, it displays a piece of the pathogen, called a processed antigen, on its membrane. In lymphatic tissues, the macrophage, with the process antigen on its surface, binds to a type of lymphocyte called a **helper T cell**. This process activates the helper T cell. The activated helper T cell then activates B cells and another type of T cell to produce antibodies. B cells make many types of antibodies.

Antibodies help to kill microorganisms in two ways. Antibodies bind to the microorganisms, making phagocytosis more likely. They can also trigger an inflammatory response.

Picture This

6. Determine By what process does the B cell divide?

____

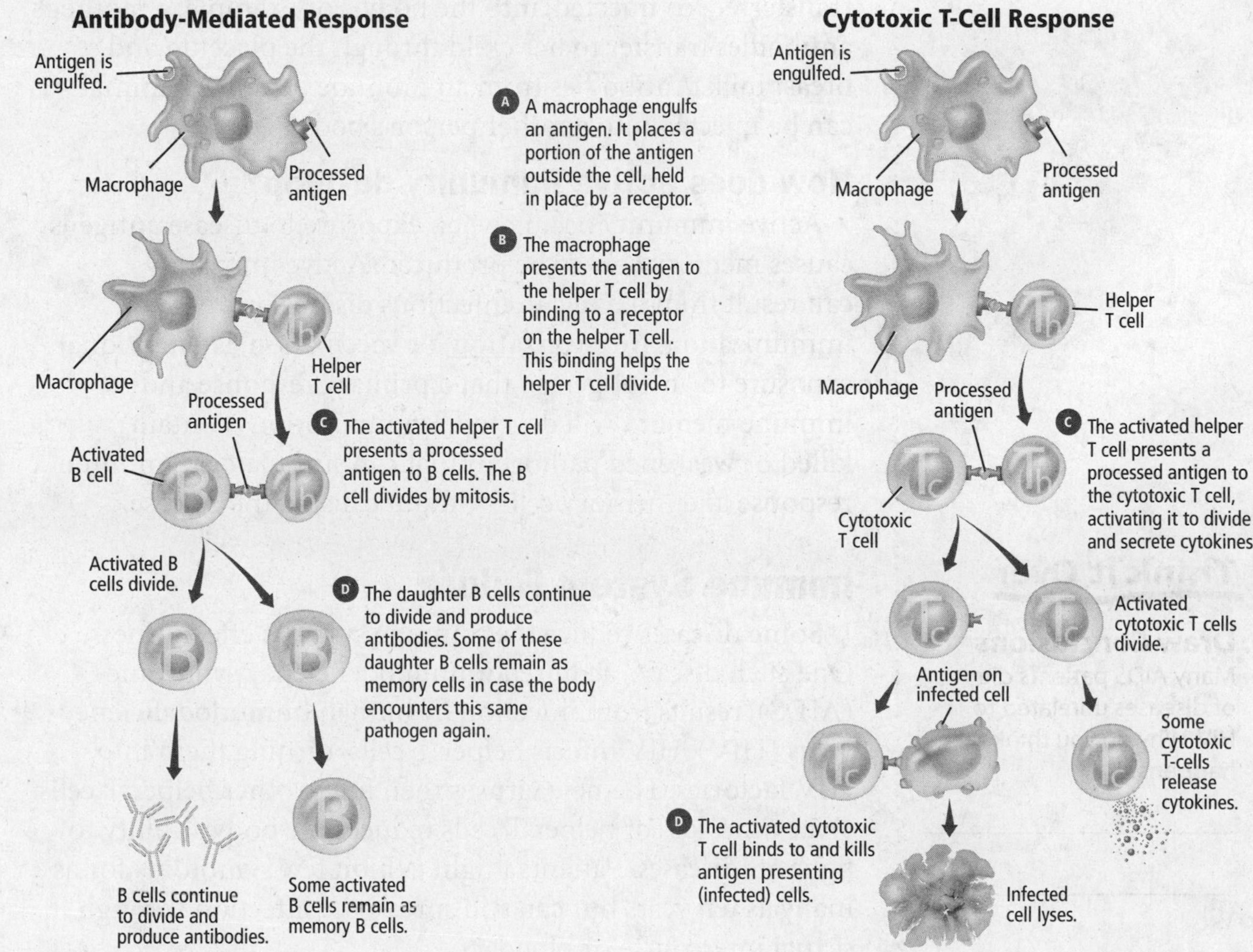

FOLDABLES

Take Notes Make a two-tab Foldable, as shown below. As you read, take notes and organize what you learn about active and passive immunity.

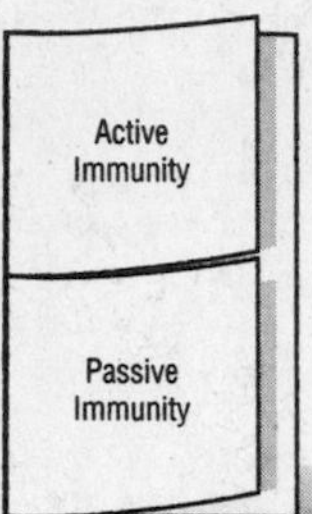

T Cell Response

Once activated, helper T cells can activate lymphocytes called cytotoxic T cells. Activated **cytotoxic T cells** destroy pathogens and release chemicals called cytokines. Cytokines stimulate cells of the immune system to divide and attract immune cells to the infected area.

Passive and Active Immunity

When a virus enters the body, nonspecific and specific immunity respond and defeat the pathogen. This is the body's first, or primary, response. A result of this response is the production of memory B and T cells. **Memory cells** are long-living cells that are exposed to the antigen during the primary immune response. If the body is later exposed to the same pathogen, memory cells respond rapidly to protect the body.

How is passive immunity produced?

Passive immunity provides temporary protection against an infection. Antibodies made by other people or animals can be transferred, or injected, into the body. For example, a mother's antibodies transfer to her child through the placenta and breast milk. Antibodies from an immune person or animal can be injected into another person's body.

How does active immunity develop?

Active immunity occurs when exposure to disease antigens causes memory cells to be produced. Active immunity can result from having an infectious disease or from immunization. **Immunization**, or vaccination, is intentional exposure to an antigen so that a primary response and immune memory will develop. Immunizations contain killed or weakened pathogens that can stimulate an immune response and memory cells without causing the disease.

Think it Over

7. Draw Conclusions Many AIDS patients die of diseases unrelated to HIV. Why do you think this happens?

____

____

____

Immune System Failure

Some diseases reduce the immune system's effectiveness. One such disease, acquired immunodeficiency syndrome (AIDS), results from infection by human immunodeficiency virus (HIV). HIV infects helper T cells, turning them into HIV factories. The new viruses then infect other helper T cells. Over time, loss of helper T cells reduces the body's ability to fight off diseases. Patients might exhibit few symptoms for as many as ten years but can still spread the infection through sexual intercourse or blood.

chapter 37 Immune System

section 3 Noninfectious Disorders

MAIN Idea

There are five categories of noninfectious diseases.

What You'll Learn

- noninfectious diseases include genetic, metabolic, and body deterioration diseases
- the role of allergens in allergies
- the difference between allergies and anaphylactic shock

Before You Read

Do you or a friend have an allergy? On the lines below, describe the symptoms. Read the section to learn the causes of common noninfectious disorders such as allergies.

____

____

Study Coach

Make Flash Cards Make a flash card for each category of disorder as you read about it. On the back of the flash card, list the characteristics of diseases in that category. Use the flash cards to review what you have learned.

Read to Learn

Genetic Disorders

Not all disorders are caused by pathogens. Some result from defective genes or chromosomes. Other disorders have both a genetic and an environmental cause. One example is coronary artery disease (CAD). Inherited genes increase the risk of developing CAD. Environmental factors, such as diet, contribute to the development of the disease.

Degenerative Diseases

<u>Degenerative diseases</u> (dih JEH nuh ruh tihv • dih ZEEZS) result from part of the body deteriorating. Degenerative arthritis and arteriosclerosis (ar tir ee oh skluh ROH sus) are two examples. Degenerative arthritis is common. Most people have it by age 70. Inherited genes can make some individuals more likely to develop degenerative diseases.

Metabolic Diseases

<u>Metabolic disease</u> results from an error in a biochemical pathway, such as an inability to regulate body processes. One example is Type II diabetes, in which cells do not recognize insulin or the pancreas does not make the proper amount of insulin. The result is too much glucose in the blood, which damages many organs. ☑

Reading Check

1. **Describe** a body process that does not function properly in Type II diabetes.

____

____

____

Cancer

Cancer is abnormal cell growth. Normally, the body can control cell division. If this control is lost, abnormal cell growth results, which can lead to tumors. The abnormal cells can interfere with normal body functions and travel throughout the body. Cancer tumors can develop in any tissue or organ. Cancer in blood cells is called leukemia. Both genes and the environment contribute to cancer. ☑

Reading Check

2. Identify Where can cancer tumors develop?

____

____

Inflammatory Diseases

Inflammatory diseases, such as allergies, result when the body produces an inflammatory response to a common substance. When the body responds to an infectious disease, the inflammatory response helps fight the disease. In inflammatory disease, the inflammatory response does not help the body.

What can cause allergies?

An **allergy** is an abnormal response to antigens in the environment. These antigens, called allergens, include plant pollens, dust, dust mites, and some foods. Symptoms may include itchy eyes, stuffy nose, sneezing, and sometimes a skin rash. The symptoms result from the chemical histamine that is released by certain white blood cells.

Anaphylactic (an uh fuh LAK tik) **shock** is a severe allergic reaction, resulting in a massive release of histamine. The smooth muscles in the bronchioles contract, restricting air flow into and out of the lungs. Severe reactions can result from bee stings, penicillin, peanuts, or latex, which is a substance used to make balloons and surgical gloves. Anaphylactic shock can be life threatening. Allergies and anaphylactic reactions are genetic.

Think it Over

3. Compare How does anaphylactic shock differ from more common allergic reactions?

____

____

____

How does the body respond in autoimmunity?

The immune system normally learns to tolerate "self" and will not attack self-proteins. However, some people develop autoimmunity (aw toh ih MYOON ih tee) and form antibodies to their own proteins, which injure their cells. In rheumatoid (roo MAH toyd) arthritis, antibodies attack the joints. In rheumatic fever, antibodies attack the valves of the heart causing the valves not to close properly. In lupus (LEW pus), antibodies attack cell nuclei, resulting in damage to many organs.